# MATLAB® FOR NEUROSCIENTISTS

## SECOND EDITION

# MATLAB® FOR NEUROSCIENTISTS

## An Introduction to Scientific Computing in MATLAB®

## SECOND EDITION

Pascal Wallisch

Michael E. Lusignan

Marc D. Benayoun

Tanya I. Baker

Adam S. Dickey

Nicholas G. Hatsopoulos

AMSTERDAM • BOSTON • HEIDELBERG • LONDON
NEW YORK • OXFORD • PARIS • SAN DIEGO
SAN FRANCISCO • SINGAPORE • SYDNEY • TOKYO
Academic Press is an imprint of Elsevier

Academic Press is an imprint of Elsevier
32 Jamestown Road, London NW1 7BY, UK
225 Wyman Street, Waltham, MA 02451, USA
525 B Street, Suite 1800, San Diego, CA 92101-4495, USA

**Notice**
No responsibility is assumed by the publisher for any injury and/or damage to
persons or property as a matter of products liability, negligence or otherwise, or from
any use or operation of any methods, products, instructions or ideas contained in
the material herein. Because of rapid advances in the medical sciences, in particular,
independent verification of diagnoses and drug dosages should be made.

**British Library Cataloguing-in-Publication Data**
A catalogue record for this book is available from the British Library

**Library of Congress Cataloging-in-Publication Data**
A catalog record for this book is available from the Library of Congress

ISBN: 978-0-12-383836-0

For information on all Academic Press publications
visit our website at elsevierdirect.com

Typeset by MPS Limited, Chennai, India
www.adi-mps.com

# Contents

# IV

# DATA MODELING WITH MATLAB

This is the space where one would – in ancient times – find what is called the "book curse". This was one of the earliest and most effective forms of copy protection. In modern times, where a free copy of a book is only a click away, the book curse is making a big comeback. However, we decided against using a book curse in this case. While we recognize that you could easily find a free copy of this work, we trust that you find this book helpful enough to support our work in turn. That's all we can hope for.

# Preface to the First Edition

I hear and I forget.
I see and I remember.
I do and I understand.

*Confucius*[1]

The creation of this book stems from a set of courses offered over the past several years in quantitative neuroscience, particularly within the graduate program in computational neuroscience at the University of Chicago. This program started in 2001 and is one of the few programs focused on computational neuroscience with a complete curriculum including courses in cellular, systems, behavioral, and cognitive neuroscience; neuronal modeling; and mathematical foundations in computational neuroscience. Many of these courses include not only lectures but also lab sessions in which students get hands-on experience using the MATLAB® software to solve various neuroscientific problems.

The content of our book is oriented along the philosophy of using MATLAB as a comprehensive platform that spans the entire cycle of experimental neuroscience: stimulus generation, data collection and experimental control, data analysis, and finally data modeling. We realize that this approach is not universally followed. Quite a number of labs

use different—and specialized—software for stimulus generation, data collection, data analysis, and data modeling, respectively. Although this alternative is a feasible strategy, it does introduce a number of problems: namely, the need to convert data between different platforms and formats and to keep up with a wide range of software packages as well as the need to learn ever-new specialized home-cooked "local" software when entering a new lab. As we have realized in our own professional life as scientists, these obstacles can be far from trivial, constitute a significant detriment to productivity and are the root cause of many a conniption.

We also believe that our comprehensive MATLAB "strategy" makes particular sense for educational purposes, as it empowers users to progressively solve a wide variety of computational problems and challenges within a single programming environment. It has the added advantage of an elegant progression within the problem space. Our experience in teaching has led us to this approach that focuses on the inherent structure of MATLAB not as a computer programming language, but rather as a tool for solving problems within neuroscience. In addition, it is well founded in our current

---

[1]In the West, this quote is commonly attributed to Confucius. However, in China itself, it is often pointed out (and it has been brought to our attention by Qian Cheng) that a very similar saying goes back to the Chinese philosopher Xunzi. While there is some controversy regarding whether similar sayings originated multiple times, there is no question that Confucius is a quote magnet. In the case of Einstein, this has been modeled. If current trends continue, it is not unlikely that over time, *all* quotes will be attributed to him. Be that as it may, we find the saying to be truthful, regardless of its source. It is an attempt at attribution, not an implicit argument from authority.

understanding of the learning process. Constant use of the information forces the repeated retrieval of the introduced concepts, which—in turn—facilitates learning (Karpicke and Roediger, 2008).

The book is structured in four parts, each with several chapters. The first part serves as a brief introduction to some of the most commonly used functions of the MATLAB software, as well as to basic programming in MATLAB. Users who are already familiar with MATLAB may skip it. It serves the important purpose of a friendly invitation to the power of the MATLAB environment. It is elementary insofar as it is necessary to have mastered the content within before progressing any further. Later parts focus on the use of MATLAB to solve computational problems in neuroscience. The second part focuses on MATLAB as a tool for the collection of data. For the sake of generality, we focus on the collection of data from human subjects in these chapters, although the user can easily adapt them for the collection of animal data as well. The third part focuses on MATLAB as a tool for data analysis and graphing. This part forms the core of the book, as this is also how MATLAB is most commonly used. In particular, we explore the analysis of a variety of datasets, including "real" data from electrophysiology as well as neuroimaging. The fourth part focuses on data modeling with MATLAB, and appendices address the philosophy of MATLAB as well as the underlying mathematics. Each chapter begins with the goals of the chapter and a brief background of the problem of interest (neuroscientific or psychological), followed by an introduction to the MATLAB concepts necessary to address the problem by breaking it down into smaller parts and providing sample code. You are invited to modify, expand, and improvise on these examples in a set of exercises. Finally, a project is assigned at the end of the chapter which requires integrating the parts into a coherent whole. Based on our experience, we believe that these chapters can serve as self-contained "lab" components of a course if this book is used in the context of teaching.

In essence, we strived to write the book that we wished to have had when first learning MATLAB ourselves, as well as the book that we would have liked to have had when teaching MATLAB to our students in the past. Our hope is that this is the very book you are holding in your hands right now.

We could have not written this book without the continuous support of a large number of friends. First and foremost, we would like to thank our families for their kind support, their endless patience, as well as their untiring encouragement. We also would like to extend thanks to our students who provided the initial impetus for this undertaking as well as for providing constant feedback on previous versions of our manuscript. Steve Shevell deserves thanks for suggesting that the project is worth pursuing in the first place. In addition, we would like to thank everyone at Elsevier who was involved in the production and development of this book—in particular our various editors, Johannes Menzel, Sarah Hajduk, Clare Caruana, Christie Jozwiak, Chuck Hutchinson, Megan Wickline, and Meg Day—their resourcefulness, professionalism and patience really did make a big difference. Curiously, there was another Meg involved with this project, specifically Meg Vulliez from The MathWorks™ book program. In addition, we would like to thank Kori Lusignan and Amber Martell for help with illustrations, and Wim van Drongelen for advice and guidance in the early stages of this project. Moreover, we thank Armen Kherlopian and Gopathy

Purushothaman, who were kind enough to provide us with valuable insights throughout our undertaking. We also would like to thank Kristine Mosier for providing the finger-tapping functional magnetic imaging data that we used in the fMRI lab, and would like to thank Aaron Suminski for his help in the post-processing of that data. Importantly, we thank everyone whom we neglected to name explicitly, but who deserves our praise. Finally, we would like to thank you, the reader, for your willingness to join us on this exciting journey. We sincerely hope that we can help you reach your desired destination.

*The authors*

# Preface to the Second Edition

The publication of the first edition of *MATLAB for Neuroscientists* was met with a reception that far exceeded our expectations, vindicating our intuition that there was an urgent need for such a text. Cynical voices often suggest that a new edition of a textbook is primarily designed to enrich the publisher. Not so in this case. While the first edition was widely adopted as a textbook as well as by individual students and investigators, several developments made it prudent to consider a second edition. First, neuroscience itself has changed, e.g., there is now an increased interest in the exploration of LFP signals. Second, MATLAB® has evolved, e.g., through the introduction of parallel computing environments. Finally, and most importantly, we received copious feedback in response to the first edition. For example, there was an overwhelming consensus that the book would benefit from an increased number of basic tutorials in the front matter. Taken together, all of this suggested to us that it might be time for an update. Deciding to release a second edition afforded us the opportunity to address these issues, and also to improve upon the first version in other ways. For instance, we were now able to introduce full-color figures throughout the book, something which we think will improve its usability considerably, given that data visualization is one of MATLAB's greatest strengths.

One thing that has not changed in the second edition is our philosophy of and focus on trying to foster behavioral change.

Unless a book somehow leads to a change in behavior (that is, the way you go about doing things), it is very likely that you will forget what you read. We know; it happened to us countless times. Sometimes, the only thing one remembers about a book is that one read it, but nothing else. That's not what this book is about. This book is about creating lasting behavioral change, specifically allowing you to use MATLAB more effectively, which in turn will (hopefully) make your research more productive. This requires more than just reading. It requires interaction with the content on a deep level. Thus, we tried to frame the content of this book to maximize the probability of meaningful engagement with the material.

The unbroken popularity of MATLAB among the neurosciences underscores the need for an accessible and up-to-date guide to its use. We hope that we succeeded in our intention to fulfill this need.

Many people helped us in our attempt to do so, and we thank them here. In addition to all the people we thanked in the first edition (on which this second edition is based), we also would like to thank April Graham, Mica Haley, Melissa Walker, and Greg Harris as well as Caroline Johnson and Melinda Rankin for their almost inexhaustible patience, kindness, and support; Donald McLaren for helping us with the neuroimaging chapter; and Qian Cheng for educational advice.

Finally, we also—and in particular—would like to thank our students and colleagues for feedback on the first edition.

In a sense, experts are the last people who should write a book like this. By their very experthood, they are often incapable of appreciating what would be helpful to someone who is not an expert and is just beginning to build cognitive structures in this domain. Therefore, feedback is absolutely crucial to give the experts information on how to make these materials more accessible to nonexperts. In a sense, writing a book like this teaches the authors—through feedback—how to write a book like this. And thus, the circle is complete.

*The authors*

# About the Authors

**Pascal Wallisch, PhD, Center for Neural Science, New York University**

Pascal received his PhD from the University of Chicago and now works as a research scientist at New York University. He is currently studying the processing of visual motion, physiological models of autism, and neurocinematics. Pascal is passionate about teaching, including teaching MATLAB as well as the communication of scientific concepts to a wider audience. He was recognized for his distinguished teaching record by the University of Chicago Booth Prize for excellence in teaching.

**Michael Lusignan, PhD, University of Chicago**

Michael received his PhD from the Committee on Computational Neuroscience at the University of Chicago, where he investigated physiological and behavioral models of song acquisition in birds. During this time, he served as a teaching assistant for a large number of courses which offered many opportunities to teach MATLAB techniques to neuroscience and other biological science students. Currently, Michael serves as a compiler and language developer for CrowdStrike, Inc., a company focusing on big data approaches to computer security.

**Marc Benayoun, MD/PhD, Emory University**

Marc is a radiology resident at Emory University. He is pursuing a combined radiology and nuclear medicine residency, with plans to complete a neuroradiology fellowship and continue as an academic neuroradiologist. Previously, he was a teaching assistant for *Mathematical Methods for the Biological Sciences*, a course taught at the University of Chicago.

Marc obtained his MD and PhD from the University of Chicago, studying stochastic models of neuronal dynamics with applications to epilepsy. He was also the recipient of the University of Chicago Booth Prize for excellence in teaching.

Marc would like to thank the Frank family for their financial support during his time at the University of Chicago.

**Tanya I. Baker, PhD, FICO**

Tanya is an analytic scientist at FICO applying neural network and other predictive analytics algorithms for fraud detection. Previously she was a junior research fellow at The Salk Institute for Biological Studies modeling large-scale neuronal population dynamics using modern statistical methods. As a post-doctoral lecturer at the University of Chicago, she developed and taught Mathematical Methods for the Biological Sciences, a new year-long course with a computer lab component. She received her Ph.D in Physics at the University of Chicago and her B.S. in Physics and Applied Mathematics at U.C.L.A.

**Adam S. Dickey, PhD, Pritzker School of Medicine, University of Chicago**

Adam is currently finishing medical school at University of Chicago and plans to complete a neurology residency. He completed his PhD in the laboratory of Dr. Nicholas G.

Hatsopoulos, looking at the encoding of corrective movements in motor cortex. Adam is interested in the clinical application of the brain-machine interfaces, particularly for patients with movement disorders.

**Nicholas G. Hatsopoulos, PhD, Department of Organismal Biology and Anatomy & Department of Neurology, University of Chicago**

Nicholas is Professor and Chairman of the graduate program on Computational Neuroscience. He teaches a course in Cognitive Neuroscience which formed the basis for some of the chapters in the book. His research focuses on how ensembles of cortical neurons work together to control, coordinate, and learn complex movements of the arm and hand. He is also developing brain/machine interfaces by which patients with severe motor disabilities could activate large groups of neurons to control external devices.

# How to Use this Book

A text of a technical nature tends to be more readily understood if its design principles are clear from the very outset. This is also the case with this book. Hence, we will use this space to briefly discuss what we had in mind when writing the chapters. Hopefully, this will improve usability and allow you to get most out of the book.

## STRUCTURAL AND CONCEPTUAL CONSIDERATIONS

A chapter typically begins with a concise overview of what material will be covered. Moreover, we usually put the chapter in the broader context of practical applications. This brief introduction is followed by a discussion of the conceptual and theoretical background of the topic in question. The heart of each chapter is a larger section in which we introduce relevant MATLAB® functions that allow you to implement methods or solve problems that tend to come up in the context of the chapter topic. This part of the chapter is enriched by small exercises and suggestions for exploration. We believe that doing the exercises is imperative to attain a sufficiently deep understanding of the function in question, while the suggestions for exploration are aimed at readers who are particularly interested in broadening their understanding of a given function. In this spirit, the exercises are usually rather specific, while the suggestions for exploration tend to be of a rather sweeping nature. This process of successive introduction and reinforcement of functions and concepts culminates in a "project," a large programming task that ties all the material covered in the book together. This will allow you to put the learned materials to immediate use in a larger goal, often utilizing "real" experimental data. Finally, we list the MATLAB functions introduced in the chapter at the very end. It almost goes without saying that you will get the most out of this book if you have a version of MATLAB open and running while going through the chapters. That way, you can just try out the functions we introduce, try out new code, etc.

Hence, we implicitly assumed this to be the case when writing the book.

Moreover, we made sure that all the code works when running the latest version of MATLAB (currently 8.1). Don't let this concern you too much, though. The vast majority of code should work if you use anything above version 7.7. We did highlight some important changes where appropriate.

## LAYOUT AND STYLE

The reader can utilize not only the conceptual structure of each chapter as outlined above, but also profit from the fact that we systematically encoded information about the function of different text parts in the layout and style of the book.

The main text is set in 10/12 Palatino-Roman. In contrast, executable code is **bolded** and offset by ≫, such as this:

> ≫ **figure**
> ≫ **subplot(2,2,1)**
> ≫ **image(test_disp)**

The idea is to type this text (without the ≫) directly into MATLAB. Moreover, functions that are first introduced at this point are **bolded** in the text. Exercises and Suggestions for exploration are set in italics and separated from the main text by boxes. When referring to directories, we alternate between the Mac (using a / slash) and PC (using a \ backslash) format of addressing. Please always use the appropriate format—slash or back-slash—for Mac or PC, respectively.

Equations are set in 10/12 Palatino-Roman. Sample solutions are in 10/12 Palatino-Bold.

## COMPANION WEB SITE

The successful completion of many chapters of this book depends on additional material (experimental data, sample solutions and other supplementary information) which is accessible from the web site that accompanies this book. For example, a database of executable code will be maintained as long as the book is in print. For information on how to access this online repository, please see page ii.

# PART I

# FUNDAMENTALS

# 1

# Introduction

Neuroscience is at a critical juncture. In the past few decades, the essentially biological nature of the field has been infused by the tools provided by mathematics. At first, the use of mathematics was mostly methodological in nature—primarily aiding the analysis of data. Soon, this influence turned conceptual, framing the very issues that characterize modern neuroscience today. Naturally, this development has not remained uncontroversial. Some neurobiologists of yore resent what they perceive to be a hostile takeover of the field, as many quantitative methods applied to neurobiology were pioneered by nonbiologists with a background in physics, engineering, mathematics, statistics, and computer science. Their concerns are not entirely without merit. For example, Hubel and Wiesel (2004) warn of the faddish nature that the idol of "computation" has taken on, even likening it to a dangerous disease that has befallen the field that we should overcome quickly in order to restore its health.

While these concerns are valid to some degree, and while excesses do happen, we strongly believe that—all in all—the effect of mathematics in the neurosciences has been very positive. Moreover, we believe that our science is and will continue to be one that is computational at its very core. The reason for this is that—as pointed out by Konrad Körding (http://www.nature.com/news/neuroscience-solving-the-brain-1.13382)—the human brain produces in 30 seconds as much data as the Hubble Space Telescope has produced in its lifetime. That is a staggering number, given that Hubble has been in operation for well over 23 years and generates more than 100 GB of data each week. Eventually, we will develop experimental methods that will fully tap this wellspring of data. We expect that computational methods to tackle this data will be developed in parallel. Put differently, not only is a computational perspective on neuroscience here to stay, we are likely only at the very beginning of this process. Historically, this notion stems in part from the influence that cognitive psychology has had in the study of the mind. Cognitive psychology and cognitive science—more generally—posited that the mind and, by extension, the brain should be viewed as information processing devices that receive inputs and transform these inputs into intermediate representations that ultimately generate observable outputs. At the same time that cognitive science was taking hold in psychology in the 1950s and 1960s, computer science was developing beyond mere number crunching and considering the possibility that intelligence could be

*MATLAB® for Neuroscientists.*
DOI: http://dx.doi.org/10.1016/B978-0-12-383836-0.00001-1

modeled computationally, leading to the birth of artificial intelligence. The information processing perspective, in turn, ultimately influenced the study of the brain, and is best exemplified by an influential book by David Marr titled *Vision*, published in 1982. In that book, Marr proposed that vision and, more generally, the brain should be studied at three levels of analysis: the computational, algorithmic, and implementational levels. The challenge at the computational level is to determine what computational problem a neuron, neural circuit, or part of the brain is solving. The algorithmic level identifies the inputs, the outputs, their representational format, and the algorithm that takes the input representation and transforms it into an output representation. Finally, the implementational level identifies the neural "hardware" and biophysical mechanisms underlying the algorithm that solves the problem. Today this perspective has permeated not only cognitive neuroscience, but also systems, cellular, and even molecular neuroscience.

Importantly, such a conceptualization of our field places chief importance on the issues surrounding scientific computing. For someone to participate in or even appreciate state of the art debates in modern neuroscience, that person has to be well-versed in the language of computation. Of course, it is the task of education—if it is to be truly liberal—to enable students to do so. Yet, this poses a quite formidable challenge. The point of a truly liberal

FIGURE 1.1   The prisoners in Plato's cave. Contemporary neuroscientists without profound scientific computing skills are arguably in a much more desperate situation, even if it doesn't feel like it.

education is to free the recipient from the most severe bondage—ignorance and accidents of birth. The situation is akin to that of the prisoners in Plato's cave (see Figure 1.1). Those prisoners are chained to rocks in a cave (in actuality, probably a stone quarry in Syracuse) and only see the shadows, never the forms. Of course, these prisoners are actually better off than the ignorant. At least they know that they are prisoners. In contrast, the shackles of ignorance often seem light, and even quite comfortable. Once freed, the recipient of a liberal education can walk out of the cave and take part in the life of the mind.

For most students interested in neuroscience, mathematics amounts to what is essentially a foreign language. Similarly, the language of scientific computing is typically as foreign to students as it is powerful. The prospects of learning both at the same time can be daunting and—at times—overwhelming. So what is a student or educator to do? To quote from Alfred North Whitehead's *Aims of Education* essay:

> There is only one subject-matter for education, and that is Life in all its manifestations. Instead of this single unity, we offer children—Algebra, from which nothing follows; Geometry, from which nothing follows; Science, from which nothing follows; History, from which nothing follows; a Couple of Languages, never mastered; and lastly, most dreary of all, Literature, represented by plays of Shakespeare, with philological notes and short analyses of plot and character to be in substance committed to memory.
>
> p. 194

Whitehead makes two points. First, teaching should not be disjointed. It is crucial to make connections between subjects. Second, teaching "inert ideas" is worse than useless; it is paralyzing. The tonic is to provide actionable information that allows the pursuit of relevant goals. This will tie the information together and make it come to life.

Immersion has been shown to be a powerful way to learn foreign languages (Genesee, 1985). Hence, it is imperative that students are using these languages as often as possible when facing a problem in the field. For immersion to work, the learning experience has to be positive, yielding useful results that solve some real or perceived problem. Unfortunately, the inherent complexity as well as the seemingly arcane formalisms that characterize both are usually very off-putting to students, requiring much effort with little tangible yield and reducing the likelihood of further voluntary immersion.

To break this catch-22, the utility of learning these languages has to be drastically increased while making the learning process more accessible and manageable at the same time, even during the learning process itself. As we alluded to previously, this is a tall order. Fortunately, there is a way out of this conundrum. Recent advances in software as well as hardware have instantiated scientific computing within the framework of a unified computational environment. One of these environments is provided by the MATLAB® software. For reasons that will become readily apparent in this book, MATLAB fulfills the requirements that are necessary to meet and overcome the challenges outlined earlier. In addition—and partly for these reasons—MATLAB has become the de facto standard of scientific computing in our field. Stated more strongly, MATLAB really has become the lingua franca that all serious students of neuroscience are expected to understand in the very near future, if not already today.

This, in turn, introduces a new—albeit more tractable—problem. How does one teach MATLAB to a useful level of proficiency without making the study of MATLAB itself an additional problem and simply another chore for students? Overcoming this problem as a key to reaching the deeper goals of fluency in mathematics and scientific computing is a crucial goal of this book. We reason that a gentle introduction to MATLAB with a special emphasis on immediate results will computationally empower you to such a degree that the practice of MATLAB becomes self-sustaining by the end of the book. We carefully picked the content such that the result constitutes a confluence of ease (gradually increasing sophistication and complexity) and relevance. We are confident that at the end of the book you will be at a level where you will be able to venture out on your own, convinced of the utility of MATLAB as a tool and of your ability to harness this power henceforth. We have tested the various parts of the contents of this book on our students, and believe that our approach has been successful. It is our sincere wish and hope that the material contained will be as beneficial to you as it was to those students.

With this in mind, we would like to outline two additional specific goals of this book. First, the material covered in the chapters to follow gives a MATLAB perspective on many topics within computational neuroscience across multiple levels of neuroscientific inquiry from decision-making and attentional mechanisms to retinal circuits and ion channels. It is well known that an active engagement with new material facilitates both understanding and long-time retention of said material. The secondary aim of this book is to acquire proficiency in programming using MATLAB while going through the chapters. If you are already proficient in MATLAB, you can go right to the chapters following the tutorial. For the rest, the tutorial chapter will provide a gentle introduction to the empowering qualities that the mastery of a language of scientific computing affords.

We take a project-based approach in each chapter so that you will be encouraged to write a MATLAB program that implements the ideas introduced in the chapter. Each chapter begins with background information related to a particular neuroscientific or psychological problem, followed by an introduction to the MATLAB concepts necessary to address that problem with sample code and output included in the text. You are invited to modify, expand, and improvise on these examples in a set of exercises. Finally, the project assignment introduced at the end of the chapter requires integrating the exercises. Most of the projects will involve genuine experimental data that are either collected as part of the project or were collected through experiments in research labs. In rare cases, we use published data from classical papers to illustrate important concepts, giving you a computational understanding of critically important research.

In addition, solutions to exercises and executable code can be found in the online repository accompanying this book (booksite.elsevier.com/9780123838360).

Finally, we would like to point out that we are well aware that there is more than one way to teach—and learn—MATLAB in a reasonably successful and efficient manner. This book represents a manifestation of our approach; it is the path we chose, for the reasons we outlined here.

# 2

# MATLAB Tutorial

## 2.1 GOAL OF THIS CHAPTER

The primary goal of this chapter is to help you to become familiar with the MATLAB® software, a powerful tool. It is particularly important to familiarize yourself with the user interface and some basic functionality of MATLAB. To this end, it is worthwhile to at least work through the examples in this chapter (actually type them in and see what happens). Of course, it is even more useful to experiment with the principles discussed in this chapter instead of just sticking to the examples. The chapter is set up in such a way that it encourages you to do this.

If desired, you can work with a partner, although it is advisable to select a partner of similar skill to avoid frustrations and maximize your learning. Advanced MATLAB users can skip this tutorial altogether, while the rest are encouraged to start at a point where they feel comfortable.

The basic structure of this tutorial is as follows: each new concept is introduced through an example, an exercise, and some suggestions on how to explore the principles that guide the implementation of the concept in MATLAB. While working through the examples and exercises is indispensable, taking the suggestions for exploration seriously is also highly recommended. It has been shown that negative examples are very conducive to learning; in other words, it is very important to find out what does not work, in addition to what does work (the examples and exercises will—we hope—work). Since there are infinite ways in which something might not work, we can't spell out exceptions explicitly here. That's why the suggestions are formulated very broadly.

## 2.2 PURPOSE AND PHILOSOPHY OF MATLAB

MATLAB is a high-performance programming environment for numerical and technical applications. The first version was written at the University of New Mexico in the 1970s. The "MATrix LABoratory" program was created by Cleve Moler to provide a simple and interactive way to write programs using the Linpack and Eispack libraries of FORTRAN subroutines for matrix manipulation. MATLAB has since evolved to become an effective and powerful tool for programming, data visualization and analysis, education, engineering and research.

The strengths of MATLAB include extensive data handling and graphics capabilities, powerful programming tools and highly advanced algorithms. Although it specializes in numerical computation, MATLAB is also capable of performing symbolic computation by having an interface with Maple (a leading symbolic mathematics computing environment). Besides fast numerics for linear algebra and the availability of a large number of domain-specific built-in functions and libraries (e.g., for statistics, optimization, image processing, neural networks), another useful feature of MATLAB is its capability to easily generate various kinds of visualizations of your data and/or simulation results.

For every MATLAB feature in general, and for graphics in particular, the usefulness of MATLAB is mainly due to the large number of built-in functions and libraries. The intention of this tutorial is not to provide a comprehensive coverage of all MATLAB features but rather to prepare you for your own exploration of its functionality. The *online help system* is an immensely powerful tool in explaining the vast collection of functions and libraries available to you, and should be the most frequently used tool when programming in MATLAB. Note that this tutorial will not cover any of the functions provided in any of the hundreds of toolboxes, since each toolbox is licensed separately. If you have additional toolboxes available to you, we recommend using the online help system to familiarize yourself with the additional functions provided. Another tool for help is the Internet. A quick online search will usually bring up numerous useful web pages designed by other MATLAB users trying to help each other out. Including on the Mathworks website itself: www.mathworks.com/matlabcentral.

As stated previously, MATLAB is essentially a tool—a sophisticated one, but a tool nevertheless. Used properly, it enables you to express and solve computational and analytic problems in a wide variety of domains. The MATLAB environment combines computation, visualization, and programming around the central concept of the matrix. Almost everything in MATLAB is represented in terms of matrices and matrix-manipulations. If you would like a refresher on matrix-manipulations, a brief overview of the main linear algebra concepts needed is given in the next chapter, Chapter 3, "Mathematics and Statistics Tutorial." We will start to explore this concept and its power later in this tutorial. For now, it is important to note that, properly learned, MATLAB will help you get your job done in a very efficient way. Giving it a serious shot is worth the effort.

## 2.2.1 Getting Started

You can start MATLAB by simply clicking on the MATLAB icon, the "L-shaped Membrane" on your desktop or taskbar. The command window will pop up, awaiting your commands and instructions.

In the context of this tutorial, all commands that are supposed to be typed into the MATLAB command window, as well as expected MATLAB responses, are typeset in

**bold**. The beginnings of these commands are indicated by the >> prompt. Press Enter at the end of this line, after typing the instructions for MATLAB. All instructions discussed in this tutorial will be in MATLAB notation, to enhance your familiarity with the MATLAB environment.

Don't be afraid as you delve into this new programming world. Help is readily at hand. Using the command **help** followed by the name of the command (for example, **help save**) in the command window gives you a brief overview on how to use the corresponding command (i.e., the command/function **save**). You can also easily access these help files for functions or commands by highlighting the command for which you need assistance in either the command window or in an M-file and right-clicking to select the Help on Selection option. Entering the commands **helpwin**, **helpdesk**, or **helpbrowser** will also open the MATLAB help browser. Another way of accessing a specific function in the help browser is to use **doc save** instead of help save. This accesses the entry of "save" in the help browser, whereas help outputs the help into the command line.

## 2.2.2 MATLAB as a Calculator

MATLAB implements and affords all the functionality that you have come to expect from a fine scientific calculator. While MATLAB can, of course, do much more than that, this is probably a good place to start. This functionality also demonstrates the basic philosophy behind this tutorial—discussing the principles behind MATLAB by showing how MATLAB can make your life easier, in this case by replicating the functionality of a scientific calculator.

**Elementary mathematical operations:** Addition, subtraction, multiplication, division. These operations are straightforward:
Addition:

```
>>2 + 3
ans =
    5
```

Subtraction:

```
>> 7 − 5
ans = 2
```

Multiplication:

```
>>17*4
ans =
   68
```

Division:

```
>> 24/7
ans =
   3.4286
```

Following are some points to note:

1. It doesn't matter how many spaces are between the numbers and operators, if only numbers and operators are involved (this does not hold for characters):

    >> 5 + 12
    ans =
       17

2. Of course, operators can be concatenated, making a statement arbitrarily complex:

    >> 2 + 3 + 4 − 7*5 + 8/9 + 1 − 5*6/3
    ans =
       − 34.1111

3. Parentheses disambiguate statements in the usual way:

    >> 5 + 3*8
    ans = 1
       29
    >> (5 + 3)*8
    ans = 1
       64

"Advanced" mathematical operators: Powers, log, exponentials, trigonometry.

Power: $x\char`\^p$ is $x$ to the power $p$:

    >> 2^3
    ans =
       8

Natural logarithm: log:

    >> log (2.7183)
    ans =
       1.0000
    >> log(1)
    ans =
       0

Exponential: $\exp(x)$ is $e^x$

    >> exp(1)
    ans =
       2.7183

Trigonometric functions; for example, sine:

    >> sin(0)
    ans =
       0
    >> sin(pi/2)
    ans =
       1
    >> sin(3/2*pi)
    ans =
       − 1

*Note:* Many of these operations are dependent on the desired accuracy. Internally, MATLAB works with 16 significant decimal digits (for floating point numbers—see Chapter 4, "Programming Tutorial"), but you can determine how many should be displayed. You do this by using the **format** command. The **format short** command displays 4 digits after the decimal point; **format long** displays 14 or 15 (depending on the version of Matlab). Example:

```
>> log(2.7183)
ans =
   1.0000
>> format long
>> log(2.7183)
ans =
   1.000006684913988
>> format short
>> log(2.7183)
ans =
   1.0000
```

As an exercise, try to "verify" numerically that $x*y = \exp(\log(x) + \log(y))$. A possible example follows:

```
>> 5*7
ans =
   35
>> exp(log(5) + log(7))
ans =
   35.0000
```

*Hint:* Keep track of the number of your parentheses. This practice will come in handy later.

One of the reasons MATLAB is a good calculator is that—on modern machines—it is very fast and has a remarkable numeric range.

For example:

```
>> 2^500
ans =
   3.2734e + 150
```

*Note:* e is scientific notation for the number of digits of a number.
$x \, e + y$ means $x*10 \char`\^ y$.
Example:

```
>> 2e3
ans =
   2000
>> 2*10^3
ans =
   2000
```

Note that in the preceding exercises MATLAB has responded to a command entered by defining a new variable *ans* and assigning to it the value of the result of the command. The variable *ans* can then be used again:

>> ans + ans
ans =
   4000

The variable *ans* has now been reassigned to the value 4000. We will explore this idea of variable assignments in more detail in the next section.

---

### EXERCISE 2.1

Try to find the numeric range of MATLAB. For which values of $x$ in $2^{\wedge}x$ does MATLAB return a numeric value? For which values does it return infinity or negative infinity, **Inf** or **− Inf**, respectively?

---

### 2.2.3 Defining Matrices

Of course, MATLAB can do much more than described in the preceding section. A central concept in this regard is that of vectors and matrices—arrays of vectors. Vectors and matrices are designated by square brackets: [ ]. Everything between the brackets is part of the vector or matrix.

A simple row vector can be defined as follows:

>> [1  2  3]
ans =
   1  2  3

It contains the elements 1, 2, and 3.
A simple matrix can be created as follows:

>> [2  2  2;  3  3  3]
ans =
   2  2  2
   3  3  3

This matrix contains two rows and three columns. When you are entering the elements of the matrix, a semicolon separates rows, whereas spaces separate the columns.

Make sure that all rows have the same number of column elements to avoid errors:

>> [2  2  2; 3  3]
??? Error using == > vertcat
CAT arguments dimensions are not consistent.

In MATLAB, the concept of a variable is closely associated with the concept of matrices. MATLAB stores matrices in variables, and variables typically manifest themselves as matrices. *Caution:* This variable is not the same as a mathematical variable, just a place in memory.

Assigning a particular matrix to a specific variable is straightforward. In practice, you do this with the equal operator ( = ). Following are some examples:

```
>> a = [1  2  3  4  5]
a =
   1  2  3  4  5
>> b = [6  7  8  9]
b =
   6  7  8  9
```

Once in memory, the matrix can be manipulated, recalled, or deleted.

The process of recalling and displaying the contents of the variable is simple. Just type its name:

```
>> a
a =
   1  2  3  4  5
>> b
b =
   6  7  8  9
```

Note:

1. Variable names are case-sensitive. Watch out what you assign and recall:
   ```
   >> A
   ??? Undefined function or variable 'A'.
   ```
   In this case, MATLAB—rightfully—complains that there is no such variable, since you haven't assigned *A* yet.
2. Variable names can be of almost arbitrary length. Try to assign meaningful variable names for matrices:
   ```
   >> uno = [1  1  1; 1  1  1; 1  1  1]
   uno =
      1  1  1
      1  1  1
      1  1  1
   >> thismatrixisreallyempty = [ ]
   thismatrixisreallyempty =
      [ ]
   ```

You can easily create some commonly used matrices by using the functions **eye**, **ones**, **zeros**, **rand**, and **randn**. The function **eye(n)** will create an *nxn* identity matrix. The function **ones(n,m)** will generate an *n* by *m* matrix whose elements are all equal to 1, and the function **zeros(n,m)** will generate an *n* by *m* matrix whose elements are all equal to 0. When you leave out the second entry, *m*, in calling those functions, they will generate square matrices of either zeros or ones. So, for example, the matrix *uno* could have been more easily created using the command **uno = ones(3).**

In a similar way, MATLAB will generate matrices of random numbers pulled from a uniform distribution between 0 and 1 through the **rand** function, and matrices of random numbers pulled from a normal distribution with zero mean and a variance of one through the **randn** function.

MATLAB uses so-called workspaces to store variables. The command **who** will allow you to see which variables are in your workspace, and the command **whos** will return additional information regarding the dimensions ("size"), size in memory ("bytes"), and type ("class") of the variables stored in the active workspace.

Now create two variables, $x$ and $y$, and assign to them the values 23 and 57, respectively:

```
>> x = 23; y = 57;
```

Note that when you add a semicolon to the end of your statement, MATLAB suppresses the display of the result of that statement. Next, create a third variable, $z$, and assign to it the value of $x + y$.

```
>> z = x + y
z = 80
```

Let's see what's in the working memory, i.e., the workspace:

```
>> who
Your variables are:
a ans b thismatrixisreallyempty uno x y z
>> whos
```

| Name | Size | Bytes | Class |
|---|---|---|---|
| a | 1 × 5 | 40 | double |
| ans | 2 × 3 | 48 | double |
| b | 1 × 4 | 32 | double |
| thismatrixisreallyempty | 0 × 0 | 0 | double |
| uno | 3 × 3 | 72 | double |
| x | 1 × 1 | 8 | double |
| y | 1 × 1 | 8 | double |
| z | 1 × 1 | 8 | double |

When you use the command **save**, all the variables in your workspace can be saved into a file. MATLAB data files have a .mat ending. The command **save** is followed by the filename and a list of the variables to be saved in the file. If no variables are listed after the filename, then the entire workspace is saved. For example,

**save my_workspace x y z**

will create a file named my_workspace.mat that contains the variables $x$, $y$, and $z$. Now rewrite that file with one that includes all the variables in the workspace. Again, you do this by omitting a list of the variables to be saved:

>> **save my_workspace**

You can now clear the workspace using the command **clear all**:

>> **clear all**
>> **who**
>> **x**
**??? Undefined function or variable 'x'.**

Note that nothing is returned by the command **who**, as is expected because all the variables and their corresponding values have been removed from memory. For the same reason, MATLAB complains that there is no variable named $x$ because it has been cleared from the workspace. You can now reload the workspace with the variables using the command **load**:

>> **load my_workspace**
>> **who**
**Your variables are:**
**a ans b thismatrixisreallyempty uno x y z**

If they are no longer needed, specific variables and their corresponding values can be removed from memory. The command **clear** followed by specific variable names will delete only those variables:

>> **clear x y z**
>> **who**
**Your variables are:**
**a ans b thismatrixisreallyempty uno**

Try using the command **help** (i.e., via **help save**, **help load**, and **help clear**) in the command window to learn about some of the additional options these functions provide.

The size of the matrix assigned to a given variable can be obtained by using the function **size**. The function **length** is also useful when only the size of the largest dimension of a matrix is desired:

>> **size(a)**
**ans =**
   **1   5**
>> **length(a)**
**ans =**
   **5**

The content of matrices and variables in your workspace can be reassigned and changed on the fly, as follows:

>> **thismatrixisreallyempty = [5]**
**thismatrixisreallyempty =**
   **5**

It is very common to have MATLAB create long vectors of incremental elements just by specifying a start and end element:

```
>> thisiscool = 4:18
thisiscool =
  4   5   6   7   8   9   10   11   12   13   14   15   16   17   18
```

The size of the increment of the vector can be changed by specifying the step size in between the start and end element:

```
>> thisiscool = 4:2:18
thisiscool =
  4   6   8   10   12   14   16   18
```

Two convenient functions that MATLAB has for creating vectors are **linspace** and **logspace**. The command **linspace($a,b,n$)** will create a vector of $n$ evenly spaced elements whose first value is $a$ and whose last value is $b$. Similarly, **logspace($a,b,n$)** will generate a vector of $n$ equally spaced elements between decades $10^a$ and $10^b$:

```
>> v = logspace(1,5,5)
v =
   10   100   1000   10000   100000
```

Transposing a matrix or a vector is quite simple: It's done with the '(apostrophe) command:

```
>> a
a =
  1   2   3   4   5
>> a'
ans =
  1
  2
  3
  4
  5
```

Variables can be copied into each other, using the = command. In this case, the right side is assigned to the left side. What was on the left side before is overwritten and lost, as shown here:

```
>> b
b =
  6   7   8   9
>> b = a
b =
  1   2   3   4   5
```

*Note:* Don't confuse the = (equal) sign with its mathematical meaning. In MATLAB, this symbol does not denote the equality of terms, but is an assignment instruction. Again, the right side of the = will be assigned to the left side, while the left side will be lost. This is the source of many errors and misunderstandings which is why this is emphasized again here. The conceptual difference is nowhere clearer than in the case of "self-assignment":

```
>> a
a =
1  2  3  4  5
>> a = a'
a =
   1
   2
   3
   4
   5
>> a
a =
   1
   2
   3
   4
   5
```

The assignment of the transpose eliminates the original row vector and renders *a* as a column vector.

This reassignment also works for elements of matrices and vectors:

```
>> a(2,1) = 9
a =
   1
   9
   3
   4
   5
```

Generally, you can access a particular element of a two-dimensional matrix with the indices *i* and *j*, where *i* denotes the row and *j* denotes the column. Specifying a single index *i* accesses the $i^{th}$ element of the array counted column-wise:

```
>> a(2)
ans =
   9
```

We will explore indexing further in Section 2.2.6.

---

**EXERCISE 2.2**

Clear the workspace. Create a variable $A$ and assign to it the following matrix value:

$$A = \begin{pmatrix} 7 & 5 \\ 2 & 3 \\ 1 & 8 \end{pmatrix}.$$

Access the element $i = 2$, $j = 1$, and change it to a number twice its original value. Create a variable $B$ and assign to it the transpose of $A$. Verify that the fifth element of the matrix $B$ counted column-wise is the same as the $i = 1$, $j = 3$ element.

---

**EXERCISE 2.3**

Using the function **linspace** generates a row vector $v1$ with seven elements which uniformly cover the interval between 0 and 1. Now generate a vector $v2$ which also covers the interval between 0 and 1, but with a fixed discretization of 0.1. Use either the function **length** or **size** to determine how many elements the vector $v2$ has. What is the value of the third element of the vector $v2$?

---

Solutions to exercises are available on the companion website.

### 2.2.4 Basic Matrix Algebra

Almost everything that you learned in the previous section on mathematical operators in MATLAB can now be applied to what you just learned about matrices and variables. In this section we explore how this synthesis is accomplished—with the necessary modifications.

First, define a simple matrix and then add 2 to all elements of the matrix, like this:

```
>> p = [1   2;   3   4]
p =
   1   2
   3   4
>> p = p + 2
p =
   3   4
   5   6
```

As a quick exercise, check whether this principle extends to the other basic arithmetic operations such as subtraction, division, or multiplication.

What if you want to add a different number to each element in the matrix? It is not inconceivable that this operation could be very useful in practice. Of course, you could do it element by element, as in the end of the preceding section. But doing this would be very tedious and time-consuming. One of the strengths of MATLAB is its matrix operations, allowing you to do many things at once.

Here, you will define a new matrix with the elements that will be added to the old matrix and assign the result to a new matrix to preserve the original matrices:

```
>> q = [2   1;   1   1]
q =
   2   1
   1   1
>> m = p + q
m =
   5   5
   6   7
```

*Note:* The number of elements has to be the same for this element-wise addition to work. Specifically, the matrices that are added to each other must have the same number of rows and columns. Otherwise, nothing is added, and the new matrix is not created. Instead, MATLAB reports an error and gives a brief hint what went wrong:

```
>> r = [2   1; 1   1; 1   1]
r =
   2   1
   1   1
   1   1
>> n = p + r
??? Error using ==> plus
Matrix dimensions must agree.
```

As a quick exercise, see whether this method of simultaneous, element-wise addition generalizes to other basic operations such as subtraction, multiplication, and division.

*Note:* It is advisable to assign a variable to the result of a computation, if the result is needed later. If this is not done, the result will be assigned to the MATLAB default variable *ans*, which is overwritten every time a new calculation without explicit assignment of a variable is performed. Hence, *ans* is at best a temporary storage.

Note that in the preceding exercise, you get consistent results for addition and subtraction, but not for multiplication and division. The reason is that * and/really symbolize a different level of operations than + or −. Specifically, they refer to matrix multiplication and division, respectively, which can be used to calculate outer products, etc. Refer to the next chapter for a refresher if necessary. If you want an analogous operation to + and − , you have to preface the * or/with a dot (.). This is known as *element-wise operations*:

```
>> p
p =
   3   4
   5   6
>> q
```

```
q =
   2  1
   1  1
>> p*q
ans =
   10   7
   16  11
>> p.*q
ans =
   6  4
   5  6
```

Due to the nature of outer products, this effect is even more dramatic if you want to multiply or divide a vector by another vector:

```
>> a = [1  2  3  4  5]
a =
   1  2  3  4  5
>> b = [5  4  5  4  5]
b =
   5  4  5  4  5
>> c = a.*b
c =
   5   8   15   16   25
>> c = a*b
??? Error using == > mtimes
Inner matrix dimensions must agree.
```

Raising a matrix to a power is similar to matrix multiplication; thus, if you wish to raise each element of a matrix to a given power, the dot (.) must be included in the command. Therefore, to generate a vector $c$ having the same length as the vector $a$, but for each element $i$ in $c$, it holds that $c(i) = [a(i)]^{\wedge}2$, you use the following command:

```
>> c = a.^2
c =
   1  4  9  16  25
```

As you might expect, there exists a function **sqrt** that will raise every element of its input to the power 0.5. Note that the omission of the dot (.) to indicate element-wise operations when it is intended is one of the most common errors when beginning to program in MATLAB. Keep this point in mind when troubleshooting your code.

Of course, you do not have to use matrix algebra to manipulate the content of a matrix. Instead, you can "manually" manipulate individual elements of the matrix. For example, if $A$ is a matrix with four rows and three columns, you can permanently add 5 to the element in the third row and second column of this matrix by using the following command:

```
>> A(3,2) = 5 + A(3,2);
```

We will explore indexing further in the next section.

Earlier, we rather casually introduced matrix operations like outer products versus element-wise operations. Now, we will briefly take the liberty to rectify this state of affairs in a systematic way. MATLAB is built around the concept of the matrix. As such, it is ideally suited to implement mathematical operations from linear algebra. MATLAB distinguishes between *matrix operations* and *array operations*. Basically, the former are the subject of linear algebra and denoted in MATLAB with symbols such as $+$, $-$, $*$, $/$, or $\wedge$. These operators typically involve the entire matrix. Array operations, are indicated by the same symbols prefaced by a dot, such as $.*$, $./$, or $.\wedge$. Array operators operate element-wise, or one by one. The rest of the sections will mostly deal with array operations. Hence, we will give the more arcane matrix operations—and the linear algebra that is tied to it—a brief introduction here. Linear algebra has many useful applications, most of which are beyond the scope of this tutorial. One of its uses is the elegant and painless (particularly with MATLAB) solution of systems of equations. Consider, for example, the system

$$x + y + 2z = 9$$
$$2x + 4y - 3z = 1$$
$$3x + 6y - 5z = 0$$

You can solve this system with the operations you learned in middle school, or you can represent the preceding system with a matrix and use a MATLAB function that produces the reduced row echelon form of A to solve it, as follows:

```
>> A = [1  1   2  9;  2  4  −3  1;  3  6  −5  0]
A =
   1  1    2  9
   2  4  − 3  1
   3  6  − 5  0
>> rref(A)
ans =
   1  0  0  1
   0  1  0  2
   0  0  1  3
```

From the preceding, it is now obvious that $x = 1$, $y = 2$, $z = 3$. As you can see, tackling algebraic problems with MATLAB is quick and painless—at least for you.

Similarly, matrix multiplication can be used for quick calculations. Suppose you sell five items, with five different prices, and you sell them in five different quantities. This can be represented in terms of matrices. The revenue can be calculated using a matrix operation:

```
>> Prices = [10  20  30  40  50];
>> Sales = [50;  30;  20;  10;  1];
>> Revenue = Prices*Sales
Revenue =
   2150
```

*Note:* Due to the way in which matrix multiplication is defined, one of the vectors (**Prices**) has to be a row vector, while the other (**Sales**) is a column vector.

---

### EXERCISE 2.4

Double-check whether the matrix multiplication accurately determined revenue.

---

### EXERCISE 2.5

Which set of array operations achieves the same effect as this simple matrix multiplication?

---

*Exploration:* As opposed to array multiplication (.*), matrix multiplication is NOT commutative. In other words, **Prices * Sales** $\neq$ **Sales * Prices**. Try it by typing the latter. What does the result represent?

---

### EXERCISE 2.6

Create a variable $C$ and assign to it a $5 \times 5$ identity matrix using the function **eye**. Create a variable $D$ and assign to it a $5 \times 5$ matrix of ones using the function **ones**. Create a third variable $E$ and assign to it the square of the sum of $C$ and $D$.

---

### EXERCISE 2.7

Clear your workspace. Create the following variables and assign to them the given matrix values (superscript $T$ indicates transpose):

(a) $x = \begin{pmatrix} 2 \\ 1 \end{pmatrix}$

(b) $y = x^T \cdot 17 \begin{pmatrix} 1 & 0 \\ 0 & 1 \end{pmatrix}$

(c) $A = \begin{pmatrix} 3 & 7 \\ 2 & 1 \end{pmatrix}$

(d) $b = yA$

(e) $c = x^T A^{-1} b^T$

(f) $E = cA^T$

---

### EXERCISE 2.8

Create a time vector $t$ that goes from 0 to 100 in increments of 5. Now create a vector $q$ whose length is that of $t$ and each element of $q$ is equal to $2 + 5$ times the corresponding element of $t$ raised to the power of 1.7.

---

## 2.2.5 Indexing

Individual elements of a matrix can be identified and manipulated by the explicit use of their index values. When indexing a matrix, $A$, you may identify an element with two numbers using the format $A(row, column)$. You could also identify an element with a single number, $A(num)$, where the elements of the matrix are counted columnwise. Let's explore this a bit through a series of exercises. First, remove all variables from the workspace (use the command **clear all**) and create a variable $A$:

$$A = \begin{pmatrix} 1 & 2 & 3 & 4 \\ 5 & 6 & 7 & 8 \\ 10 & 20 & 30 & 40 \\ 50 & 60 & 70 & 80 \end{pmatrix}.$$

```
>> clear all
>> A = [1  2  3  4;  5  6  7  8;  10  20  30  40;  50  60  70  80];
```

Now assign the value 23 to each entry in the first row:

```
>> A(1,:) = 23
A =
    23   23   23   23
     5    6    7    8
    10   20   30   40
    50   60   70   80
```

The colon (:) in the col position indicates all column values. Similarly, you can assign the value 23 to each entry in the first column:

```
>> A(:,1) = 23
A =
    23   23   23   23
    23    6    7    8
    23   20   30   40
    23   60   70   80
```

Suppose you didn't know the index values for the elements that you wanted to change. For example, presume you wanted to assign the value 57 to each entry of $A$ that is equal or larger than 7 in the second row. What are the column indices for the elements of the second row of the matrix $A$ [i.e., $A(2,:)$] which satisfy the criteria to change? For this task, the **find** function comes in handy:

```
>> find(A(2,:) >= 7)
ans =
     1   3   4
```

Thus, the following command will produce the desired result:

```
>> A(2,find(A(2,:) >= 7)) = 57
A =
   23   23   23   23
   57    6   57   57
   23   20   30   40
   23   60   70   80
```

To further illustrate the use of the function **find** and indexing, consider the following task. Assign the value 7 to each entry in the fourth column of the matrix $A$ that is equal or larger than 40 and lower than 60. For this example, it is clearer to split this operation into two lines:

```
>> i = find((A(:,4) >= 40)&(A(:,4) < 60))
i =
   2
   3
>> A(i,4) = 7
A =
   23   23   23   23
   57    6   57    7
   23   20   30    7
   23   60   70   80
```

Back to a nice and simple task, assign the value 15 to the entry in the third row, second column:

```
>> A(3,2) = 15
A =
   23   23   23   23
   57    6   57    7
   23   15   30    7
   23   60   70   80
```

Similarly, you could have used the command **A(7) = 15**. If you try entering the command **find(A == 15)**, you will get the answer 7. The reason is that MATLAB stores the elements of a matrix column after column, so 15 is stored in the seventh element of the matrix when counted this way. Had you entered the command **[r,c] = find(A == 15);** you would see that $r$ is now assigned the row index value and $c$ the column index value of the element whose value is 15; that is, $r = 3$, $c = 2$.

```
>> [r,c] = find(A == 15)
r =
   3
c =
   2
```

The **find** function is often used with relational and logical operators. We used a few of these in the preceding examples and will summarize them all here. The relational operators are as follows:

$==$ (equal to)
$\sim=$ (not equal to)
$<$ (less than)
$>$ (greater than)
$<=$ (less than or equal to)
$>=$ (greater than or equal to)

MATLAB also uses the following syntax for logical operators:

**&** (AND)
**|** (OR)
**$\sim$** (NOT)
**xor** (EXCLUSIVE OR)
**any** (true if any element is nonzero)
**all** (true if all elements are nonzero).

---

### EXERCISE 2.9

Find the row and column indices of the matrix elements of *A* whose values are less than 20. Set all elements of the third row equal to the value 17. Assign the value 2 to each of the last three entries in the second column.

---

## 2.3 GRAPHICS AND VISUALIZATION

Whereas we re-created the functionality of a scientific calculator in the previous sections, here we will explore MATLAB as a graphing calculator. As you will see, visualization of data and data structures is one of the great strengths of MATLAB.

### 2.3.1 Basic Visualization

In this section, it will be particularly valuable to experiment with possibilities other than the ones suggested in the examples, since the examples can cover only a very small number of possibilities that will have a profound impact on the graphs produced.

For aesthetic purposes, start with a trigonometric function, which was introduced before—sine. First, generate a vector *x*, take the sine of that vector, and then plot the result:

```
>> x = 0:10
x =
   0  1  2  3  4  5  6  7  8  9  10
>> y = sin(x)
```

y =

   0  0.8415  0.9093  0.1411  − 0.7568  − 0.9589  − 0.2794  0.6570  0.9894  0.4121  − 0.5440
>> plot(x,y)

The result of this series of commands will look something like Figure 2.1.

A quick result was reached, but the graphic produced is admittedly rather crude, albeit sinusoidal in nature. Note the values on the x-axis (0 to 10), as desired, and the values on the y-axis, between −1 and 1, as it's supposed to be, for a sine function. The problem seems to be with sampling. So let's redraw the sine wave with a finer mesh.

Recall that a third parameter in the quick generation of vectors indicates the step size. If nothing is indicated, MATLAB assumes 1 by default. This time, you will make the mesh 10 times finer, with a step size of 0.1.

---

**EXERCISE 2.10**

Use >> x = 0:0.1:10 to create the finer mesh.

---

Notice that MATLAB displays a long series of incremental elements in a vector that is 101 elements long. MATLAB did exactly what you told it to do, but you don't necessarily want to see all that. Recall that the ; (semicolon) command at the end of a command suppresses the "echo," the display of all elements of the vector, while the vector is still created in memory. You can operate on it and display it later, like any other vector.

So try this:

>> x = 0:0.1:10;
>> y = sin(x);
>> plot(x,y)

This yields something like that shown in Figure 2.2, which is arguably much smoother.

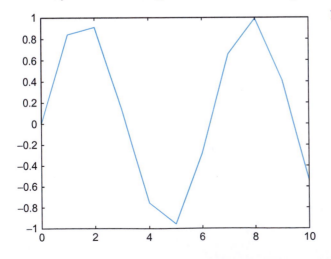

FIGURE 2.1  Crude sinusoid.

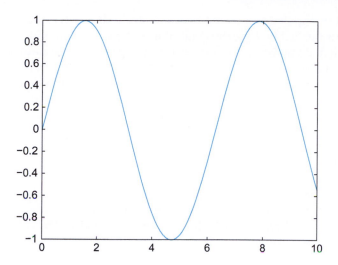

FIGURE 2.2 Smooth sinusoid.

---

**EXERCISE 2.11**

Plot the sine wave on the interval from 0 to 20, in 0.1 steps.

---

Upon completing Exercise 2.11, enter the following commands:

```
>> hold on
>> z = cos(x);
>> plot(x,z,'color','k')
```

The result should look something like that shown in Figure 2.3.

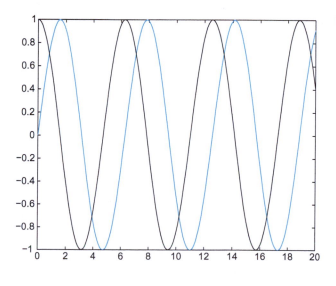

FIGURE 2.3 Sine vs. cosine.

Now you have two plots on the canvas, the sine and cosine from 0 to 20, in different colors. The command **hold on** is responsible for the fact that the drawing of the sine wave didn't just vanish when you drew the cosine function. If you want to erase your drawing board, type **hold off** and start from scratch. Alternatively, you can draw to a new figure, by typing **figure**, but be careful. Only a limited number of figures can be opened, since every single one will draw on the memory the resources of your computer. Under normal circumstances, you should not have more than about 30 figures open—if that. The command **close all** closes all the figures.

---

### EXERCISE 2.12

Draw different functions with different colors into the same figure. Things will start to look interesting very soon. MATLAB can draw lines in a large number of colors, eight of which are predefined: *r* for red, *k* for black, *w* for white, *g* for green, *b* for blue, *y* for yellow, *c* for cyan, and *m* for magenta.

---

Give your drawing an appropriate name. Type something like the following:

>> **title('My trigonometric functions')**

Now watch the title appear in the top of the figure.

Of course, you don't just want to draw lines. Say there is an election and you want to quickly visualize the results. You could create a quick matrix with the hypothetical results for the respective candidates and then make a bar graph, like this:

>> **results = [55   30   10   5]**
**results =**
    55   30   10   5
>> **bar(results)**

The result should look something like that shown in Figure 2.4.

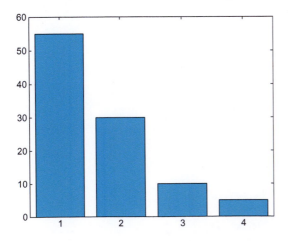

FIGURE 2.4   Lowering the bar.

---

**EXERCISE 2.13**

To get control over the properties of your graph, you will have to assign a handle to the drawing object. This can be an arbitrary variable, for example, $h$:

```
>> h = bar(results)
h = 298.0027
```

```
>> set(h,'linewidth', 3)
>> set(h,'FaceColor', [1   1   1])
```

The result should be white bars with thick lines. Try **get(**$h$**)** to see more properties of the bar graph. Then try manipulating them with **set(**$h$, **'Propertyname',Propertyvalue)**.

---

Finally, let's consider histograms. Say you have a suspicion that the random number generator of MATLAB is not working that well. You can test this hunch by visual inspection.

First, you generate a large number of random numbers from a normal distribution, say 100,000. Don't forget the ; (semicolon). Then you draw a histogram with 100 bins, and you're done. Try this, for example:

```
>> suspicious = randn(100000,1);
>> figure
>> hist(suspicious, 100)
```

The result should look something like that shown in Figure 2.5. No systematic deviations from the normal distribution are readily apparent. Of course, statistical tests could yield a more conclusive evaluation of your hypothesis.

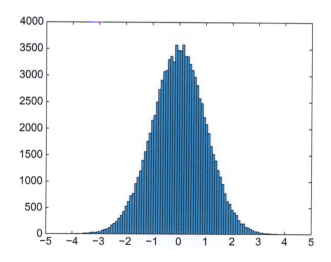

**FIGURE 2.5**  Gaussian normal distribution.

---

**EXERCISE 2.14**

You might want to run this test a couple of times to convince yourself that the deviations from a normal distribution are truly random and inconsistent from trial to trial.

---

A final remark on the display outputs: most of the commands that affect the display of the output are permanent. In other words, the output stays like that until another command down the road changes it again. Examples are the **hold** command and **format** command. Typing **hold** will hold plot and allow something else to be plotted on it. Typing **hold** again toggles hold and releases the plot. This is similarly true for the **format** commands, which keep the format of the output in a certain way.

We have thus far introduced only a small number of the many visualization tools that give MATLAB its strength. In addition to the functions **plot**, **bar**, and **hist**, you can explore other plotting commands and get a feel for more display options by viewing the help files for the following plotting commands that you might find useful: **loglog**, **semilogx**, **semilogy**, **stairs**, and **pie**.

Want to know what your data sounds like? MATLAB can send your data to the computer's speakers, allowing you to visually manipulate your data and *listen* to it at the same time. To hear an example, load the built-in chirp.mat data file by typing **load chirp**. Use **plot(y)** to see these data and **sound(y)** to listen to the data.

We will cover more advanced plotting methods in the following section as well as in future chapters.

## 2.4 FUNCTION AND SCRIPTS

Until now, we have driven MATLAB by typing commands directly in the command window. This is fine for simple tasks, but for more complex ones you can store the typed input commands into a file and tell MATLAB to get its input commands from the file. Such files must have the extension .m and are thus called *M-files*. If an M-file contains statements just as you would type them into the MATLAB command window, they are called *scripts*. If, however, they accept input arguments and produce output arguments, they are called *functions*.

The primary goal of this section is to help you become familiar with M-files within MATLAB. M-files are the primary vehicle to implement programming in MATLAB. So while the previous sections showed how MATLAB can double as a scientific calculator and as a calculator with graphing functions, this section will introduce you to a high-level programming language that should be able to satisfy most of your programming needs if you are a casual user. It will become apparent in this section of the tutorial how MATLAB can aid the researcher in all stages of an experiment or study. By no means is this tutorial the last word on M-files and programming. Later we will elaborate on all concepts introduced in this section—particularly in terms of improving efficiency and performance of the programs you are writing. One final goal of this tutorial is to demonstrate the remarkable versatility of MATLAB—and don't worry, we'll move on to neuroscience-heavy topics soon enough.

### 2.4.1 Scripts

Scripts typically contain a sequence of commands to be executed by MATLAB when the filename of the M-file is typed into the command window.

M-files are at the heart of programming in MATLAB. Hence, most of our future examples will take place in the context of M-files. You can work on M-files with the M-file editor, which comes with MATLAB. To go to the editor, open the File menu (top left), select New, and then select M-File (see Figure 2.6). In recent versions of MATLAB, there is now a distinction between "Script" and "Function." Open a new "Script" here. The most recent versions of MATLAB don't even have a File menu any more, but rather a number of icons in the home tab. They are still on the upper left. The functionality of the menu—and more—has now been put into a toolbar. The layout of the user interface will likely continue to change in future versions of MATLAB, so do not get too attached to it. However, the core functionality can be expected to say the same. Figure 2.6 shows a screenshot from the distant past. That's why it is in black and white.

The first thing to do after a new M-file is created is to name it. For this purpose, you have to save it to the hard disk. There are several ways of doing this. The most common is to click the editor's File menu and then click Save As. Recent versions of Matlab feature a toolbar that replaces the menu, see figure 2.7. To save, click on the floppy disk (never mind that no one has used a floppy disk in over a decade and that the pictured 3.5″ disk wasn't actually floppy). You can then save the file with a certain name. Using **myfirstmfile.m** would probably be appropriate for the occasion.

As a script, an M-file is just a repository for a sequence of commands that you want to execute repeatedly. Putting them in an M-file can save you the effort of typing the commands anew every time. This is the first purpose for which you will use M-files.

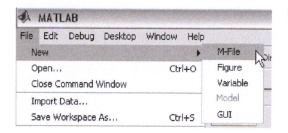

FIGURE 2.6   Creating a new M-File.

FIGURE 2.7

```
1 -    figure
2 -    x = 0:0.1:10;
3 -    y = sin (x);
4 -    plot (x,y)
5
```

FIGURE 2.8   The editor.

Type the commands into your M-file editor as they appear in Figure 2.8. Make sure to save your work after you are done (by pressing Ctrl + S), if you already named it. If you now type **myfirstmfile** into the MATLAB command window (not the editor), this sequence will be executed by MATLAB. You can do this repeatedly, tweaking things as you go along (don't forget to save).

### 2.4.2 Functions

We have already been using many of the functions built into MATLAB, such as **sin** and **plot**. You can extend the MATLAB language by writing your own functions. MATLAB has a very specific syntax for declaring and defining a function. Function M-files must start with the word **function**, followed by the output variable(s) within square brackets, the equal sign, the name of the function, and the input variable(s) in parenthesis, in that order. Functions do not have to have input or output arguments. The following is a valid function definition line for a function named **flower** that has three inputs, *a*, *b*, and *c*, and two outputs, *out_1* and *out_2*:

**function [out_1, out_2] = flower(a,b,c)**

To demonstrate this further, you will write a function named **triple** that computes and returns three times the input, i.e., **triple(2) = 6, triple(3) = 9**, etc. First, type the following two lines into an M-file and save it as **triple.m**:

**function r = triple(i)**
**r = 3*i;**

If you want to avoid confusion, it is strongly advised to match the name of the M-file with the name of the function. The input to the function is *i* and the output is *r*. You can now test this function:

**>> a = triple(7)**
**a =**
   **21**
**>> b = triple([10 20 30])**
**b =**
   **30   60   90**

*Note:* This function is trivial. Here, however, you should learn to apply the syntax only for defining and calling functions. Also note that function variables do not appear in the main workspace; rather their scope is limited to themselves. For instance, you do not have access to the variable "r" from the main workspace.

## 2.4.3 Control Structures

Of course, what you just saw is only the most primitive way of using M-files. M-files will allow you to harness the full power of MATLAB as a programming language. For this to happen, you need to familiarize yourself with loops and conditionals—in other words, with statements that allow you to control the flow of your program.

*Loops:* The two most common statements to implement loops in MATLAB are **for** and **while**.

The structure of all loops is as follows (in terms of a **while** loop):

**while** certain-conditions-are-true
Statements
. . .
Statements
**end**

All statements between **while** and **end** are executed repeatedly until the conditions that are checked in **while** are no longer the case. This is best demonstrated with a simple example: open a new M-file in the editor and give it a name. Then type the following code and save. Finally, type the name of the M-file in the command window (not the editor) to execute the loop.

This is a good place to introduce comments. As your programs become more complex, it is highly recommended that you add comments. After a week or two, you will not necessarily remember what a particular variable represents or how a particular computation was done. Comments can help, and MATLAB affords functionality for it. Specifically, it ignores everything that is written after the percent sign (%) in a line. In the editor itself, comments are represented in green. So here is the program that you should write now, implementing a simple **while** loop. If you want to, you can save yourself the comments (everything after % in each line). We placed them here to explain the program flow (what the program will do) to you:

```
%A simple counter

ii = 1          %Initializing our counter as 1

while ii < 6    %While ii is smaller than 6, statements are executed

 ii = ii + 1    %Incrementing the counter and displaying new value

end             %Ending the loop, it contains only one statement
```

What happened after you executed the program? Did it count from 1 to 6?

Note that we use the variable ii, not i as a counter here. This has several major advantages. The first one is that "i" and "j" are already predefined in MATLAB. They both represent the imaginary part of a complex number (1i). Try it. Clear the workspace if you defined i and j before (we did above, in the indexing section), then type i and j, respectively. It is generally a bad idea to overwrite something that is already predefined in MATLAB and can lead to puzzling and counterintuitive error messages later on. We did it in the section on indexing for didactic purposes only (mathematicians like to use i and j as indices). Another reason is that people tend to confuse i and 1 when coding. This is empirically true. So a for loop that would be defined as for i = 1:10 is often written as i = i:10. It might sound preposterous, but happens surprisingly often. Good luck finding the error later on. It is better to use for ii = 1:10 instead. One is much less likely to make this mistake. Programming is challenging work, as one has to constantly juggle objectives (what the code is supposed to do) with implementation (how the code does it). Keeping all of that in working memory is hard and the literature is full of examples of how quickly performance can break down, particularly if there are added problems like sleep deprivation or distractions. Better to play it safe and stay away from i and j altogether. This goes for all functions. A prominent example is "size." People often call the variable with which they represent the size of something "size," e.g. the size of the population under study. If you do this, you basically broke your code if it invokes "size" to refer to the function at some other point in the code. MATLAB will throw an error message at that point and it will surprise you. So use more specific names, like pop_size for your variables instead. We use ii instead i in the same spirit.

---

### EXERCISE 2.15

Let your program count from 50 to 1050. Then redo this with a for loop for practice.

If you execute this program on a slow machine, chances are that this operation will take a while.

---

### EXERCISE 2.16

Let your program count from 1 to 1,000,000.

If you did everything right, you will be sitting for at least a minute, watching the numbers go by. While we set up this exercise deliberately, chances are that you will underestimate the time it takes to execute a given loop sometime in the future. Instead of just biding your time, you have several options at your disposal to terminate runaway computations. The most benign of these options is pressing Ctrl + C in the MATLAB command window. That shortcut should stop a process that hasn't yet completely taken over the resources of your machine. Try it.

---

*Note:* The display of the numbers takes most of the time. The computation itself is relatively quick. Make the following modifications to your program; then save and run it:

**%A silent counter**

**ii = 1          %Initializing our counter as 1**

```
while ii < 1000000
    ii = ii + 1;   %Incrementing the counter without displaying new value
end                %Ending the loop, it contains only one statement
ii                 %Displaying the final value of ii
```

*Note:* One of the most typical ways to get logical errors in complex programs is to forget to initialize the counter (after doing something else with the variable). This is particularly likely if you reuse the same few variable names (*ii, jj,* etc.) throughout the program. In this case, it would not execute the loop, since the conditions are not met. Hence, you should make sure to always initialize the variables that you use in the loop *before* the loop. As a cautionary exercise, reduce your program to the following:

```
%A simple counter, without initialisation
while ii < 1000000     %While ii is smaller than 1M, statements are executed
    ii = ii + 1        %Incrementing the counter and displaying new value
end                    %Ending the loop, it contains only one statement
```

Save and run this new program. If you ran one of the previous versions, nothing will happen. The reason is that the loop won't be entered because the condition is not met; *i* is already larger than 1,000,000 before the first loop is executed.

Of course, the most common way to get runaway computations is to create infinite loops—in other words, loops with conditions that are always true after they are entered. If that is the case, they will never be exited. A simple case of such an infinite loop is a modified version of the initial loop program—one without an increment of the counter; hence, *ii* will always be smaller than the conditional value and never exit.

Try this, save, and run:

```
%An infinite loop
ii = 1           %Initializing our counter as 1
while ii < 6     %While ii is smaller than 6, statements are executed
    ii = ii      %NOT incrementing the counter, yet displaying its value
end              %Ending the loop, it contains only one statement
```

If you're lucky, you can also exit this process by pressing Ctrl + C. If you're not quick enough or if the process already consumed too many resources—this is particularly likely for loops with many statements, not necessarily this one—your best bet is to summon the Task Manager by pressing Ctrl + Alt + Delete simultaneously in Windows (for a Mac, the corresponding key press is Command + Option + Escape to call the Force Quit menu). There, you can kill your running version of MATLAB. The drawbacks of this method are that you have to restart MATLAB and your unsaved work will be lost. So beware the infinite loop.

*If statements:* In a way, **if** statements are pure conditionals. Statements within **if** statements are either executed once (if the conditions are met) or not (if they are not met). Their syntax is similar to loops:

**if** these-conditions-are-met
Execute-these-Statements
**else**
Execute-these-Statements
**end**

It is hard to create a good example consisting solely of **if** statements. They are typically used in conjunction with loops: the program loops through several cases, and when it hits a special case, the **if** statement kicks in and does something special. We will see instances of this in later examples. For now, it is enough to note the syntax.

### Fun with loops—How to make an American quilt

This is a rather baroque but nevertheless valid exercise on how to simply save time writing all the statements explicitly by using nested loops. If you want to, you can try replicating all the effects without the use of loops. It's definitely possible—just very tedious.

Open a new window in the editor, name it, type the following statements (without comments if you prefer), save it, and see what happens when you run it:

```
figure                      %Open a new figure
x = 0:0.1:20;               %Have an x-vector with 201 elements
y = sin(x);                 %Take the sine of x, put it in y
k = 1;                      %Initialize our counter variable k with 1
while k < 3;                %For k = 1 and 2
QUILT1(1,:) = x;            %Put x into row 1 of the matrix QUILT1
QUILT2(1,:) = y;            %Put y into row 1 of the matrix QUILT2
QUILT1(2,:) = x;            %Put x into row 2 of the matrix QUILT1
QUILT2(2,:) = −y;           %Put −y into row 2 of the matrix QUILT2
QUILT1(3,:) = −x;           %Put −x into row 3 of the matrix QUILT1
QUILT2(3,:) = y;            %Put y into row 3 of the matrix QUILT2
QUILT1(4,:) = −x;           %Put − x into row 4 of the matrix QUILT1
QUILT2(4,:) = −y;           %Put −y into row 4 of the matrix QUILT2
hold on                     %Always plot into the same figure
for ii = 1:4                %A nested loop, with ii as counter, from 1 to 4
    plot(QUILT1(ii,:),QUILT2(ii,:))  %Plot the iith row of QUILT1 vs. QUILT2
```

```
pause                    %Waiting for user input (key press)
end                      %End of ii-loop
for ii = 1:4             %Another nested loop, with ii as counter, from 1 to 4
   plot(QUILT2(ii,:),QUILT1(ii,:))  %Plot the iith row of QUILT2 vs. QUILT1
   pause                 %Waiting for user input (key press)
end                      %End of ii-loop
y = y + 19;              %Incrementing y by 19 (for every increment of k)
k = k + 1;               %Incrementing k by 1
end                      %End of k-loop
```

*Note:* This program is the first time we use the **pause** function. If the program pauses after encountering a **pause** statement, press a key to continue until the program is done. This is also the first time that we wrote a program that depends on user input—albeit in a very small and limited form—to execute its flow. We will expand on this later.

*Note:* This program used both **for** and **while** loops. The **for** loops increment their counter automatically, whereas the **while** loops must have their counter incremented explicitly.

Now that you know what the program does and how it operates, you might want to take out the two **pause** functions to complete the following exercises more smoothly.

---

### EXERCISE 2.17

What happens if you allow the conditional for $k$ to assume values larger than 1 or 2?

---

### EXERCISE 2.18

Do you know why the program increments $y$ by 19 at the end of the $k$ loop? What happens if you make that increment smaller or larger than 19?

---

### EXERCISE 2.19

Do you remember how to color your quilt? Try it.

### 2.4.4 Advanced Plotting

We introduced basic plotting of two-dimensional figures previously. This time, our plotting section will deal with subplots and three-dimensional figures. Subplots are an efficient way to present data. You probably have seen the use of the subplot function in published papers. The syntax of the subplot command is simply **subplot(*a,b,c*)**, where *a* is the number of rows the subplots are arranged in, *b* is the number of columns, and *c* is the particular subplot you are drawing to. It's probably best to illustrate this command in terms of an example. This requires you to open a new program, name it, etc.

Then type the following:

**figure**               %**Open a new figure**

**for ii = 1:9**          %**Start loop, have counter ii run from 1 to 9**

  **subplot(3,3,ii)**   %**Draw into the subplot ii, arranged in 3 rows, 3 columns**

  **h = bar(1,1);**     %**This is just going to fill the plot with a uniform color**

  **set(h,'FaceColor',[0 0 ii/9]);**   %**Draw each in a slightly different color**

**end**                  %**End loop**

This program will draw nine colored squares in subplots in a single figure, specifically, different shades of blue (from dark blue to light blue) and should look something like Figure 2.9.

*Note:* The three numbers within the square brackets in the **set(h,'FaceColor',[0 0 ii/9]);** statement represent the red, green, and blue color components of the bar that is plotted. Each color component can take on a value between 0 and 1. A bar whose color components are

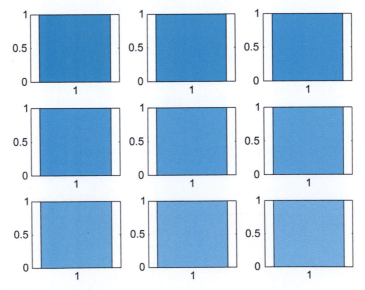

**FIGURE 2.9**  Color gradient subplots.

[0 0 0] is displayed black and [1 1 1] is white. By setting the color components of the pixels of your image to different combinations of values, you can create virtually any color you desire.

---

### EXERCISE 2.20

Make the blocks go from black to red instead of black to blue.

---

### EXERCISE 2.21

Make the blocks go from black to white (via gray). Try 49 shades.

---

### SUGGESTION FOR EXPLORATION

Can you create more complex gradations? It is possible, given this simple program and your recently established knowledge about RGB values in MATLAB as a basis.

---

Three-dimensional plotting is a simple extension of two-dimensional plotting. To appreciate this, we will introduce a classic example: drawing a two-dimensional exponential function. The two most common three-dimensional plotting functions in MATLAB are probably **surf** and **mesh**. They operate on a grid. Magnitudes of values are represented by different heights and colors. These concepts are probably best understood through an example.

Open a new program in the MATLAB editor, name it, and type the following; then save and run the program:

```
a = −2:0.2:2;           %Creating a vector a with 21 elements
[x, y] = meshgrid(a, a); %Creating x and y as a meshgrid of a
z = exp (− x.^2 − y.^2); %Take the 2-dimensional exponential of x and y
figure                  %Open a new figure
subplot(1,2,1)          %Create a left subplot
mesh(z)                 %Draw a wire mesh plot of the data in z
subplot(1,2,2)          %Create a right subplot
surf(z)                 %Draw a surface plot of the data in z
```

After running this program, you probably need to maximize the figure to be able to see it properly. To do this, click the maximize icon in the upper right of your figure (see Figure 2.10; or if using a Mac, click on the green button in the upper left corner). Both the left and right figures illustrate the same data, but in different manners. On the left is a wire mesh; on the right, a surface plot.

If you did everything right, you should see something like that shown in Figure 2.11.

---

**EXERCISE 2.22**

Improve the resolution of the meshgrid. Then redraw.

---

FIGURE 2.10   Maximizing a figure.

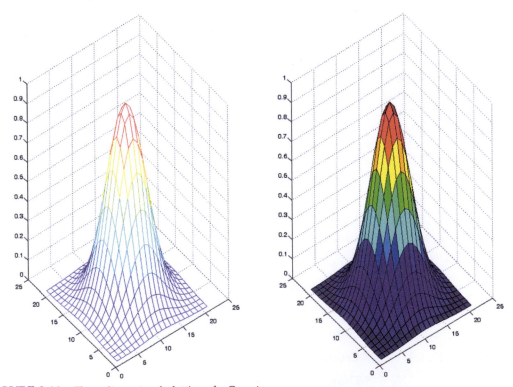

FIGURE 2.11   Three-dimensional plotting of a Gaussian.

---

**EXERCISE 2.23**

Can you improve the look of your figure? Try shading it in different ways by using the following:

**shading interp**

Now try the following:

**colormap hot**

---

**SUGGESTION FOR EXPLORATION**

As you can see, meshgrid is extremely powerful. With its help, you can visualize any quantity as a function of several independent variables. This capability is at the very heart of what makes MATLAB so powerful and appealing. Some say that one is not using MATLAB unless one is using meshgrid. While this statement is certainly rather strong, it does capture the central importance of the meshgrid function. We recommend trying to visualize a large number of functions to try and get a good handle on it. Start with something simple, such as a variable that is the result of the addition of a sine wave and a quadratic function. Use **meshgrid**, then **surf** to visualize it. This makes for a lot of very appealing graphs.

---

## 2.4.5 Interactive Programs

Many programs that are actually useful crucially depend on user input. This input comes mostly in one of two forms: from the mouse or from the keyboard. We will explore both forms in this section.

First, create a program that allows you to draw lines. Open a new program in the editor, write the following code, then save and run the program:

```
figure              %Opens a new figure
hold on;            %Make sure to draw all lines in same figure
xlim([0 1])         %Establish x-limits
ylim([0 1])         %Establish y-limits
for ii = 1:5        %Start for-loop. Allow to draw 5 lines
a = ginput(2);      %Get user input for 2 points
plot(a(:,1),a(:,2));  %Draw the line
end                 %End the loop
```

The program will open a new figure and then allow you to draw five lines. When the cross-hairs appear, click the start point of your line and then on the end point of your line. Repeat until you're done. The result should look something like that shown in Figure 2.12.

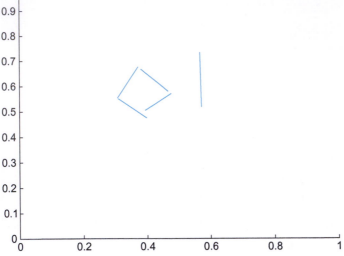

FIGURE 2.12    The luck of the draw.

---

### EXERCISE 2.24

Allow the program's user to draw 10 lines, instead of five.

---

### EXERCISE 2.25

Allow the user to draw "lines" that are defined by three points instead of two.

---

Remember to use **close all** if you opened too many figures.

Most user input will likely come from the keyboard, not the mouse. So let's create a little program that exemplifies user input very well. In effect, we are striving to re-create the "sugar factory" experiment by Berry and Broadbent (1984). In this experiment, research participant were told that they are the manager of a sugar factory and instructed to keep sugar output at 12,000 tons per month. They were also told that their main instrument of steering the output is to determine the number of workers per month. The experiment showed that participants are particularly bad at controlling recursive systems. Try this exercise on a friend or classmate (after you're done programming it). Each month, you ask the participant to indicate the number of workers, and each month, you give feedback on the production so far.

Here is the code:

```
P = [ ];                %Assigning an empty matrix. Making P empty.
a0 = 6000;              %a0 is 6000;
m0 = 0;                 %m0 is 0;
w0 = 300;               %w0 is 300;
P(1,:) = [m0, w0, a0];  %First production values
figure                  %Open a new figure
plot(0,a0,'.', 'markersize', 24);   %Plot the first value
hold on;                %Make sure that the plot is held
xlim([0 25])            %Indicate the right x-limits
ii = 1;                 %Initialize our counter
while ii < 25           %The participant is in charge for 24 months = 2 years.
P                       %Show the production values thus far
a = input('How many workers this month?')   %Get the user input
b = 20 * a - a0         %This is the engine. Determines how much sugar is produced
a0 = b;                 %Assign a new a0
plot (ii,a0,'.', 'markersize', 24);   %Plot it in the big figure
P(ii + 1,:) = [ii, a, b];  %Assign a new row in the P(roduction) matrix
plot (P(:,1),P(:,3),'color','k');   %Connect the dots
ii = ii + 1;            %Increment counter
end                     %End loop
```

The result (of a successful participant) should look something like that shown in Figure 2.13.

---

**EXERCISE 2.26**

Add more components to the production term, like a trend that tends to increase production over time (efficiency) or decrease production over time (attrition).

---

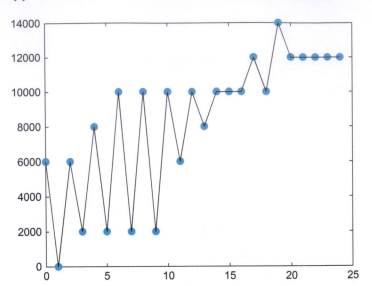

FIGURE 2.13   Game over.

---

**EXERCISE 2.27**

Add another plot (a subplot) that tracks the development of the workforce (in addition to the development of production; refer to Figure 2.13).

---

## 2.5 DATA ANALYSIS

Section 2.4.5 described a good way to get data into MATLAB: via user input. Conversely, this section is concerned with data analysis after you already have data. One of the primary uses of MATLAB in experimental neuroscience is the analysis of data.

### 2.5.1 Importing and Storing Data

Of course, data analysis is fun only if you already have large amounts of data. Cases in which you will have to manually enter the data before analyzing them will (we hope) be rare. For this scenario, suppose that you are in the marketing department of a major motion picture studio. You just produced a series of movies and asked people how they like these movies.

Specifically, the movies are *Matrix I*, *Matrix II: Matrix Reloaded*, and *Matrix III: Matrix Revolutions*. You asked 1603 people how much they liked any of these movies. They were instructed to use a nine-point scale (0 for awful, 4 for great and everything in between, in 0.5 steps). Also, they were instructed to abstain from giving a rating if they hadn't seen the movie. Now you will construct a program that analyzes these data, piece by piece. So open a new program in the editor and then add commands as we add them in our discussion here.

**Data import**: Download the data from the companion website to a suitable directory in your hard disk. Try using a directory that MATLAB will read without additional specifications of a path (file location) in the following code. First, import the data into MATLAB. To do this, add the following pieces of code to your new analysis program:

```
%Data import

M1 = xlsread('Matrix1.xls')  %Importing data for Matrix I

M2 = xlsread('Matrix2.xls')  %Importing data for Matrix II
```

These commands will create two matrices, M1 and M2, containing the ratings of the participants for the respective movies (contained in Excel files). Type **M1** and **M2** to inspect the matrices. You can also click on them in the workspace to get a spreadsheet view. One of the things that you will notice quickly is cells that contain "NaN." These are responders that didn't see the movie or didn't give a rating for this movie for other reasons. In any case, you don't have ratings for these and MATLAB indicates "NaN," which means "not a number"—or empty, in our case. The problem is that retaining this entry will defeat all attempts at data analysis if you don't get rid of these missing values. For example, try a correlation:

```
>> corrcoef(M1,M2)
ans =
   NaN   NaN
   NaN   NaN
```

You want to know how much an average person likes *Matrix II* if he or she saw *Matrix I* and vice versa. Correlating the two matrices is a good start to answering this question. However, the correlation function (**corrcoef**) in MATLAB assumes two vectors that consist only of numbers, not NaNs. A single NaN somewhere in the two vectors will render the entire answer NaN. This result is not very useful. So the first thing you need to do is to prune the data and retain only those people that gave ratings for both movies.

**Data pruning:** There are many ways of pruning data, and the way that we're suggesting here is certainly not the most efficient one. It does, however, require the least amount of introduction of new concepts and is based on what you already know, namely loops. As a side note, loops are generally slow (compared to matrix operations); therefore, it is almost always more efficient to substitute the loop with such an operation, particularly when calculating things that take too long with loops. We'll discuss this issue more later. For now, you should be fine if you add the following code to the program you already started:

```
%Data pruning
Movies = [ ];               %New Movies variable. Initializing it as empty.
temp = [M1 M2];             %Create a joint temporary Matrix, consisting of two long vectors
k = 1;                      %Initializing index k as 1
for ii = 1:length(temp)    %Could have said 1603, this is flexible. Start ii loop
if isnan(temp(ii,1)) == 0 & isnan(temp(ii,2)) == 0    %If both are numbers (=valid)
   Movies(k,:) = temp(ii,:);   %Fill with valid entries only
```

```
   k = k + 1;              %Update k index only in this case
   end                     %End if clause
end                        %End for loop
```

The **isnan** function tests the elements of its input. It returns 1 if the element is not a number and returns 0 if the element is a number. By inspecting M1, you can verify visually that M1(2,1) is a number but that M1(3,1) is not. So you can test the function by typing the following in the command window:

```
>> isnan(M1(2,1))
ans =
   0
>> isnan(M1(3,1))
ans =
   1
```

Recall that **&** is the MATLAB symbol for logical AND. The symbol for logical OR is |. So you are effectively telling MATLAB in the **if** statement that you want to execute the statements it contains only if both vectors contain numbers at that row using **isnan** in combination with **&**.

---

### EXERCISE 2.28

What would have happened if you had made everything contingent on the index *ii*, instead of declaring another specialized and independent index *k*? Would the program have worked?

---

It's time to look at the result. In fact, it seems to have worked: There is a new matrix, "Movies," which is 805 entries long. In other words, about half the people in the survey report to have seen both movies.

After these preliminaries (data import and data pruning), you're ready to move to data analysis and the presentation of the results. The next step is to calculate the correlation you were looking for before, so add that to the code:

**corrcoef(Movies(:,1),Movies(:,2)) %Correlation between Matrix I and Matrix II**

The correlation is 0.503. That's not substantial, but not bad, either. The good news is that it's positive (if you like one, you tend to like the other) and that it's moderately large (definitely not 0). To get a better idea of what the correlation means, use a scatterplot to visualize it:

**figure %Create a new figure**
**plot(Movies(:,1), Movies(:,2),'.', 'markersize', 24) %Plot ratings vs. each other**

The result looks something like that shown in Figure 2.14.

The problem is that the space is very coarse. You have only nine steps per dimension—or 81 cells overall. Since you have 805 ratings it is not surprising that almost every cell is taken by at least one rating. This plot is clearly not satisfactory. We will improve on it later. The white space on the top left of the figure is, however, significant. It means that there was no one in the sample who disliked the first *Matrix* movie but liked the second one. The opposite seems to be very common.

Let's look at this in more detail and add the following line to the code:

**averages = mean(Movies) %Take the average of the Movie matrix**

**mean** is a MATLAB function that takes the average of a vector. It is not the opposite of the "nice" function, which is undefined. The **averages** variable contains both means.

As it turns out, the average rating for *Matrix I* is 3.26 (out of 4), while the average rating for *Matrix II* is only about 2.28. Figure 2.14 makes sense in light of these data. This can be further impressively illustrated in a bar graph, as shown in Figure 2.15.

However, this graph doesn't tell about the variance among the means. Let's rectify this in a quick histogram. Now add the following code:

```
figure                  %Open new figure
subplot(1,2,1)          %Open new subplot
hold on;                %Hold the plot
hist(Movies(:,1),9)     %Matrix I data. 9 bins is enough, since we only have 9 ratings
histfit(Movies(:,1),9)  %Let's fit a gaussian
xlim([0 4]);            %Let's make sure that plotting range is fine
title('Matrix I')       %Add a title
subplot(1,2,2)          %Open new subplot
```

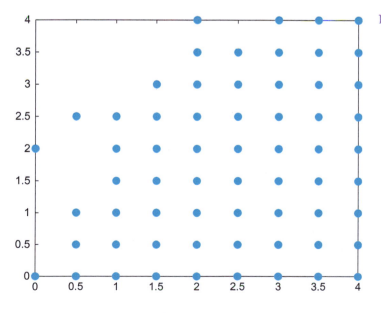

FIGURE 2.14    Low resolution.

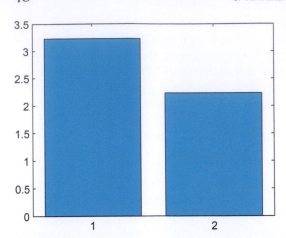

FIGURE 2.15    Means.

```
hold on;                  %Hold the plot
hist(Movies(:,2),9)       %Matrix II data. 9 bins is enough, since we only have 9 ratings
histfit(Movies(:,2),9)    %Let's fit a gaussian
xlim([0 4]);              %Let's make sure that plotting range is fine
title('Matrix II: reloaded') %Add a title
```

As you can see in Figure 2.16, it looks as though almost everyone really liked the first *Matrix* movie, but the second one was just okay (with a wide spread of opinion). Plus, fewer people actually report having seen the second movie.

The last thing to do—for now—is to fix the scatterplot that you obtained in Figure 2.14. You will do that by using what you learned about surface plots, keeping in mind that you will have only a very coarse plot ($9 \times 9$ cells).

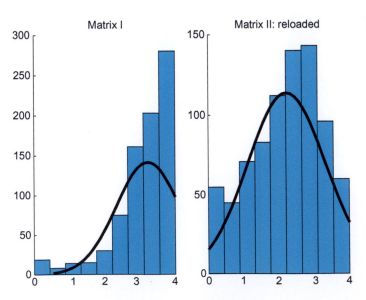

FIGURE 2.16    Variation.

Nevertheless, add the following code to the program:

**MT1 = (Movies(:,1)*2) + 1; %** Assign a temporary matrix, multiplying ratings by 2 to get
**MT2 = (Movies(:,2)*2) + 1; %** integral steps and adding 1 matrix indices start w/1, not 0.
**c = zeros(9,9); %** Creates a matrix "c" filled with zeros with the indicated dimensions
**ii = 1;      %** Initialize index
**for ii = 1:length(Movies) %** Start ii loop. This loop fills c matrix with movie rating counts
**c(10-MT1(ii,1),MT2(ii,1)) = c(10-MT1(ii,1),MT2(ii,1)) + 1; %** Adding one in the cell count
**end %** End loop
**figure %** New figure
**surf(c) %** Create a surface
**shading interp %** Interpolate the shading
**xlabel('Ratings for matrix I') %** Label for the x-axis
**ylabel('Ratings for matrix II: reloaded') %** Label for the y-axis
**zlabel('Frequency') %** Get in the habit of labeling your axes.

The result looks rather appealing—something like that shown in Figure 2.17. It gives much more information than the simple scatterplot shown previously—namely, how often a given cell was filled and how often a given combination of ratings was given.

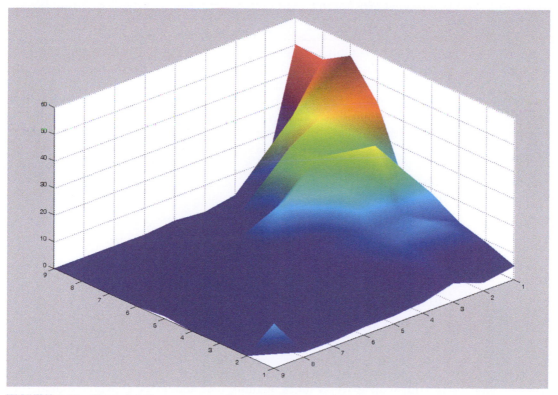

FIGURE 2.17   The real deal.

---

**EXERCISE 2.29**

Import the data for the third *Matrix* movie, prune it, and include it in the analysis. In particular, explore the relations between *Matrix I* and *Matrix III* and between *Matrix II* and *Matrix III*. The plots between *Matrix II* and *Matrix III* are particularly nice.

---

Can you now predict how much someone will like *Matrix II*, given how much he or she liked *Matrix I*? It looks as though you can. But the relationship is much stronger for *Matrix II−III*.

## 2.6  A WORD ON FUNCTION HANDLES

Before we conclude, it is worthwhile to mention function handles, as you will likely need them—either in your own code or when interpreting the code of others.

In this tutorial, we talked a lot about functions. Mostly, we did so in the context of the arguments they take. Up to this point, the arguments have been numbers—sometimes individual numbers, sometimes sequences of numbers—but they were always numbers.

However, there are quite a few functions in MATLAB that expect other *functions* as part of their input arguments. This concept will take a while to get used to if it is unfamiliar from your previous programming experience, but once you have used it a couple of times, the power and flexibility of this hierarchical nestedness will be obvious.

There are several ways to pass a function as an argument to another function. A straightforward and commonly used approach is to declare a function handle. Let's explore this concept in the light of specific examples. Say you would like to evaluate the sine function at different points. As you saw previously, you could do this by just typing

**sin(*x*)**

where *x* is the value of interest.

For example, type

**sin([0 pi/2 pi 3/2\*pi 2\*pi])**

to evaluate the sine function at some significant points of interest.

Predictably, the result is

**ans =**
   0   1.0000   0.0000   −1.0000   −0.0000

Now, you can do this with function handles. To do so, type

**h = @sin**

You now have a function handle *h* in your workspace. It represents the sine function. As you can see in your workspace, it takes memory and should be considered analogous to other handles that you have already encountered, namely figure handles.

The function **feval** evaluates a function at certain values. It expects a function as its first input and the values to-be-evaluated as the second. For example, typing

**feval(h,[0 pi/2 pi 3/2\*pi 2\*pi])**

yields

**ans =**
  0   1.0000   0.0000   − 1.0000   − 0.0000

Comparing this with the previous result illustrates that passing the function handle worked as expected.

You might wonder what the big deal is. It is arguably as easy—if not easier—to just type the values directly into the **sin** function than to formally declare a function handle.

Of course, you would be right to be skeptical. However—at the very least—you will save time typing when you use the same function over and over again—given that you use function handles that are shorter than the function itself. Moreover, you can create more succinct code, which is always a concern as your programs get longer and more intricate.

More importantly, there are functions that actually do useful stuff with function handles. For example, **fplot** plots a given function over a specified range. Typing

**fplot(h,[0 2\*pi])**

should give you a result that looks something like that shown in Figure 2.18.

Now let's consider another function that expects a function as input. The function **quad** performs numeric integration of a given function over a certain interval. You need a way to tell **quad** which function you want to integrate. This makes **quad** very powerful

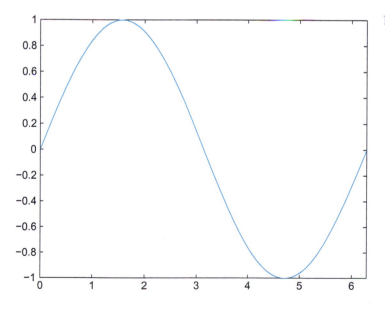

FIGURE 2.18   fplot in action.

because it can integrate any number of functions (as opposed to your writing a whole library of specific integrated functions).

Now integrate the sine function numerically. Conveniently, you already have the function handle $h$ in memory. Then type

>> **quad(h,0,pi)**
ans =
    **2.0000**
>> **quad(h,0,2*pi)**
ans =
    **0**
>> **quad(h,0,pi/2)**
ans =
    **1.0000**

After visually inspecting the graph in Figure 2.18 and recalling high school calculus, you can appreciate that the **quad** function works on the function handle as intended.

In addition, you can not only tag pre-existing MATLAB functions, but also declare your own functions and tag them with a function handle, as follows:

>> **q = @(x) x.^5 − 9.* x^4 + 8 .* x^3 − 2.* x.^2 + x + 500;**

Now you have a rather imposing polynomial all wrapped up and neatly tucked away in the function handle $q$. You can do whatever you want with it. For example, you could plot it as follows:

>> **fplot(q,[0 10])**

The result is shown in Figure 2.19.

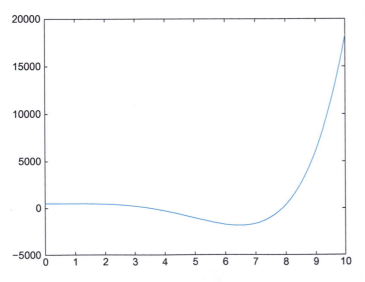

FIGURE 2.19 A polynomial in $q$, plotted from 0-10.

---

### EXERCISE 2.30

Try integrating a value of the polynomial. Does the result make sense?

---

### EXERCISE 2.31

Do everything you just did, but using your own functions and function handles.

Try declaring your own functions and evaluating them, e.g. "vice" or "virtue".

---

### SUGGESTION FOR EXPLORATION

Find another function that takes a function handle as input by using the

MATLAB help function. See what it does.

---

Finally, you can save your function handles as a workspace. This way, you can build your own library of functions for specific purposes.

As usual, there are many ways to do the same thing in MATLAB. As should be clear by now, function handles are a convenient and robust way to pass functions to other functions that expect functions as an input.

## 2.7 THE FUNCTION BROWSER

Since release 2008b (7.7), MATLAB contains a function browser. This helps the user to quickly find—and appropriately use—MATLAB functions. The introduction of this feature is timely. MATLAB now contains thousands of functions, most of which are rarely used. Moreover, the number of functions is still growing at a rapid pace, particularly with the introduction of new toolboxes. Finally, the syntax and usage of any given function may change in subtle ways from one version to the next.

In other words—and to summarize—even experts can't be expected to be aware of all available MATLAB functions as well as their current usage and correct syntax. A crude but workable solution up to this date has been to constantly keep the MATLAB "Help Navigator" open at all times. This approach has several tangible drawbacks. First, it takes up valuable screen real estate. Second, it necessitates switching back and forth between what are essentially different programs, breaking up the workflow. Finally, the Help Navigator window requires lots of clicking, copying and pasting and the like. It is not as well integrated in the MATLAB software as one would otherwise like.

The new "function browser" is designed to do away with these drawbacks. It is directly integrated into MATLAB. You can now see this in the form of a little *fx* that is placed just to the left of the command prompt, at the far left edge of the command window. Clicking on it (or pressing Shift and F1 at the same time) opens up the browser. Importantly, the functions are grouped in hierarchical categories, allowing you to find particular functions even if you are not aware of their name (such as plotting functions). The hierarchical trees can be rather deep, first distinguishing between MATLAB and its Toolboxes, then between different function types (e.g., Mathematics vs. Graphics) and then particular subfields thereof. Of course, the function browser also allows to search for functions by name. Type something in the search function field provides a quick list of functions that match the string that was inputted in the field. The list of functions also gives a very succinct but appropriate short description of what the function does. Hovering over a given entry with the cursor brings up a popup window with a more elaborate description of the function and its usage.

Finally, the function browser allows to drag and drop a given function from the browser into the command window.

Figure 2.20 illustrates the use of the function browser for a function introduced in this chapter, **isnan**.

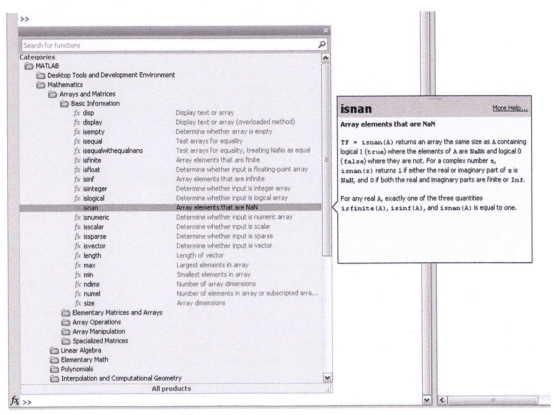

FIGURE 2.20    The function browser.

## 2.8 SUMMARY

This tutorial introduced you to the functionality and power of MATLAB. MATLAB contains a large number of diverse operators and functions, covering virtually all applied mathematics, with particularly powerful functions in calculus and linear algebra. If you would like to explore these functions, the MATLAB **help** function provides an excellent starting point. You can summon the help with the **help** command. Of course, you will encounter many useful functions in the sections to follow. Before we move on, a brief word on errors. You have probably encountered your fair share of errors at this point. Try to embrace them. Errors are not mistakes. Errors are just MATLAB's way of saying that it did not understand a particular input. It is doing you a favor by pointing out what is wrong (even if some error messages can be quite cryptic). While this can be a trying experience, error messages help to improve the code. The same is not the case for mistakes. If you make a mistake in the program (if there is logical problem), MATLAB won't say anything, but the program won't be doing what you think it is doing. This is a real problem. To put this differently: Mistakes are errors that are not caught. Having MATLAB throw errors is frustrating, but it is better than the alternative. Speaking of frustration...

Try not to get too frustrated with MATLAB while learning the program and working on the exercises. If things get rough and the commands you entered don't produce the expected results, know that MATLAB is able to provide much needed humor and a succinct answer to why that is. Just type in the command **why**.

## MATLAB FUNCTIONS, COMMANDS, AND OPERATORS COVERED IN THIS CHAPTER

| | | |
|---|---|---|
| **help** | load | cos |
| **helpwin** | **clear** | close |
| **helpdesk** | **length** | title |
| **helpbrowser** | **size** | set |
| **+** | **linspace** | **FaceColor** |
| **−** | **logspace** | **linewidth** |
| **\*** | **'** | **rref** |
| **/** | **./** | **loglog** |
| **()** | **.\*** | **semilogx** |
| **^** | **.^** | **semilogy** |
| **log** | **find** | **stairs** |
| **exp** | **==** | **pie** |

| sin | ~= | sound |
|---|---|---|
| pi | < | function |
| format | > | for |
| e | <= | while |
| [ ] | >= | end |
| : | & | % |
| ; | \| | if |
| = | ~ | else |
| eye | xor | pause |
| ones | any | subplot |
| zeros | all | surf |
| rand | plot | mesh |
| randn | bar | meshgrid |
| who | hist | shading |
| whos | figure | colormap |
| save | hold | xlim |
| ylim | isnan | @ |
| ginput | histfit | quad |
| markersize | xlabel | why |
| corrcoef | feval | doc |
| xlsread | fplot | |

# Mathematics and Statistics Tutorial

## 3.1 INTRODUCTION

Not everyone who intends to start practicing the neurosciences can be expected to do so with a perfect knowledge of mathematics. As a matter of fact, due to the inherently interdisciplinary nature of the field, it would be quite surprising if this were the case. Educational backgrounds are diverse, and not everyone is privileged enough to hail from a "math-heavy" one as afforded by physics, computer science, or engineering (to be sure, plenty do, but they tend to be similarly challenged by a lack of prior exposure to biological concepts). This state of affairs can usually be traced to the quality and style of the typical high school and college education in mathematics, which tends to be heavy on proofs and rote memorization of formulae, but falls short on good explanations that could foster conceptual understanding, visualization, and establishing a working knowledge that allows problem solving. In reality, the only thingk most people actually learn (in terms of long term retention) from their high school education in mathematics is that the field is deeply foreign and full of alien and intimidating topics that can trigger deep-seated insecurities. But people do learn, so most usually stay clear of math-heavy fields after an initial negative exposure if they can help it, further solidifying the deficiency. Worse than just the absence of knowledge, many people are actively avoiding math. In our information-based society, few admissions of ignorance are received with such impunity and, indeed, acclaim as that of "not getting math." Math phobia has swept wide parts of the population and is flaunted as a badge of honor. The biological sciences are not immune from this; citations of a paper drop 35% for every additional equation per page (Fawcett et al., 2009). Yet a solid and workable knowledge of some key mathematical concepts is absolutely indispensable if one is to follow and partake in contemporary neuroscience research. There is no question that not overcoming the acquired fear of math will be severely limiting if not debilitating to the budding researcher, a state of affairs that will only get more severe as the mathematization of neuroscience progresses relentlessly. Such self-limitation is needless, and it is a shame that droves of budding researchers trying to uncover answers to questions they care about passionately find themselves in this situation without any fault of their own.

Thus, the purpose of this tutorial is largely therapeutic in nature. We will focus on introducing a few key concepts in linear algebra and statistics that are central to neuroscience research. We will do so in the most gentle and affirming way possible. In the process, the reader will (hopefully) realize how MATLAB® itself can be used to help overcome math anxiety. One piece of advice upfront: If you know that looking at an equation raises your blood pressure, there is a straightforward trick to calm the nerves. Simply *translate* the equation into a series of MATLAB commands. Every equation ultimately corresponds to a couple of lines of code. Once you get familiar with MATLAB, even the most intimidating looking equation will lose its sting.

In addition to this primary goal, we will also set up the mathematical groundwork for the math that is used in the rest of the chapters, so that there are no bad surprises later on.

If you feel already sufficiently steeled in the art and practice of mathematics, you can safely skip this tutorial. If you are on the fence, you probably need the reminder (in spite of the central importance of explanations, math is effectively a motor skill; it benefits tremendously from practice).

We explicitly focus on a *gentle introduction* here, as it serves our purposes. If you are in need of a more rigorous or comprehensive treatment, we refer you to *Mathematics for Neuroscientists* by Gabbiani and Cox. If you want to see what math education *could* be like, centered on great explanations that build intuition, we recommend *Math, Better Explained* by Kalid Azad.

## 3.2 LINEAR ALGEBRA

Linear algebra is as fitting a topic as any with which to start this tutorial. As it so happens, the central concept of linear algebra, the matrix, is also the principal data structure underlying MATLAB itself. MATLAB is at its best when it comes to the manipulation of matrices. Linear algebra is, broadly speaking, the study of matrix manipulations.

But what is a matrix and why is it so central? Didn't we get in enough trouble when we started to mix the alphabet into equations back in middle school? What does the concept buy us? Why is it a suitable representation, and of what exactly?

We will discuss these issues in turn.

### 3.2.1 Matrices, Vectors, and Arrays

To avoid confusion, we need to clarify some concepts and the terms we use to reference these concepts. In linear algebra, the term *scalar* refers to a nondimensional quantity, whereas *values* commonly refers to vectors, matrices, or arrays. Informally, the terms matrix, vector, and array are sometimes used interchangeably, but more formally, an array is a set of numbers organized by a finite number of fixed sized dimensions. Within MATLAB, the term *array* can also denote a data structure, a set of numeric values. However, in this tutorial, we will use "array" in its mathematical sense.

A matrix is a two-dimensional array of numbers or variables. Matrices are usually depicted as a rectangular group of numbers, with rows and columns corresponding to the two dimensions. If there are more than two dimensions, we would call it a "tensor," but let's not get into that at this point. The sizes of the two dimensions of a matrix are often written as $m \times n$, where $m$ indicates the number of rows and $n$ indicates the number of columns. Here is an example of a $2 \times 3$ matrix, $A$:

$$A = \begin{pmatrix} 2 & 4 & 8 \\ 1 & 7 & 3 \end{pmatrix}$$

In MATLAB, we use square brackets for defining a matrix. The following MATLAB code defines a matrix $A$ with the above values. The matrix name is usually capitalized.

```
>> A = [2  4  8; 1  7  3]
A =
   2  4  8
   1  7  3
```

The entire content of the matrix is contained by the square brackets, and a semicolon is used to separate the rows.

In contrast to a matrix, a vector is a one-dimensional array of numbers or variables. An individual row or column of a matrix can be identified as a row vector or a column vector. Here is an example row vector $B$ from matrix $A$, above:

$$B = (2 \quad 4 \quad 8)$$

Just like matrices, vectors are also entered into MATLAB with square brackets.

```
B = [2  4  8];
```

Note that no semicolons appeared within the square brackets for the definition of $B$. This was because $B$ has only a single row.

You can refer to a particular element in a matrix by its row and column placement. So, for the matrix $A$, the element in the first row and third column is the number 8. These two values identifying an element within the matrix are called *indices*. Likewise, an element of a vector can be identified with a single index.

In MATLAB, indices can be specified using parentheses to select elements from matrices or vectors. For example,

```
A(1,3)
```

A vector would need only one index.

```
B(2)
```

Some matrices are special and can be categorized further. We will refer to these definitions in the following sections.

*Square* matrices are those matrices where both dimensions are equal. Square matrices in which only the values along the main diagonal are non-zero are called *diagonal* matrices. Here is an example:

$$C = \begin{pmatrix} 3 & 0 & 0 \\ 0 & 7 & 0 \\ 0 & 0 & 1 \end{pmatrix}$$

Finally, diagonal matrices where all the non-zero values are 1 are termed *identity* matrices. The capital letter *I* is usually reserved in linear algebra for representing identity matrices. Here is an example of a $4 \times 4$ identity matrix:

$$I = \begin{pmatrix} 1 & 0 & 0 & 0 \\ 0 & 1 & 0 & 0 \\ 0 & 0 & 1 & 0 \\ 0 & 0 & 0 & 1 \end{pmatrix}$$

Often in linear algebra, an identity matrix is referred to as "the identity matrix," with dimension inferred from context, and subsequent sections will adhere to this convention. MATLAB has a special function for creating an identity matrix of any desired size: **eye(n)**, where *n* is the dimension desired.

```
>> eye(3)
ans =
   1  0  0
   0  1  0
   0  0  1
```

There is a reason MATLAB has its own function for the identity matrix. It plays a central role in linear algebra, as will become clear in the rest of this tutorial.

### 3.2.2 Transposition

One common operation on matrices is transposition. Transposition flips rows and columns; each row of the original matrix becomes the corresponding column of the new matrix. In mathematical notation, transposition is usually indicated with the superscript *t*. Here is how we would write the transposition of the matrix *A* defined in the previous section.

$$A^t = \begin{pmatrix} 1 & 2 \\ 7 & 4 \\ 3 & 8 \end{pmatrix}$$

You carry out this operation in MATLAB by using the punctuation ' after the matrix name: In this case, *A'*.

```
>> A'
ans =
   1  2
   7  4
   3  8
```

These preliminaries might seem excessive, but a precise nomenclature of operations matters a lot in linear algebra. It will soon become obvious why.

### 3.2.3 Addition

Addition is an operation that is defined for two matrices or two vectors of the same dimensionality. Adding matrices algebraically is adding corresponding components to form a new matrix. Thus, each of the two matrices or two vectors being added must contain only elements that correspond with those in the other. Because addition is defined only for cases where the two values being added have the same dimensionality, cases where the dimensionality differ, such as adding $A$ from the previous section with its transpose, would be termed *undefined* or *meaningless*. But you don't actually have to worry about this. MATLAB will literally not let you add matrices with differing dimensionalities; it will complain that there has been an error and that "Matrix dimensions must agree," all red and bothered (unless you changed the color preferences).

Here is an example of matrix addition. We define the matrix $F$ as

$$F = \begin{pmatrix} 10 & 20 & 30 \\ 5 & 10 & 15 \end{pmatrix}$$

$$A + F = \begin{pmatrix} 2 & 4 & 8 \\ 1 & 7 & 3 \end{pmatrix} + \begin{pmatrix} 10 & 20 & 30 \\ 5 & 10 & 15 \end{pmatrix} = \begin{pmatrix} 12 & 24 & 38 \\ 6 & 17 & 18 \end{pmatrix}$$

---

**EXERCISE 3.1**

Add $A + A$ in MATLAB.
Add $A + F$ in MATLAB.
Add $F + F'$ in MATLAB.

---

### 3.2.4 Scalar Multiplication

If this is starting to look to you like we are retracing our steps from elementary school, you would be right. All operations you learned in first grade for actual numbers have their corresponding operation for arrays in linear algebra (except for transpose; that wouldn't make any sense for scalars, as each scalar is its own transpose, so we mercifully skipped that in elementary school; now you know). Note that if you add $A$ to itself, as in Exercise 3.1, the resulting matrix has values double to those of the corresponding values in $A$. This suggests a simple definition for the scalar multiplication of a matrix. Indeed, when a matrix is multiplied by a scalar value, *each* element of the matrix is simply multiplied by that number.

$$5A = 5 \begin{pmatrix} 2 & 4 & 8 \\ 1 & 7 & 3 \end{pmatrix} = \begin{pmatrix} 5 \cdot 2 & 5 \cdot 4 & 5 \cdot 8 \\ 5 \cdot 1 & 5 \cdot 7 & 5 \cdot 3 \end{pmatrix} = \begin{pmatrix} 10 & 20 & 40 \\ 5 & 35 & 15 \end{pmatrix}$$

In MATLAB, a scalar multiplication is performed with an asterisk, if one of the multiplicants is a scalar number.

```
>> 5*A
ans =
    10   20   40
     5   35   15
```

---

**EXERCISE 3.2**

---

Evaluate 7F in MATLAB, using the matrices A and F defined in the previous section. Evaluate 2A + 3F in MATLAB.

---

### 3.2.5 Matrix Multiplication

So far, so simple. But this is the precise point where things get hairy and the majority of students get lost with linear algebra. This is because matrix multiplication is the first point where the analogy to elementary school math starts to break down. As you already learned elementary school math, this highly practiced cognitive template will start to interfere with learning this crucial step. We urge you to pay extreme attention to matrix multiplication and practice it as much as you can to override your strong cognitive priors. As most of linear algebra crucially hinges on matrix multiplication, this dire warning is not overstated. This is the point where you most likely will get lost, if you get lost. So proceed with the utmost care.

Multiplication can also be defined for two matrices or for two vectors. When you multiply two matrices together, $AB$, each element of the resulting matrix, $C$, is the sum of the corresponding row elements of A times the corresponding column elements of B. In other words, all elements of C may be obtained by using the following simple but perhaps counterintuitive rule (we are not big fans of rote memorization, but it pays to memorize this one by heart; otherwise, it will haunt you forever):

The element in row $i$ and column $j$ of the product matrix $AB$ is equal to the row $i$ of $A$ times the column $j$ of $B$, added.

Here is an example with two square matrices C and D.

$$CD = \begin{pmatrix} 1 & 2 \\ 3 & 4 \end{pmatrix} \begin{pmatrix} 5 & 6 \\ 7 & 8 \end{pmatrix} = \begin{pmatrix} 1\cdot 5 + 2\cdot 7 & 1\cdot 6 + 2\cdot 8 \\ 3\cdot 5 + 4\cdot 7 & 3\cdot 6 + 4\cdot 8 \end{pmatrix}$$

$$CD = \begin{pmatrix} 19 & 22 \\ 43 & 50 \end{pmatrix}$$

This definition constrains the dimensionality of the two matrices or vectors in a matrix multiplication. For two matrices A and B, the number of columns in A must match the number of rows in B for the product AB to be defined. Also, the dimensions of the product are $m \times n$, where $m$ is the number of rows in A and $n$ is the number of columns in B. If you try to multiply "incompatible" matrices (in terms of dimensionality), MATLAB won't let you do it and will inform you of this fact.

Matrix multiplication can occur between a vector and a matrix, provided that both meet the dimensionality constraints.

$$AB = \begin{pmatrix} 2 & 4 & 8 \\ 1 & 7 & 3 \end{pmatrix} \begin{pmatrix} 5 \\ 6 \\ 3 \end{pmatrix} = \begin{pmatrix} 2.5 + 4.6 + 8.3 \\ 1.5 + 7.6 + 3.3 \end{pmatrix} = \begin{pmatrix} 58 \\ 56 \end{pmatrix}$$

Observe that $AB$ is *not* the same as $BA$, throwing off elementary school intuitions. Unlike scalar multiplication, matrix multiplication is *not commutative*; in general, matrices do *not commute* under multiplication. In the mathematical sense, commuting has nothing to do with traveling to your place of business. It simply means, as stated above, that $AB$ is not the same as $BA$. It is extremely important to keep this property in mind when manipulating non-scalar values in algebraic equations.

In MATLAB, matrix multiplication also uses the asterisk like scalar multiplication, but both multiplicants are now non-scalars:

```
>> B = [1;6;3];
>> A*B
ans =
   58
   56
```

---

### EXERCISE 3.3

Verify the matrix product $CD$ above in MATLAB.

---

Much like in non-matrix multiplication where every number $N$ has a reciprocal such that $N \cdot \frac{1}{N} = 1$, matrix multiplication defines the concept of an inverse. However, unlike with scalar multiplication, only some matrices have inverses. We were not kidding when we mentioned that matrix multiplication is where the vanilla world of elementary school scalar multiplication is shattered.

The inverse of a matrix $D$, $D^{-1}$, is the matrix that, when multiplied with the original matrix, equals the identity matrix:

$$DD^{-1} = I$$

Note that this definition requires that the matrix $D$ be square. This falls out from the constraints of matrix multiplication and the definition of the identity matrix as a square matrix.

So, for example, if we define $D$ as

$$D = \begin{pmatrix} 2 & 3 \\ 5 & 7 \end{pmatrix}$$

then its inverse $D^{-1}$ is

$$D^{-1} = \begin{pmatrix} -7 & 3 \\ 5 & -2 \end{pmatrix}$$

---

### EXERCISE 3.4

Use MATLAB to demonstrate in an example that the matrix product $DD^{-1}$ is indeed the identity matrix.

---

For your convenience, MATLAB provides a function **inv(A)**, which calculates the inverse of a matrix.

>> **inv(D)**
ans =
   −7   3
    5   −2

As mentioned above, only square matrices have a defined inverse. Even among square matrices, not all have inverses. The MATLAB function **inv()** returns **Inf** in such cases. For instance, the matrix $X$ below looks completely innocent, but alas, it does not have an inverse.

>> **X = [2   3; 1/3   1/2]**
X =
   2     3
   0.3333   0.5
>> **inv(X)**
**Warning: Matrix is singular to working precision.**
ans =
   Inf   Inf
   Inf   Inf

As MATLAB's warning implies, such square matrices without inverses are termed *singular*. We will discuss criteria for assessing when a matrix has a defined inverse when we discuss determinants, in Section 3.2.6.

### 3.2.6 Geometrical Interpretation of Matrix Multiplication

In addition to linear *algebra*, there is also a corresponding *geometrical* interpretation of matrix-vector multiplication that can be extremely useful. First, see what happens when a vector is multiplied by a scalar. Suppose that

$$B = \begin{pmatrix} 3 \\ 4 \end{pmatrix}$$

You can plot the vector $B$ on the Cartesian plane if you assume that the $x$-component of the vector is the element in the first row and the $y$-component of the vector is the element in the second row. In such cases, we can define unit vectors $\hat{x} = \begin{pmatrix} 1 \\ 0 \end{pmatrix}$ and $\hat{y} = \begin{pmatrix} 0 \\ 1 \end{pmatrix}$. Therefore, the vector $B$ can be written in terms of simple and elementary unit-length *component vectors* as $B = 3\hat{x} + 4\hat{y} = 3\begin{pmatrix} 1 \\ 0 \end{pmatrix} + 4\begin{pmatrix} 0 \\ 1 \end{pmatrix}$. This can be readily seen by substituting the definitions for $\hat{x}$ and $\hat{y}$ into the equation for $B$:

$$B = 3\hat{x} + 4\hat{y}$$

$$B = 3\begin{pmatrix} 1 \\ 0 \end{pmatrix} + 4\begin{pmatrix} 0 \\ 1 \end{pmatrix}$$

$$B = \begin{pmatrix} 3 \\ 0 \end{pmatrix} + \begin{pmatrix} 0 \\ 4 \end{pmatrix}$$

$$B = \begin{pmatrix} 3 \\ 4 \end{pmatrix}$$

This decomposition in terms can be demonstrated in MATLAB as well.

```
>> x = [1; 0];
>> y = [0; 1];
>> 3*x + 4*y
ans =
    3
    4
>> B = 3*x + 4*y
B =
    3
    4
```

This results in the graph shown in Figure 3.1.

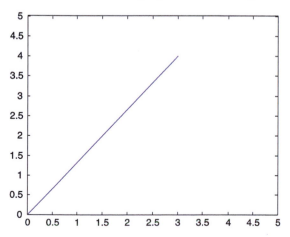

FIGURE 3.1    This figure shows the vector $B$ plotted in the $x,y$ coordinate space.

Next, you can multiply the vector $B$ by a scalar, 2, to get:

>> 2*B
ans =
    6
    8

If you plot this new vector alongside $B$, then you get the graph shown in Figure 3.2.

Notice that multiplying a vector by a scalar changes only its length. It does not change the direction of the vector. Now see what happens when a vector is multiplied by a matrix.

$$A = \begin{vmatrix} 1 & 1 \\ 4 & 1 \end{vmatrix}$$

---

**EXERCISE 3.5**

What is the product $A$ times $B$? Use MATLAB to calculate this product. Is it the same as $B$ times $A$?

---

Since the matrix $A$ is square, the product of $A$ and $B$ has the same dimensions as the vector $B$ (in this case, both are $2 \times 1$). Therefore, you can plot the vectors $A * B$ and $B$ on the same graph to obtain the result shown in Figure 3.3.

Here, you can see that multiplication of vector $B$ by the matrix $A$ has resulted in *rotating B* counterclockwise and stretching it out. Now, try another example, where $A$ is the same, but:

$$B = \begin{pmatrix} 1 \\ 2 \end{pmatrix}$$

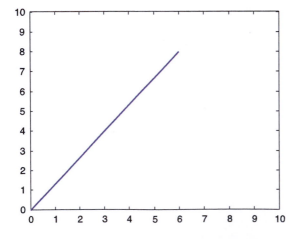

FIGURE 3.2   This graph shows the result of multiplying the vector $B$ by a scalar (the value 2).

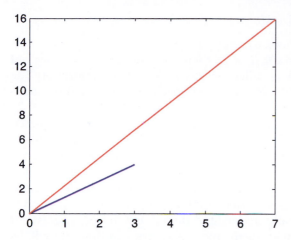

**FIGURE 3.3** Multiplying the vector $B$ (blue) by the matrix produces the rotated and rescaled vector in red.

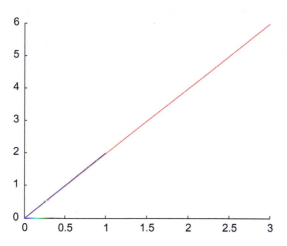

**FIGURE 3.4** Multiplying the matrix $A$ by the vector $B = (1\ 2)$ (in blue) produces a vector with the same direction but different magnitude (in red).

---

### EXERCISE 3.6

What is the product $A$ times $B$? Plot the product in MATLAB and $B$ on the same graph. What do you notice about the direction of the product, relative to the vector $B$? Can you express the product in terms of $B$ alone?

---

If you plot $B$ and $AB$ on the same graph, then you get the result shown in Figure 3.4. In this case, multiplication of the vector $B$ by the matrix $A$ is equivalent to multiplication of $B$ by a scalar—in this case 3. It turns out that this scenario is a general one. For many square matrices $A$, there exist corresponding vectors $B$ such that

$$AB = \lambda B$$

where $\lambda$ is a scalar constant. (So, in the previous example, $\lambda = 3$.)

Geometrically, this means that for a given matrix $A$, there is a vector $B$ that does not rotate when multiplied by $A$. The scalar $\lambda$ is called an *eigenvalue* of the matrix $A$. The invariant vector $B$ is called an *eigenvector* of the matrix $A$, and each eigenvector $B$ is associated with a particular eigenvalue $\lambda$.

While we mentioned this concept in passing, it pays to do a full stop and truly appreciate the concept. It is even more fundamental to linear algebra than matrix multiplication itself. There are quite a few people who spend their days calculating eigenvalues of systems (represented by matrices). Even worse, there are plenty of intellectual posers who fail this basic academic shibboleth; they tellingly refer to them as "Igon values" instead (Gladwell, 2009). You don't want to be that guy. Seriously, throwing buzzwords around is all fun and games, but we want you to actually understand them. Only if you understand these concepts can you meaningfully work with them, and we assure you that you will be dealing with eigenvalues and eigenvectors as long as you do linear algebra. And you will probably be doing linear algebra as long as you are doing science. As for how long you want to do science, that's up to you.

Back to eigenvectors, there is actually a relatively simple visual interpretation. Imagine rotating a globe around its axis (or imagine the actual planet earth spinning around its axis on a daily basis). The values on this axis are rotation invariant: They do not change when the system is rotated. You can imagine that these are special values and it is important to know them, as they characterize in a way what the system as a whole (the spinning earth) is doing. If the system was doing something else, the values would be different.

We do believe that in addition to this visual you get the best appreciation for eigenvectors and eigenvalues not by reading or writing about them, but by working with them, which is exactly what we will do in the next few sections, where we will discuss how to determine eigenvectors/eigenvalue pairs for square matrices and their applications. However, before we can cover eigenvectors and eigenvalues, we need to discuss the determinants of matrices.

### 3.2.7 The Determinant

As discussed earlier, only some square matrices actually have defined inverses. The *determinant* is a value (defined only for square matrices) that aids in determining whether a matrix has a defined inverse or not. In addition, the determinant aids in identifying whether a matrix has eigenvectors as well.

The definition of the determinant for larger matrices is complex, and, for completeness, we refer the reader to a suitable reference. For $2 \times 2$ matrices, however, the determinant is a relatively simple expression. Defining the matrix $A$ as

$$A = \begin{pmatrix} a & b \\ c & d \end{pmatrix}$$

the determinant is defined as $ad - bc$, multiplying then subtracting the values on the two diagonals of the matrix. The determinant is written in linear algebra as det() around a matrix or as vertical bars. The following equation shows the two notations and the value of the determinant for $2 \times 2$ square matrices.

$$\det \begin{pmatrix} a & b \\ c & d \end{pmatrix} = \begin{vmatrix} a & b \\ c & d \end{vmatrix} = ad - bc$$

MATLAB provides a function **det()** that calculates the determinant of a matrix for you.

---

**EXERCISE 3.7**

Use MATLAB to calculate the determinant of the matrix $D = \begin{pmatrix} 2 & 3 \\ 5 & 7 \end{pmatrix}$.

---

Now you know how to calculate it. But why would you want to? The value of the determinant can be used to determine (hence the name: It actually determines a range of other matrix properties as well) whether a square matrix has a defined inverse. A square matrix has a defined inverse if and only if the determinant of that matrix is nonzero (As we saw earlier that such matrices with zero valued determinants and no inverse are called singular.) This can be seen by attempting to determine the value of a matrix inverse analytically.

Let the matrices $A$ and $A^{-1}$ be defined as

$$A = \begin{pmatrix} a & b \\ c & d \end{pmatrix}, \qquad A^{-1} = \begin{pmatrix} e & f \\ g & h \end{pmatrix}$$

For $A^{-1}$ to be the inverse of $A$, $AA^{-1}$ must equal the identity matrix. This generates a set of equations in the elements of both matrices:

$ae + bg = 1$
$fa + bh = 0$
$ce + gd = 0$
$fc + hd = 1$

Ideally, we would want to represent the elements of $A^{-1}$ in terms of $a,b,c,d$; the elements of $A$. Starting with $ce + gd = 0$, the value of $e$ is

$$ce + gd = 0$$
$$ce = -gd$$
$$e = -\frac{d}{c}g$$

Substituting this value for $e$ into the equation $ae + bg = 1$ results in an equation using only one term from $A^{-1}$, $g$:

$$ae + bg = 1$$
$$a\left(-\frac{d}{c}g\right) + bg = 1$$

I. FUNDAMENTALS

$$\left(-\frac{ad}{c} + b\right)g = 1$$

$$\left(-\frac{ad}{c} + \frac{bc}{c}\right)g = 1$$

$$\left(\frac{bc - ad}{c}\right)g = 1$$

$$g = \frac{c}{bc - ad}$$

This yields an expression for the element $g$ of $A^{-1}$ solely in terms of elements of the original matrix $A$, and this process can be repeated for the other three elements of $A^{-1}$. However, in rearranging terms, the equation was divided by the term $bc - ad$, implying that this term must not be zero (as no one can divide by zero). You may recognize this term as the negative of the $2 \times 2$ determinant defined above, $bc - ad$. Thus, if the determinant is zero, the system of equations identified by the inverse has no solution. QED.

### 3.2.8 Eigenvalues and Eigenvectors

Recall that finding the eigenvalues and corresponding eigenvectors of a square matrix $A$ is equivalent to solving for scalar $\lambda$ and vector $B$ such that

$$AB = \lambda B$$

One valid but obviously degenerate solution to this equation is the zero vector,

$$B = \begin{pmatrix} 0 \\ 0 \end{pmatrix},$$

regardless of the matrix $A$, as long as $A$ is a $2 \times 2$ matrix. The zero-vector solution is called the *trivial solution* and will not be of interest here (it is rather of interest to philosophical discussions of mathematical conceptualizations of death). Thus, to limit solutions to the non-trivial solutions, we will require that any solutions for $B$ not be zero vectors.

The eigenvector equation is $AB = \lambda IB$, where $I$ is the identity matrix. This can be rearranged as:

$$AB = \lambda IB$$
$$AB - \lambda IB = 0$$
$$(A - \lambda I)B = 0$$

If the matrix $(A - \lambda I)$ has an inverse, then multiplying through this equation by the inverse gives:

$$(A - \lambda I)^{-1}(A - \lambda I)B = 0$$

Because a matrix multiplied by its inverse is the identity, this would imply that

$$(A - \lambda I)^{-1}(A - \lambda I)B = 0$$
$$IB = 0$$
$$IB = 0$$

which is exactly the trivial solution that you do NOT want, which means that $(A - \lambda I)$ must not have an inverse if nontrivial solutions for $B$ exist. Recall from the previous section that a matrix has no inverse if its determinant equals zero. So, for $(A - \lambda I)$, nontrivial solutions for $B$ exist only if $\det(A - \lambda I) = 0$. QED, yet again. This equation is called the *characteristic equation* of the matrix $A$. It is the only equation you need to calculate the eigenvalues and eigenvectors of a matrix.

Through the characteristic equation, we can solve for the eigenvalues of $A$, $\lambda$, and then we can use the values of $\lambda$ to determine the corresponding eigenvectors of $A$.

We use this now in an example calculation of eigenvectors and eigenvalues for the matrix

$$A = \begin{pmatrix} 1 & 1 \\ 4 & 1 \end{pmatrix}$$

$$A - \lambda I = \begin{pmatrix} 1 & 1 \\ 4 & 1 \end{pmatrix} - \lambda \begin{pmatrix} 1 & 0 \\ 0 & 1 \end{pmatrix}$$

$$A - \lambda I = \begin{pmatrix} 1 - \lambda & 1 \\ 4 & 1 - \lambda \end{pmatrix}$$

$$\det(A - \lambda I) = \begin{vmatrix} 1 - \lambda & 1 \\ 4 & 1 - 2 \end{vmatrix} = 0$$

$$(1 - \lambda)^2 - 4 = 0$$

You can solve the quadratic equation for $\lambda$ to get $\lambda = \{-1, 3\}$. These are the eigenvalues of the matrix $A$! You can solve for the corresponding eigenvectors as follows. For $\lambda = 3$, the equation becomes

$$AB = 3B$$

Substitute $A$ into the preceding equation and let:

$$B = \begin{pmatrix} x \\ y \end{pmatrix}$$

The preceding equation becomes:

$$\begin{pmatrix} 1 & 1 \\ 4 & 1 \end{pmatrix}\begin{pmatrix} x \\ y \end{pmatrix} = 3\begin{pmatrix} x \\ y \end{pmatrix} \Rightarrow \begin{pmatrix} x & + & y \\ 4x & + & y \end{pmatrix} = \begin{pmatrix} 3x \\ 3y \end{pmatrix}$$

Solving the system of equations gives $y = 2x$. Thus, any $B$ such that

$$B = \begin{pmatrix} x \\ 2x \end{pmatrix}$$

is an eigenvector of the matrix $A$ corresponding to the eigenvalue $\lambda = 3$. Note that this demonstrates that any vector with the same orientation will be invariant to changes in orientation imposed by multiplication by $A$ (recall the spinning globe).

---

### EXERCISE 3.8

Find the eigenvector of $A$ corresponding to the other eigenvalue, $\lambda = -1$.

---

In MATLAB, the command [V,D] = **eig**(A) will return two matrices: $D$ and $V$. The elements of the diagonal matrix $D$ are the eigenvalues of the square matrix $A$. The columns of the matrix $V$ are the corresponding eigenvectors.

---

### EXERCISE 3.9

Use the MATLAB function eig() to calculate the eigenvalues and eigenvectors of matrix $A$. Provided that eigenvectors and eigenvalues exist for some matrix $B$, the relationship $BV = VD$ holds for the matrices returned by eig(B). Demonstrate that this is the case for the matrix $A$ defined earlier. This relationship will be explored in depth in the next section.

---

## 3.2.9 Applications of Eigenvectors: Eigendecomposition

Here we will describe a powerful theorem called the *eigendecomposition theorem*. This theorem states:

For any $n \times n$ matrix A with distinct eigenvalues you can write:

$$A = VDV^{-1}$$

where $V$ is the square matrix whose columns are the eigenvectors of $A$, and $D$ is the square diagonal matrix formed by placing the eigenvalues of $A$ along the primary diagonal of $D$.

Powerful stuff indeed. Note that the matrices $V$ and $D$ are exactly those matrices returned by the MATLAB function eig()!

This theorem allows any matrix $A$ with distinct eigenvalues to be decomposed into a diagonal matrix. This decomposition is especially useful in cases where a matrix needs to be raised to a power:

$$A^N = (VDV^{-1})^N$$
$$A^N = (VDV^{-1})(VDV^{-1})(VDV^{-1})\ldots(VDV^{-1})$$
$$A^N = VDV^{-1}VDV^{-1}VDV^{-1}\ldots VDV^{-1}$$

Note here that each pair $VV^{-1}$ between diagonal matrices $D$ is equivalent to the identity matrix and thus drops out of the equation.

$$A^N = VDDD\ldots DV^{-1}$$
$$A^N = VD^NV^{-1}$$

Only the diagonal matrix $D$ is raised to the power $N$. Recall that a diagonal matrix raised to a power $N$ is exceptionally easily calculated (raise each element of the diagonal to the power $N$). Thus, raising a matrix with distinct eigenvalues to a large power becomes a far simpler calculation.

---

**EXERCISE 3.10**

---

Use the eigendecomposition theorem to calculate $A^4$, where $A$ is the matrix defined in the previous section. Verify that this is equal to $A^4$ by calculating the value in MATLAB.

---

## 3.2.10 Applications of Eigenvectors: PCA

The eigendecomposition theorem can be used in many remarkable ways. In this section, we will explore one application, principal component analysis (which we will revisit with practical examples in Chapter 17 of this book). Principal component analysis provides a means of identifying the independent axes responsible for major sources of variability in a multivariate sample. Once these axes are identified, the axes can be used for classification and simplification of the data. For instance, if two dimensions capture all the variability inherent in 200 dimensions, the data can be simplified to the "loading" of the data on those two dimensions. This will become clearer later on.

Let X be a set of data represented as an $m \times n$ matrix X, where m is the number of data points and n is the number of dimensions in the data set. For this data, we can calculate an $n \times n$ covariance matrix $\Sigma$. According to the eigendecomposition theorem, we can represent $\Sigma$ as $\Sigma = VDV^{-1}$. Under this reformulation, the eigenvectors form a new set of axes that indicate independent directions of variance. One can see this by a rearrangement of the equation:

$$\Sigma = VDV^{-1}$$
$$\Sigma V = VDV^{-1}V$$

$$\Sigma V = VD$$

$$V^{-1}\Sigma V = V^{-1}VD$$

$$V^{-1}\Sigma V = D$$

Thus, through the rotation and scaling of the eigenvector matrix, the original covariance matrix can be transformed into a diagonal covariance matrix, eliminating covariance altogether.

Equally significantly, if the eigenvectors are normalized, then the eigenvalues indicate the relative contribution of each of the eigenvectors to the covariance matrix. For large datasets, the relative weights provided by the eigenvalues can be used to reduce the dimensions of the data.

We will use an example.

Load the data file **data.mat**.

You'll notice that the included matrix is a $50 \times 3$ matrix of data, corresponding to 50 samples of a three-valued vector quantity. Because this data has more than two dimensions, visualizing this data is fairly difficult. We will use PCA to remap the data to new axes that better represent the variance of the data.

First, we can use the MATLAB **cov()** function to generate the covariance matrix:

```
>> cov(M)
C =
    0.8874   1.772   0.054
    1.772    3.544   0.114
    0.0542   0.114   0.676
```

Next, we will calculate the eigenvectors of C:

```
>> [V, D] = eig(C);
>> D
D =
    4.435   0              0
    0       1.601e − 03    0
    0       0              6.720e − 01
```

The value of $D$ shows the three eigenvalues. Note that one of the eigenvalues is far smaller than the other two ($1.601e - 03$). This indicates that the corresponding eigenvector (column 2 of $V$) only weakly contributes to the covariance matrix. As a demonstration of this, we will apply the eigenvector matrix $V$ to the original data and examine the data:

```
>> V_inv = inv(V);
>> m_rot = V_inv * m;
>> var(m_rot(:, 1))
ans =
    4.346
>> var(m_rot(:, 2))
ans =
    0.00157
```

```
>> var(m_rot(:, 3))
ans =
    0.659
```

Note that these variances of the modified data set match the eigenvalues of the original covariance matrix. Just as important, the variance corresponding to the second axis is substantially smaller than the other two. Because of this discrepancy, we can omit this axis in the rotated data set while still preserving the variance of the original data.

## 3.3  PROBABILITY AND STATISTICS

### 3.3.1  Introduction

The intent of this section is a brief, rapid introduction to probability and statistics and their use in MATLAB. This primer cannot hope to replace a good elementary statistics sequence. That said, those readers with a less extensive background in statistics may find this section useful. This primer expects a basic understanding of calculus and a passing familiarity with MATLAB, such as what might be expected by having gone through the introductory chapters of this very book.

### 3.3.2  Random Variables

Much of probability is built upon the concept of a random variable. A random variable is a variable that can take any one of a number of defined values and whose actual value is determined solely by chance. As a simple example, we will define a random variable $X$ to represent the outcome of a flip of a coin, where the value 1 denotes an outcome of "heads" and a value of 0 signifies the outcome "tails." With a fair coin, the probability of heads or tails is equal.

Usually, we will represent the probability of an outcome as a rational number, often a fraction. As a fraction, the numerator represents the number of outcomes that yield the event, and the denominator represents the total number of outcomes in the system. So, in the case of random variable $X$, the probability of a tails event is $1/2$. There are two possible outcomes, and the event of getting "tails" is the result of only one. Similarly, the probability of a heads event is also $1/2$. Together, the probability of a heads event or a tails event occurring is $2/2$ or 1. This should make sense, as flipping an idealized coin should yield one or the other.

This property of probabilities, summing to one, is a general one, and in a formal treatment of probability is usually defined axiomatically. This usually includes three axioms:

1. Probability is always nonnegative.
2. The probabilities of all possible events sum to one.
3. The probability of any of multiple mutually exclusive (nonoverlapping) events is the sum of the individual event probabilities.

These three are also known as the "Kolmogorov axioms" and form the traditional axiomatic foundation of probability theory.

---

### EXERCISE 3.11

Let $Y$ be a random variable $Y$ whose result is the roll of a six-sided die. What are the outcomes? What is the probability of a 5? What is the probability of an even number? What is the probability of rolling a number from 1 to 6?

---

We can generalize the coin flip example by allowing the probabilities of the two outcomes to differ from $1/2$. Under such a generalization, a random variable having two possible outcomes is called a **Bernoulli** random variable. Unlike the case of $X$, where we modeled a coin flip, a Bernoulli random variable does not necessarily have equal probabilities for the two outcomes. The probabilities for both must, however, sum to one.

Given a Bernoulli random variable $Y$, with probability $p$ of outcome 1 and probability $(1-p)$ of outcome 0, we denote the probability of an outcome 1 of $Y$ as $\Pr(Y=1)$. Here, $\Pr(Y=1)$ would be equal to $p$. We can define a function $f(y) = \Pr(Y=y)$ such that the value of $f(y)$ is the probability of value $y$ of random value $Y$. In other words,

$$f(y; p) = \begin{cases} p, & 1 \\ (1-p), & 0 \\ 0, & y \notin \{1, 0\} \end{cases}$$

This function is termed the *probability mass function* (PMF) of random variable $Y$. It literally outlines where the mass of the probability of the variable lies.

Thus far, our example has focused on a single Bernoulli variable representing a single binary outcome. As a more complex example, we can flip a coin multiple times and count the number of heads. This can be represented as a sum of Bernoulli random variables. We can also define a new type of random variable, a binomial random variable, to represent this scenario.

Formally, given a series of $n$ Bernoulli random variables $X_0, X_1 \ldots X_n$, all with equal probability $p$ of outcome 1 ($\Pr(X_0=1) = \Pr(X_1=1) = \ldots \Pr(X_n=1) = p$), a binomial random variable $Y$ represents the total number of positive (i.e., 1 valued) outcomes. We say here that $n$ represents the number of *trials*. Since each trial must have either a positive or negative (i.e., zero-valued) outcome, the total number of positive and negative outcomes must equal the number of trials, and the number of negative (i.e., zero-valued) outcomes is $n - Y$.

Much like with Bernoulli random variables, we can define a probability mass function. However, because a binomial random variable has more outcomes, this case is more complex. Take a series of three coin tosses and a random variable $Y$ representing the total number of heads the series of flips (see Table 3.1).

There are now eight possible outcomes. This can be calculated quickly from $2^3 = 8$. (Each flip occurs independently of the others and doubles the number of outcomes in the series. Thus, with 3 flips in the series, the total number of outcomes is $2 \times 2 \times 2 = 8$.) Of these 8, only one outcome involves 0 or 3 heads. 1 or 2 heads both involve 3 outcomes. With this information and from this table, we can construct a probability mass function for $Y$.

**TABLE 3.1** This table shows the possible outcomes in a three coin flip experiment and the total number of heads in each outcome.

| Flips | Number of Heads |
|-------|-----------------|
| TTT | 0 |
| TTH | 1 |
| THT | 1 |
| THH | 2 |
| HTT | 1 |
| HTH | 2 |
| HHT | 2 |
| HHH | 3 |

$$f(y) = \begin{cases} 1/8, & y = 0 \\ 3/8, & y = 1 \\ 3/8, & y = 2 \\ 1/8, & y = 3 \end{cases}$$

Values below 0 or above 3 were omitted as they are necessarily 0 (these already sum up to 1). In general, the probability mass function for a binomial random variable can be calculated from the formula

$$f(k; n, p) = \binom{n}{k} p^k (1-p)^{n-k}$$

Here, $n$ is the number of trials, $p$ is the probability of a positive outcome on any one trial, and $k$ is the number of successes or nonzero-valued results. The notation $\binom{n}{k}$ might look scary due to its unfamiliarity, but it is simply the number of combinations, also called the binomial coefficient, and it can be calculated from $\binom{n}{k} = \frac{n!}{k!(n-k)!}$. The MATLAB function C = **nchoosek(n, k)** will calculate the number of combinations automatically for you:

```
>> C = nchoosek(3, 1)
C =
    3
```

---

**EXERCISE 3.12**

Use MATLAB and the formula above to find a probability mass function for a four coin toss example. Verify these values by enumerating the possible outcomes.

---

For more complex distributions, a bar graph provides an excellent tool for visualizing a probability mass function. Figure 3.5 depicts the probability mass function for the probability of the total number of heads for eight coin flips. With the probabilities in the vector p, the command

>> **bar(0:8, p)**

produced the plot shown in Figure 3.5. (The expression 0:8 generated the markers for each bar at the bottom of the plot, denoting the number of successes.)

As $n$ increases, you may notice that the probability mass function acquires a bulging shape, where success counts near the middle of the possible range have much higher probabilities relative to the probabilities of all heads or all tails. *Descriptive statistics* provides a number of standard terms that we can use to characterize the distribution.

We can describe the *central tendency* of the distribution. We will discuss three common ways of describing the central tendency of a distribution. The first is the mode. The mode is defined as the most probable outcome in the distribution. From a bar graph depicting a probability mass function, the mode is the outcome with the highest probability. The second central tendency is the median. The median is defined as the outcome corresponding to the point where the probability masses above or below the outcome are equal. This can also be stated as the outcome for which the cumulative probability is equal to or exceeds 0.5.

The final central tendency that we will discuss is the mean. Occasionally, the term expected value is also used to indicate the mean *expected value*. This term suggests an interpretation for the mean, given a binomial random variable $Y$ drawn from a known distribution, what value should you expect? We can define the expected value of a function $f(x)$ relative to a distribution for a random variable $X$ as

$$E_X[f(x)] = \sum_{x \in X} \Pr(X = x) f(x)$$

or, the expected value of a function $f(x)$ is the value of the function at $x$ multiplied by the probability of $x$, summed over all values $x$ for the random variable $X$. The mean is defined as the expected value of the function $f(x) = x$. So, the mean of the three coin toss example discussed previously is

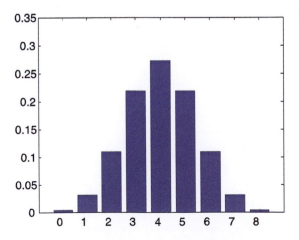

FIGURE 3.5   Bar graph

$$E_X[x] = \sum_{x \in X} x \mathrm{Pr}(X = x)$$

$$E_X[x] = 0 \cdot 1/8 + 1 \cdot 1/8 + 1 \cdot 3/8 + 2 \cdot 3/8 + 3 \cdot 1/8$$

$$E_X[x] = 3/2$$

Clearly, the mean is not a valid number of successes (we cannot have a fraction of a coin flip). The mean is not guaranteed to be a valid count for the variable in question. Nonetheless, the empirical average of many trials or random variables will grow ever closer to the true mean of distribution, assuming that all trials originate from the same underlying distribution and are statistically independent. This property, that the empirical average over many trials will approach the expected value, is called the Central Limit Theorem.

We can easily demonstrate that the mean for any Bernoulli random variable with probability $p$ of outcome 1 is $p$:

$$E_X[x] = \sum_{x \in X} x \mathrm{Pr}(X = x)$$

$$E_X[x] = (1)_p + (0)(1 - p) = p$$

As mentioned earlier, the mean provides a measure of the central tendency. What the mean doesn't provide is a measure of the dispersion of the data. One group of data may be widely dispersed, and another may be tightly clustered, but both may have very similar means.

Unfortunately, we cannot simply use the sum of the differences between the random variable and the mean as a measure of dispersion, as the positive and negative deflections around the mean tend to cancel each other out, tending toward an expected value of zero:

$$E_X[(x - \bar{x})] = \sum_{x \in X} (x - \bar{x}) \mathrm{Pr}(X = x)$$

$$E_X[(x - \bar{x})] = \sum_{x \in X} (x \mathrm{Pr}(X = x) - \bar{x} \mathrm{Pr}(X = x))$$

$$E_X[(x - \bar{x})] = \sum_{x \in X} x \mathrm{Pr}(X = x) - \sum_{x \in X} \bar{x} \mathrm{Pr}(X = x)$$

$$E_X[(x - \bar{x})] = \bar{x} - \sum_{x \in X} \bar{x} \mathrm{Pr}(X = x)$$

$$E_X[(x - \bar{x})] = \bar{x} - \bar{x} \sum_{x \in X} \mathrm{Pr}(X = x)$$

$$E_X[(x - \bar{x})] = \bar{x} - \bar{x}(1) = 0$$

Instead, we can calculate the expected value of *square* of this difference (as they don't cancel out), called the *variance*:

$$Var(x) = E_X[(x - \bar{x})^2] = \sum_{x \in X} (x - \bar{x})^2 \mathrm{Pr}(X = x)$$

$$Var(x) = \sum_{x \in X} (x^2 - 2x\bar{x} + \bar{x}^2) \mathrm{Pr}(X = x)$$

$$Var(x) = \sum_{x \in X} x^2 \mathrm{Pr}(X = x) - \sum_{x \in X} 2x\bar{x} \mathrm{Pr}(X = x) + \sum_{x \in X} \bar{x}^2 \mathrm{Pr}(X = x)$$

$$Var(x) = \sum_{x \in X} x^2 Pr(X = x) - 2\bar{x} \sum_{x \in X} x Pr(X = x) + \bar{x}^2 \sum_{x \in X} Pr(X = x)$$

$$Var(x) = \sum_{x \in X} x^2 Pr(X = x) - 2\bar{x}(\bar{x}) + \bar{x}^2 (1)$$

$$Var(x) = E_X[x^2] - \bar{x}^2$$

Thus, the variance is the difference between the expected value of the random variable squared and the square of the mean. Unlike the expectation of the difference between the random variable and the mean, the variance is rarely zero. Because the variance is in units equivalent to the square of the random variable, we will often use the *standard deviation* instead, which is defined as the square root of the variance. The Greek letter sigma is often used to represent standard deviation:

$$\sigma_x = \sqrt{Var(x)}$$

By definition, sigma squared represents the variance of the random variable.

### 3.3.2.1 Sample Estimates of Population Parameters

Often when dealing with real data the actual distribution of values will not be known. After collecting a set of data, one can look at the empirical distribution of the collected data. Is that distribution necessarily an instantiation of the exact distribution of the population from which the data was collected? Since each data point is a random variable in a sample that is often considerably smaller than the population it was drawn from, the empirical distribution of data will never match exactly, but (for most distributions) increasing numbers of samples will allow a better approximation of the distribution of the (usually much larger) population.

When we discuss sample estimates of a distribution, we use a slightly different notation from the notation we are used to for the actual distribution itself. The sample mean of a set of random variables is denoted as $\bar{x}$ rather than $\mu$. This mean of the sample forms an estimate of the mean of the actual distribution and is calculated from a sample by

$$\bar{x} = \frac{1}{N} \sum_{n}^{N} x_n$$

or the sum of the values divided by the number of values. The estimate of the standard deviation is written as $s$ instead of $\sigma$ and is calculated as

$$s = \sqrt{\frac{\sum_{n}^{N} (x - \bar{x})^2}{N - 1}}$$

The estimated variance is the square of this quantity, and is written as $s^2$. You may note the factor of $N - 1$ rather than $N$ as with the mean. This results from the use of the empirically calculated mean in the estimate for the standard deviation. Because of this factor, this value is often called the *unbiased* estimate of the standard deviation, as one degree of freedom is lost. Note that in practice you will almost always deal with sample estimates of these parameters, as you have access to a sample of data (sampled from the population at large, but not equal to that population).

MATLAB provides the functions **mean()**, **var()**, and **std()** to estimate the mean, variance, and standard deviation of a sample.

```
>> x = [2  3  5  9  5  6  7];
>> mean(x)
ans =
   5.2857
>> var(x)
ans =
   5.5714
```

---

**EXERCISE 3.12**

The MATLAB function **normrnd**(mu, sigma) generates random values that vary according to a normal distribution with mean mu and standard deviation sigma. We will discuss the properties of the normal distribution later on. For the purposes of this exercise, generate a good number of random values with **normrnd**(30, 10), and calculate empirical estimates of the mean and standard deviation. How do they compare to the known values (mean = 30, sigma = 10)?

---

### 3.3.2.2 Joint and Conditional Probabilities

At this point, we've explored single variable distributions fairly extensively. While many phenomena can be modeled quite effectively with just a single independent variable, this is not always the case.

Take for example the following scenario: Instead of one, you are now rolling two ordinary six-sided dice on each trial. The total can be modeled as a single independent variable, but it may be simpler in certain cases to treat the dice as two separate random variables $X$ and $Y$. Both variables would be from the same uniform discrete distribution, so we say these are identically distributed.

With two random variables, we can discuss the probabilities of outcomes across both of them at the same time. For example, $P(x<2, y<3)$ is the probability that the result of the first die is less than 2 and the result of the second is less than 3. This can be computed by enumerating all the possibilities and determining the fraction that complies. In this case, the outcomes are (1, 1) and (1, 2), so there are two possible outcomes that meet the criteria. There are 36 possible outcomes, so the probability is 2/36. This probability over multiple variables is called the *joint* probability over $X$ and $Y$.

You may have noticed a relationship between the probabilities of each individual case and the overall probability. In the context of multiple random variables, probabilities of individual random variables are termed *marginal*, for historical reasons (there is nothing inherently marginal about it; they were computed—manually—by writing probability sums in the margins of a probability table). Thus, the probability $P(x<2)$ in this context is a marginal probability. For P($x$<2), there is one outcome out of six (for $X$ alone, only the first die is considered). Likewise, there are two outcomes where the die roll is less than 3, so P($y$<3) is 2/6. The product of these two probabilities is equivalent to the joint: $P(X, Y) = P(X)P(Y)$. This general relationship, where the joint probability is the product of the marginal probabilities, defines *statistical independence*. In other words, the outcome of $X$ and $Y$ do *not* depend on one another. As we've defined our problem, we can say that $X$ and $Y$ are independent and identically distributed (very frequently abbreviated as i.i.d.).

Many systems of variables will not be independent. If we take for example the previous system in the context of a random variable $Z$ representing the total of $X$ and $Y$, the joint probability is clearly not just the product of the individual probabilities. For example, take the events of rolling a 12 total, and rolling a 6 on the first die (before looking at the outcome of the second). In this case, P($x = 6, z = 12$) is 1/36. However, the probability of rolling a 6 is 1/6 and the probability of rolling a 12 on the two dice is 1/36. Thus, the variables $Z$ and $X$ are not independent: the overall outcome rather strongly depends on what was rolled on the first die.

Given the relationship between $X$ and $Z$, we may want to express probabilities of certain events conditioned on other events having occurred. For example, if the first die comes up 6, what is the probability that the total will be 12? (Hopefully, it is clear that the probability of that total is 1/6, the probability of rolling a 6 with the remaining die. This holds because the two die themselves are independent. This might seem confusing, but it is important to keep in mind separately.) We can use a conditional probability to express such cases. A conditional probability takes the form P($A|B$), which is the probability of outcome $A$, given that $B$ has occurred. So, the previous example of rolling a 12 given one die is already 6 can be written as P($z = 12|x = 6$).

The joint, conditional, and marginal probabilities have the relationship

$$P(A|B)P(B) = P(A,B)$$

Thus, the joint probability of events $A$ and $B$ happening is equal to the conditional probability of $A$ occurring if $B$ occurs multiplied by the probability of $B$ occurring. We can verify this in the case of the two-die scenario.

$$P(z = 12|x = 6) \ P(x = 6) = P(z = 12, x = 6)$$

We know that the probability of P($z = 12, x = 6$) is 1/36, from the above. P($x = 6$) is 1/6 (remember, this is the probability of the first die coming up 6 considered entirely on its own). As discussed above, P($z = 12|x = 6$) is also 1/6. Thus, the relationship holds true. It is important to remember that this relationship holds even when the random variables are not independent.

---

**EXERCISE 3.13**

Calculate the conditional probabilities:

P($z < 6|x = 2$)

P($z = 12|x = 5$)

P($z = 10|x > 4$)

---

A serious drawback of using dice examples in most introductory treatments of probability is questionable relevance. It is understandable why these examples are used so often; a tightly constrained problem allows for establishing the concepts with great mathematical precision. However, most of us are hopefully not spending the majority of our days throwing dice. Introducing real world examples is dicey, as most real world examples map onto such fundamental concepts in multiple ways. No matter which topic you pick, this holds; you will imply a relationship (sometimes a causal relationship) between variables that can a priori be conceptualized as independent. But that is what science is all about: finding these relationships.

Ultimately, a conditional probability indicates that additional information is known about the problem space. Science establishes this information; society uses it beneficially. For instance, an insurance company might assess your risk of dying within the next 10 years differently if they knew that you are overweight, smoke, and don't exercise. In a way, all of life is about harvesting the information expressed in conditional probabilities, and optimizing one's outcomes in the face of uncertainty. It is now cliché that half of marriages end in divorce. It is less well-known that there are pretty solid conditional probabilities involved; the outcome strongly depends on educational and financial status as well as number of previous sexual partners. Put differently, while the probability of divorce for any couple selected randomly is around 0.5, the conditional probability can be far from 0.5 under certain conditions.

Another common situation arises in a medical context. For instance, the probability of a baby to have Down's syndrome is about 1/700. However, the conditional probability is as high as 1/20, given that the mother is over 45 years old. Similarly, the conditional probability that a child will develop autism is four times higher than the unconditional probability if it is known that the child is male.

We can also expand our understanding of expectation and variance to account for multiple random variables. Conditional expectation follows in a straightforward way from conditional probabilities. So, to use the previous two-die example, the expectation of the sum $Z$ is 7:

$$E_z[z] = \sum_{z \in Z} zP(z = Z) = 7$$

We can calculate the conditional expectation of the sum $Z$ given that the first die roll is a 6 in a similar manner:

$$E_{(Z|X=6)}[z] = \sum_{z \in Z} zP(z = Z|X = 6)$$

$$E_{(Z|X=6)}[z] = 2\,P(Z = 2|X = 6) + 3\,P(Z = 3|X = 6) + 12\,P(Z = 12|X = 6)$$

For any values of $Z$ less than 7, the conditional probability is zero, and the corresponding terms drop out, leaving

$$E_{(Z|X=6)}[z] = 7\,P(Z = 7|X = 6) + 8\,P(Z = 8|X = 6) + 12\,P(Z = 12|X = 6)$$

With one die known, only one outcome of the six possible can produce each sum, so

$$E_{(Z|X=6)}[z] = [7 + 8 + 9 + 10 + 11 + 12]\frac{1}{6} = \frac{57}{6} = 9.5$$

For systems with multiple random variables, a single variance does not sufficiently describe dispersion. The term covariance describes variance that occurs together between random variables, due to statistical dependence. Covariance between two variables is defined as

$$\text{cov}(x, y) = E_{(X,Y)}\big[(x - E_X[x])(y - E_Y[y])\big]$$

over the two variable expectation. Covariance is often written as $\sigma_{xy}$. When calculated between a variable and itself, the covariance is equivalent to the standard variance, sometimes written as $\sigma_{xx}$. A nonzero covariance indicates some interdependence between the two variables. Two entirely independent variables should have a covariance of zero.

### 3.3.3 The Poisson Distribution

The Poisson distribution is used to describe phenomena that are comparatively rare. In other words, a Poisson random variable will relatively accurately describe a phenomenon if there are few "successes" (positive outcomes) over many trials.

The Poisson distribution has a single parameter, $\lambda$. For a Poisson distribution modeling a binomial phenomenon, $\lambda$ can be taken as an approximation of $np$.

---

**EXERCISE 3.14**

Write a MATLAB function to calculate the probability of $k$ successes for a Poisson distribution with parameter lambda. Compare a binomial distribution with parameters $n = 10$ and $p = 0.01$ with the equivalent Poisson distribution.

---

Aside from use as an approximation for the binomial distribution, the Poisson distribution has another common interpretation. For an infrequently occurring event, the parameter lambda can be viewed as the mean rate, or $\lambda = nT$, where $n$ is the mean events per unit time, and $T$ is the number of time units. In such a case, a Poisson distribution with the appropriate parameter $\lambda$ will approximate the distribution of events over time or the number of events in an interval.

Events whose occurrence follows a Poisson distribution have another interesting property. Given a series of Poisson distributed independent random variables $X_1, X_2, X_3, \ldots X_n$ and their corresponding arrival times $T_1, T_2, T_3, \ldots T_n$, we can calculate the distribution of the corresponding inter-event intervals.

Let $N(t)$ equal the number of events at some time $t$, where $P(k = N(t))$ follows a Poisson process with parameter $\lambda$. Then, the probability of the nth event occurring at

$$P(T_k > t) = P(N(t) < k)$$
$$P(T_1 > t) = P(N(t) < 1)$$
$$P(T_1 > t) = P(N(t) = 0)$$

$$P(N(t) = 0) = \frac{e^{-\lambda t}(\lambda t)^0}{0!} = e^{-\lambda t}$$

Likewise, we can calculate a distribution for the inter-event intervals. Here, the probability of an interval between two successive events $T_k$ and $T_{k-1}$ being larger than some time $t$ is the same as the probability of exactly $k-1$ events occurring within the interval 0 to $T_{k-1} + t$. This can be expressed as follows:

$$P(T_k - T_{k-1} > t) = P(N(T_{k-1} + t) = k - 1)$$

The probability of $k-1$ events during the interval $T_{k+1} + t$ is equivalent to the probability of no events during the interval 0 to $t$.

$$P(N(T_{k-1} + t) = k - 1) = P(N(T_{k-1}) = k - 1)P(N(t) = 0)$$
$$P(N(T_{k-1} + t) = k - 1) = P(N(t) = 0)$$

By definition, $k-1$ events occurred during the interval 0 to $T_{k-1}$. Thus, if any events occur during the interval $T_{k-1}$ to $T_{k-1} + t$, this would mean that the number of events in the interval $T_{k-1} + t$ would not be $k-1$ but greater than $k-1$.

This implies that the intervals are distributed in a "memoryless" fashion. In other words, for an ongoing process following a Poisson distribution of events, the distribution of waiting time to the next event does not change over time. Put differently, knowing when the last event occurred does not give you any information about when to expect the next one.

The distribution of intervals that we have derived here is called the exponential distribution. This is our first example of a continuous distribution. Unlike with discrete distributions, calculating the exact probability of a single value in a continuous distribution is not feasible, as it is always zero. To understand why, one can use the previous example of the exponential distribution of time intervals for a Poisson process. The probability of a specific value, say 3 seconds, would correspond to the probability of the time interval being **exactly** equal to 3 seconds. Since this would exclude any interval even infinitesimally close to 3, this will be vanishingly small regardless of the number chosen. Therefore, when working with continuous probabilities, we compute the probability of a variable falling within a range of values.

Because of this fundamental difference, continuous distributions do not have a probability mass function like discrete distribution. Continuous distributions are defined in terms of cumulative distribution functions. The derivation of the exponential distribution above provides an excellent example. Above, the probability of an inter-event interval $T$ being greater than some value $t$ was found to be equal to $P(T_1 > t) = e^{-\lambda t}$. Usually, the cumulative distribution function $F$ is defined as the probability of a random variable being less than a given value. Following those conventions, the cumulative distribution function $F$ for the exponential distribution can be defined as

$$P(T < t) = 1 - e^{-\lambda t}$$

(This falls out of the requirement that the sum of probability be equal to one. If $P(T < t) + P(T > t) = 1$, then $P(T < t) = 1 - P(T > t)$.)

To determine the probability of a random variable falling within a specific range, we can subtract ranges. For example, the probability of a random variable $T$ falling between $t_1$ and $t_2$ can be expressed in terms of each value alone:

$$P(t_1 < T < t_2) = P(T < t_2) - P(T < t_1)$$

This holds true because the probability of the random variable $T$ being less than $t_2$ includes all cases where the random variable is less than $t_1$. So, the probability mass corresponding to $P(T < t_1)$ must be subtracted out.

This can be more clearly understood by introducing the probability density function, a function $f(x)$ such that the cumulative distribution function $F(x) = \int f(x)dx$. For a probability expression with a density function $f(x)$, the probability $P(X < x) = \int_0^x f(x)dx$. Likewise, the range $P(a < X < b) = \int_a^b f(x)dx$.

---

### EXERCISE 3.15

Demonstrate through derivation that the probability density function $f(x)$ for the exponential distribution is $f(x) = \lambda e^{-\lambda x}$. Remember that the cumulative density function is defined in terms of the probability density function as $F(x) = \int_{-\infty}^x f(t)dt$ and the cumulative density function for the exponential function as given above. (*Note*: Because the exponential distribution is only defined for non-negative numbers, the lower bound of the integral can be set at 0.)

---

Continuous distributions also have expectations like discrete distributions. Instead of summing over all probabilities, the expectation is defined in terms of an integral over the probability density function. For a probability density function $f(x)$, the expectation of the function $g(x)$ is defined as

$$E[g(x)] = \int_{-\infty}^{\infty} f(x)g(x)dx$$

With this, we can calculate the mean and variance for the exponential distribution:

$$E[x] = \int_{-\infty}^{\infty} xf(x)dx$$

$$E[x] = \int_{0}^{\infty} x(\lambda e^{-\lambda x})dx$$

$$E[x] = \lambda \int_{0}^{\infty} xe^{-\lambda x}dx$$

$$E[x] = \lambda \left[ -x\frac{1}{\lambda}e^{-\lambda x} - \frac{1}{\lambda^2}e^{-\lambda x} \right]_{0}^{\infty}$$

$$E[x] = \left[ -xe^{-\lambda x} - \frac{1}{\lambda}e^{-\lambda x} \right]_{0}^{\infty}$$

$$E[x] = 0 + 0 + 0\frac{1}{\lambda} = \frac{1}{\lambda}$$

$$Var(x) = \int_{-\infty}^{\infty} (x-\mu)^2 f(x)dx$$

$$Var(x) = \int_{-\infty}^{\infty} (x-\mu)^2 \lambda e^{-\lambda x} dx$$

$$Var(x) = \int_{-\infty}^{\infty} x^2 \lambda e^{-\lambda x} + \mu^2 \lambda e^{-\lambda x} - 2\lambda \mu e^{-\lambda x} dx$$

$$Var(x) = \lambda \int_{0}^{\infty} x^2 e^{-\lambda x} + \mu^2 \lambda \int_{0}^{\infty} e^{-\lambda x} dx - 2\mu\lambda \int_{0}^{\infty} x e^{-\lambda x} dx$$

$$Var(x) = \lambda \left( \frac{2}{\lambda^2} \right) + \frac{1}{\lambda^2} - 2\frac{1}{\lambda} = \frac{1}{\lambda^2}$$

## 3.3.4 Normal Distribution

The last classical distribution that we will discuss here is the normal distribution. There are many more, some of which will be visited in later chapters for more specialized purposes. Because its cumulative distribution function is not solvable analytically, the normal distribution is usually defined in terms of its probability density function,

$$f(x; \mu, \sigma) = \frac{1}{\sigma\sqrt{2\pi}} e^{-\frac{(x-\mu)^2}{2\sigma^2}}$$

Here, the parameters $\mu$ and $\sigma$ define the mean and standard deviation of the distribution. A normal distribution with $\mu = 0$ and $\sigma = 1$ is called a standard distribution.

The cumulative distribution function of the normal distribution is the integral over all values $x$:

$$F(x; \mu, \sigma) = \int_{-\infty}^{x} \frac{1}{\sigma\sqrt{2\pi}} e^{-\frac{(a-\mu)^2}{2\sigma^2}} da$$

The cumulative distribution function of the standard distribution is often denoted as $\Phi(x)$. This cumulative distribution function is often defined in terms of another special function whose form is very similar to the integral over the probability density. This function is commonly called the error function $\mathrm{erf}(x) = \frac{2}{\sqrt{\pi}} \int_0^x e^{-t^2} dt$. So, in terms of the error function, the cumulative distribution function of the standard distribution is $\Phi(x) = \frac{1}{2} + \frac{1}{2}\mathrm{erf}\left(\frac{x}{\sqrt{2}}\right)$.

MATLAB defines both a cumulative distribution function, **normcdf(x, mu, sigma)**, and the error function, **erf(x)**, for the normal distribution.

```
>> normcdf(0.6, 0, 1)
ans =
   0.7257
>> 0.5 + 0.5*erf(0.6/2^0.5)
ans =
   0.7257
```

The normal distribution approximates how many phenomena vary. In particular, the normal distribution is useful in understanding error.

### 3.3.5 Confidence Values

The normal distribution is particularly useful because of the central limit theorem. Given $N$ independent, identically distributed random variables with mean $\mu$ and variance $\sigma^2$, the central limit theorem asserts that the distribution of the mean of the random variables will converge to a normal distribution with mean $\mu$ and variance $\frac{\sigma^2}{N}$. The dependency of the variance on the number of variables (in this case, samples) is particularly important. As we will see, this result is especially relevant to estimating distributions from samples.

---

**EXERCISE 3.16**

Here we will explore how the precision of a mean estimate varies with the sample count.

```
>> figure
>> samples = [];
>> N = [1:15];
>> for n = 1:15
samples(n) = var(mean(randn(2^N(n),
100)));
end
>> scatter(N, samples)
```

The use of **randn(2^N(n), 100)** here selects $2^N \times 100$ samples from a standard normal distribution. The intent is to simulate picking $2^N$ samples 100 times in order to estimate the variance of the distribution of the means. **mean()** calculates the sample means, returning a vector of length 100, and **var()** estimates the variance of the distribution of means.

You should see a figure like

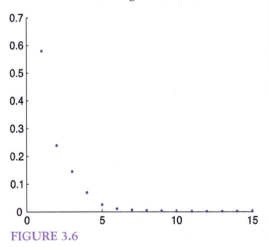

FIGURE 3.6

How does variance vary with sample count? How many more samples are required to halve the variance in the estimate of the mean?

---

The term standard error of the mean (or just standard error) is defined as

$$SE_{\bar{x}} = \frac{s}{\sqrt{n}}$$

where $s$ is the estimate of the standard deviation and $n$ is the number of samples in the estimate.

Often, the error will be expressed as confidence intervals around the mean. A confidence interval is expressed as the interval surrounding the mean within which an estimate of the mean should fall with a certain probability (often 90% or 95%). For a 95% confidence interval, this is approximately 1.96 times the standard error on either side of the mean estimate.

For example, let's assume we have a normally distributed population whose actual mean is 25 and whose variance is 5. We can collect a sample of 10 as follows.

```
>> sample = normrnd(25, sqrt(5), [1 10]);
>> mean(sample)
ans =
   25.5606
>> se = std(sample)/sqrt(10);
>> se * 1.96
ans =
   1.4133
```

In this case, the 95% confidence interval around the mean estimate 25.5606 would be 24.1473 to 26.9739.

For values other than 95%, we can calculate the factor of the standard error directly using the MATLAB function **erfinv()**. **erfinv()** calculates the *inverse* of the error function discussed earlier. To determine the factor to replace the 1.96, you will need to calculate $\sqrt{2}\mathrm{erf}^{-1}(p)$, where $p$ is the confidence interval probability. It is important to note that this assumes normally distributed values.

```
>> 2^0.5 * erfinv(0.95)
ans =
   1.9600
>> 2^0.5 * erfinv(0.90)
ans =
   1.6449
```

## 3.3.6 Significance Testing

Hand in hand with the idea of a confidence interval is significance testing. Take a known distribution: a normal distribution with a mean of 15 and a standard deviation of 3. A sample of five values has a mean of 11. Is this sample likely to be drawn from the same population? How about a sample of 100 values with the same mean? There is always a chance that the sample *was* drawn from the distribution. The question is, with which probability? Significance testing provides systematic methods for answering such questions.

Returning to the original question posed about the estimated sample mean of 11, we can describe this in a probabilistic way; in fact, we can rephrase this probabilistically in at least **two** ways. First, we can ask how probable a mean of 11 or lower might be, or

$$P(\overline{x} < 11)$$

Secondly, we can ask how probable a mean at least as extreme as 6 might be, or

$$P(|\overline{x} - \mu| > (15 - 11))$$

Classical statistics distinguishes these two refinements of our original questions as a hypothesis test: we use the properties of the distribution to test a *null hypothesis* (often written as $H_0$) that the five item sample is drawn from the known distribution. An extremely low probability of such an extreme result would argue against the null hypothesis or, alternatively, for rejecting the null hypothesis.

Typically, a maximum threshold for the probability is chosen, called the significance level. Common values are 1%, 5%, and occasionally 10%. Outcomes with probabilities below the significance level are termed statistically significant at the corresponding level, and usually strongly argue for rejecting the null hypothesis. It is important to keep in mind that hypothesis testing is only evidence for or against rejecting the null hypothesis. For example, a significance level of 5% indicates that only one out of every twenty repetitions would produce a result as extreme. If an experiment is repeated 20 times, on average, the outcome would be statistically significant once. All that the *p* value gives you is the probability that such data so extreme (or more extreme) could happen by chance, assuming that the null hypothesis is true. By this logic, if the significance level is not met, it does *not* mean that the null hypothesis is true, just that we failed to reject it at this significance level.

Often, a significance level is selected prior to analysis or even the collection of data, and a significance of 5% is especially common. The selection of a significance level requires a tradeoff between two types of error. Choosing a less stringent significance level increases the risk of interpreting a result as indicating that the null hypothesis should be rejected when it's not actually false (this is classically called a Type I error and refers to spurious findings). Alternatively, a more stringent significance level enforces a more severe threshold for the rejection of the null hypothesis, but setting too low a significance level can miss rejection when the null hypothesis is actually false (a classical Type II error: missing differences that are really there). How one should pick the significance level depends on the relative value of the outcomes in a given practical case: how serious is it to miss real effects versus how serious it is to claim the existence of effects that are not really there. More than the brief treatment of Type I/II errors here is beyond the scope of this primer, and the reader is referred to a more detailed reference for an in-depth discussion.

Significance tests can be classified as *one-tailed* or *two-tailed* hypothesis tests. The origin of these names can be easily understood from an illustration of the expected distribution of sample means. From the central limit theorem, as discussed in the last section, we know that sample means from the known distribution should vary with a normal distribution whose mean matches that of the underlying distribution and whose standard deviation is $\sigma/\sqrt{N}$.

Figure 3.7 shows the expected PDF (remember, probability distribution function) for sample means for samples with five elements. Shaded is the probability of the sample mean

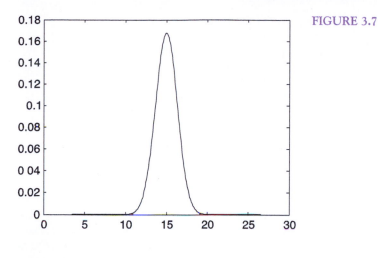

FIGURE 3.7

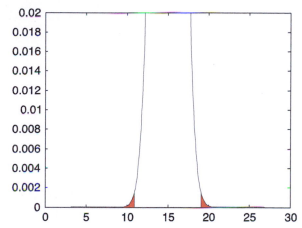

FIGURE 3.8

having a value at least as extreme as the collection value. Figure 3.8 shows the shaded portion of the PDF in greater detail. (Noting that $P(|\bar{x} - \mu| > (15 - 11)) = P(\bar{x} > 19) + P(\bar{x} < 11)$ may help in understanding Figure 3.8.)

Figure 3.9 shows the same PDF with the probability of the sample mean being less than 11 shaded. In this case, the shaded probability covers only one of the two ends of the PDF. This is a one-tailed test. Likewise, the previous case covering both tails of the PDF is called a two-tailed test.

Using significance testing correctly requires determining whether the question at hand involves a one-tailed or two-tailed test. Here we are interested in ascertaining whether the measured sample comes from the known distribution. Understanding the extreme nature of the sample mean is what we're interested in, so a two-tailed test is most appropriate.

Since the sample mean should be distributed according to a known normal distribution, we can calculate the two-tailed probability using the MATLAB function **normcdf**. Recall

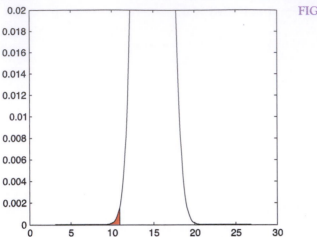

FIGURE 3.9

that **normcdf**(x, mu, sigma) returns the cumulative distribution function at $x$ for a normal distribution with $\mu = $ mu and $\sigma = $ sigma.

First, $P(\bar{x} \leq 11)$,

```
>> mu = 15;
>> sigma = 3;
>> N = 5;
>> s = sigma/sqrt(N);
>> p_tail_one = normcdf(11, mu, s)
p_tail_one =
   0.0014
```

Next, we can calculate $P(\bar{x} > 19)$. **normcdf()** can be used for this as well, but if the same procedure is followed with 19 substituted for 11, we will calculate $P(\bar{x} \leq 19)$:

```
>> p = normcdf(19, mu, s)
p =
   0.9986
```

Note that the value here is substantially larger than the probability of the first tail. To calculate $P(\bar{x} > 19)$, we can use the equality $P(\bar{x} > 19) + P(\bar{x} \leq 19) = 1$:

```
>> p_tail_two = 1 - p
p_tail_two =
   0.00143
```

This matches the probability mass of the first tail, as one might expect from Figure 3.9. The sum of the two is the probability that the sample mean is at least as extreme as the estimated mean here:

```
>> p = p_tail_one + p_tail_two
p =
   0.0029
```

So, roughly 3% of the time, the empirical mean of five samples from the known distribution would be at least as extreme as 11. At the 1% significance level, this would not provide sufficient support for rejecting the null hypothesis, but it would at the 5% level.

---

### EXERCISE 3.17

Does the same conclusion hold for a larger sample (the second example of 100 samples with the same mean of 11)? Determine the probability using normcdf().

---

#### 3.3.6.1 Student's t Distribution

Imagine an electrophysiology experiment attempting to determine whether a single neuron responds to a stimulus. In trials without the stimulus, you see firing rates as in the second column of Table 3.2. The third column of the table shows the firing rates in trials with the stimulus. Obviously, we're interested in whether the stimulus alters the firing rate. This question can be rephrased as a statistical test: what is the probability that the two distributions are the same or, rather, that two these samples were drawn from the same distribution?

**TABLE 3.2**

| Trial | Rate without stimulus | Rate with stimulus |
|-------|----------------------|--------------------|
| 1 | 54.5 | 67.1 |
| 2 | 43.5 | 63.8 |
| 3 | 36.5 | 73.5 |
| 4 | 48.7 | 57.2 |
| 5 | 41.8 | 31.0 |
| 6 | 52.6 | 54.2 |
| 7 | 28.7 | 33.1 |
| 8 | 57.1 | 117.0 |
| 9 | 40.5 | 71.4 |
| 10 | 48.2 | 133.8 |
| 11 | 57.3 | 60.0 |
| 12 | 50.8 | 41.1 |
| 13 | 62.5 | 93.0 |
| 14 | 30.8 | 33.5 |
| 15 | 28.9 | 52.0 |

Unfortunately, since we don't know either of the distributions from which the firing counts are sampled, we can't use the procedure we used in the previous section. This will often be the case in real scenarios. Fortunately, a wide variety of tests for general purpose significance testing have been defined. For example, Student's *t* test is appropriate here.

Student's (actually William S. Gosset, but being in the employ of Guinness, he had to publish under a pseudonym) *t* test (Student, 1908) is useful in a number of statistical scenarios. Here, we will use a paired *t* test. In this experimental paradigm, the measurements of firing rate with and without the stimulus are not independent (the experiment measures the same cell under two different conditions). We can pair the measurements before and during stimulus presentation, and use the *t* test to determine whether the data is significantly different.

Under the paired *t* test, we need to calculate a *t* statistic from

$$t = \frac{\overline{x}_D}{s_D/\sqrt{N}}$$

where $\overline{x}_D$ is the mean of the difference between elements of each pair, $S_D$ is the estimated standard deviation of the differences between elements of each pair, and $N$ is the number of pairs. Then, we need to use a Student's *t* distribution in much the same way we used a normal distribution in the prior section. It is worth nothing that the *t* distribution is very similar to the normal distribution anyway, just with heavier tails to account for unknown population variance with small sample sizes.

MATLAB offers a number of functions to simplify this process. The function **ttest()** has two forms that are particularly used. **ttest(x,y)** performs a paired *t* test. This would be easiest if we have vectors x and y for the two samples.

```
>> x = [54.5 43.5 36.5 48.7 41.8 52 6 28.7 57.1 40.5 48.2 57.3 50.8 62.5 30.8 28.9];
>> y = [67.1 63.8 73.5 57.2 31.0 54.2 33.1 117.0 71.4 133.8 60.0 41.1 93.0 33.5 52.0];
>> ttest(x, y)
ans =
   1
```

Without specifying return parameters, **ttest()** performs a test and returns whether the null hypothesis should be rejected at the default significance level (5%). Thus, the data supports rejecting the null hypothesis that the stimulus has no effect on the firing rate. (In other words, the data suggests that the stimulus has an effect on the firing rate at the 5% significance level.) We can supply a different significance level as a final parameter to **ttest()**:

```
>> ttest(x, y, 0.01)
ans =
   0
```

This implies that the data does not support rejection of the null hypothesis at the 1% level of significance. Depending on the criteria of the experiment, significance at the 5% level may be sufficient, or a lack of significance at the 1% level may suggest that the experiment had insufficient "power" to detect an effect at this level, and more data should be collected to yield this power.

We can obtain the exact probability of the result or one more extreme by supplying a second return parameter:

```
>> [h, p] = ttest(x, y)
h =
    1
p =
    0.0106
```

If we have the differences between the sample pairs already calculated, we could also use another form of the **ttest** function. With a single vector, **ttest**(x) tests against a mean of 0. This is appropriate when we just have the differences between the pairs, not the actual values.

```
>> d = x − y;
>> ttest(d)
ans =
    1
```

### 3.3.6.2 ANOVA Testing

Student's $t$ test covers a number of hypothesis scenarios for testing the results of a single factor (one independent variable) and between pairs of samples. Multiple samples or multiple experimental factors create a scenario that is difficult for a single $t$ test to handle. To use a $t$ test under such circumstances, we would need a separate $t$ test for each distinct pair in the experiment. For a five factor experiment, this would require $2^5 = 32$ separate tests! Note that on a significance level of 5%, we would expect 1 in 20 differences to test positively just by chance, even in the absence of any real experimental effects, so we would also have to adjust our significance level for multiple comparisons. If this sounds like a recipe for disaster, it is. An alternative approach for the statistical treatment of experimental data from experiments with more than one independent variable is the so-called "analysis of variance."

An analysis of variance (ANOVA) allows us to ask the probability that a group of samples all originate from the same larger population without the inflated risk of a Type I error as with multiple $t$ tests. As an example, we'll expand our hypothetical experiment to three different stimuli. For simplicity, we will call them A, B, and C (see Table 3.3).

First, we need to calculate the variance across groups.

```
>> stim_a = [39.2  45.7  45.9  42.8  60.2  50.7  39.9  50.8  43.0  55.9];
>> stim_b = [43.2  56.7  32.8  61.2  54.6  44.6  53.2  43.3  35.1  53.7];
>> stim_c = [66.5  54.5  62.6  45.6  46.8  34.9  53.3  60.1  69.7  61.0];
```

This requires calculating the mean for each stimulus

```
>> mean_a = mean(stim_a);
>> mean_b = mean(stim_b);
>> mean_c = mean(stim_c);
>> means = [mean_a mean_b mean_c];
```

TABLE 3.3

| Trial | Stimulus A | Stimulus B | Stimulus C |
|-------|-----------|-----------|-----------|
| 1 | 39.2 | 43.2 | 66.5 |
| 2 | 45.7 | 56.7 | 54.5 |
| 3 | 45.9 | 32.8 | 62.6 |
| 4 | 42.8 | 61.2 | 45.6 |
| 5 | 60.2 | 54.6 | 46.8 |
| 6 | 50.7 | 44.6 | 34.9 |
| 7 | 39.9 | 53.2 | 53.3 |
| 8 | 50.8 | 43.3 | 60.1 |
| 9 | 43.0 | 35.1 | 69.7 |
| 10 | 55.9 | 53.7 | 61.0 |

the overall mean

```
>> mean_overall = mean([mean_a mean_b mean_c]);
```

and the weighted sum of the squares of differences between the overall group mean and the mean of each group

```
>> sum_between = sum(10*(means - mean_overall).^2);
>> between_mean = sum_between/2;
```

Next, we need to calculate the variance within each group. We sum the squares of the differences between each measurement and its difference from the mean of its corresponding group:

```
>> sum_within = sum((stim_a - mean_a).^2 + (stim_b - mean_b).^2 + (stim_c -
mean_c).^2);
>> within_mean = sum_within/(3*9);
```

We use these two values, the mean square difference between groups and the mean square difference within groups, to calculate the test statistic, $f$:

```
>> f_value = between_mean/within_mean
f _value = 2.483
```

What does this tell us? If all three sets of data follow the same distribution, the distribution of $f$ should follow an $F$ distribution, which has two parameters. A discussion of the analytic form of the $F$ distribution is beyond the scope of this text. For the purposes of our use of the $F$ distribution, the two parameters are equivalent to the degrees of freedom in our data set: the number of stimuli minus one ($3 - 1 = 2$) and the total count of data points, minus the number of stimuli ($30 - 3 = 27$).

In this case, we need to determine the cumulative probability function for the $F$ distribution given our $f$ value and the two parameters. The **fcdf()** function will calculate this value.

>> p = fcdf(f_value, 2, 27)
  p = 0.651

This $p$ value, 0.651, indicates a little over 65% of the time samples consistent with the null hypothesis (i.e., all distributed similarly) would yield an $f$ value at least this extreme. Thus, we cannot discount the null hypothesis here.

### 3.3.7 Linear Regression

Assume two variables $x$ and $y$, with an expected linear relationship between them such that $y = \alpha + \beta x$. Under this relationship, we will call the $y$ the dependent variable and $x$ the independent variable. Let's say that we have collected sample pairs $(X_n, Y_n)$ and want to estimate the parameters $\alpha$ (constant offset) and $\beta$ (slope) so that we can express $y$ as a function of $x$.

This estimation procedure is called linear regression (think "prediction"). Here, we are *regressing* $x$ onto $y$. Under traditional terminology, as the independent variable, $x$ is the regressor (plural: regressors). We can approach the estimation problem as a search for values of $\alpha, \beta$ (we will call these estimates $\hat{\alpha}, \hat{\beta}$) that minimize the distance between the predicted value $\hat{y}_n = \hat{a} + \hat{\beta} X_n$ and the actual measured value $Y_n$. The distance for each measured pair can be represented as

$$S = \sum_{i=1}^{N} (y_i - (\alpha + \beta x_i))^2$$

$$S = \sum_{i=1}^{N} (y_i - \alpha - \beta x_i)^2$$

Then, the estimates $\hat{\alpha}, \hat{\beta}$ can be calculated from the partial derivatives of $S$:

$$\frac{\partial S}{\partial \alpha} = \sum_{i=1}^{N} 2(-1)(y_i - \hat{\alpha} - \hat{\beta} x_i) = 0$$

$$\sum_{i=1}^{N} y_i - \sum_{i=1}^{N} \hat{\alpha} - \sum_{i=1}^{N} \hat{\beta} x_i = 0$$

$$\sum_{i=1}^{N} \hat{\alpha} = \sum_{i=1}^{N} y_i - \sum_{i=1}^{N} \hat{\beta} x_i$$

$$N\hat{\alpha} = \sum_{i=1}^{N} y_i - \sum_{i=1}^{N} \hat{\beta} x_i$$

$$\hat{\alpha} = \frac{1}{N} \left( \sum_{i=1}^{N} y_i - \hat{\beta} \sum_{i=1}^{N} x_i \right)$$

The estimate $\hat{\beta}$ can be calculated in the same manner:

$$\frac{\partial S}{\partial \beta} = \sum_{i=1}^{N} 2(-x_i)(y_i - \hat{\alpha} - \hat{\beta}x_i) = 0$$

$$\sum_{i=1}^{N}(x_i y_i - x_i \hat{\alpha} - \hat{\beta} x_1^2) = 0$$

$$\hat{\beta}\sum_{i=1}^{N} x_i^2 = \sum_{i=1}^{N} x_i y_i - \sum_{i=1}^{N} x_i \hat{\alpha}$$

$$\hat{\beta} = \frac{\displaystyle\sum_{i=1}^{N} x_i y_i - \hat{\alpha}\sum_{i=1}^{N} x_i}{\displaystyle\sum_{i=1}^{N} x_i^2}$$

This has yielded a pair of equations in $\hat{\alpha}, \hat{\beta}$. The equation for the estimate of $\hat{\alpha}$ can be substituted in and a closed form solution for $\hat{\beta}$ obtained:

$$\hat{\beta} = \frac{\displaystyle\sum_{i=1}^{N} x_i y_i - \left(\frac{1}{N}\sum_{i=1}^{N} y_i - \hat{\beta}\sum_{i=1}^{N} x_i\right)\sum_{i=1}^{N} x_i}{\displaystyle\sum_{i=1}^{N} x_i^2}$$

$$\hat{\beta}\left(1 - \frac{\left(\displaystyle\sum_{i=1}^{N} x_i\right)^2}{\displaystyle\sum_{i=1}^{N} x_i^2}\right) = \frac{\displaystyle\sum_{i=1}^{N} x_i y_i - \frac{1}{N}\sum_{i=1}^{N} y_i \sum_{i=1}^{N} x_i}{\displaystyle\sum_{i=1}^{N} x_i^2}$$

$$\hat{\beta} = \frac{\displaystyle\sum_{i=1}^{N} x_i y_i - \frac{1}{N}\sum_{i=1}^{N} y_i \sum_{i=1}^{N} x_i}{\displaystyle\sum_{i=1}^{N} x_i^2 - \left(\sum_{i=1}^{N} x_i\right)^2}$$

This method of linear regression is called least squares optimization, because the estimates originate from optimizing (here, minimizing) the sum of the squared distance.

---

### EXERCISE 3.18

Write a MATLAB function that calculates estimates for $\hat{\alpha}, \hat{\beta}$, given two vectors of data: [a,b] = least_squares(x, y). Test against the following observations, where $x$ is the independent variable, and $y$ is dependent.

$x = [-0.454, 4.68, 6.93, 7.43, 4.58, 6.40, 6.04, 0.846, 3.49, 4.53]$

$y = [-4.10, 46.8, 69.0, 74.7, 47.5, 63.8, 61.6, 7.76, 0.739, 45.3]$

---

Once an estimate is calculated, one should look at the *residuals*, the difference between the values predicted by the estimated parameters in the regression equation and the measured values. Assuming that the estimates for $\hat{\alpha}, \hat{\beta}$ are stored within MATLAB variables $a$ and $b$ respectively, the residuals for a single variable can be calculated with

```
>> r = y - b * x - a;
```

After any linear regression, it is important to review the residuals. In a perfectly linear relationship between dependent variables and regressors, the residuals will be randomly distributed (and ideally small). This implies that the difference between the prediction and measured value is primarily the result of error and not an additional nonlinear relationship.

After fitting, we can also look at the coefficient of determination, $r^2$.

$$r^2 = 1 - \frac{\sum_i (y_i - f_i)^2}{\sum_i (y - \bar{y})^2}$$

Here the numerator is the sum of the squares of the residuals, and the denominator is the sum of the squares of the difference between the dependent variable and its mean, essentially N times the var(y). Thus, given estimated parameters in MATLAB variables $a, b$, we can calculate $r^2$ with

```
>> r_2 = 1 - (sum((y - b * x - a).^2) / sum((y - mean(y)).^2));
```

The coefficient of determination ranges from 0 to 1, with values near 1 indicating a better fit to the data. One interpretation of $r^2$ is the proportion of variance explained by the model. Thus, a value closer to 1 indicates that most of the variance in the dependent variables originate in the variation of the regressors, as propagated through the model. A lower $r^2$ implies that the dependent variables have some variance unaccounted for by the model.

Instead of calculating the estimates and coefficient of determination by hand, we can use the MATLAB function **regress()** in the Statistics Toolbox.

**[b,bint] = regress(y, x)**

**regress** can perform multivariate linear regression, so for a single variable, $x$ should be an $N \times 2$ matrix, where $N$ is the number of observations. The first column data are the single variate observations, and the second column data consists of ones to indicate the constant offset. **regress()** returns a vector b with the estimates for the coefficients, $\hat{\alpha}, \hat{\beta}$. **regress()** also returns 95% confidence intervals for the coefficients in bint. **regress()** will also return the residuals if a third return parameter, $r$, is included.

### 3.3.8 Introduction to Bayesian Reasoning

Briefly put, Bayes' theorem allows you to invert conditional probabilities if the base rates (or priors) are known. Illustrating the perils of eponyms, Thomas Bayes, a Presbyterian minister who first worked out the basic idea (but never published it), would probably be surprised by the attribution, in light of his other work.

When would one want to invert conditional probabilities? Surprisingly often. As a matter of fact, an awareness of Bayes' theorem is likely much more useful in contributing to everyday wisdom in decision making than, say, calculus (which doesn't stop the professionals from making up contrived examples anyway). This is due to the fact that it is often much easier to measure the conditional probability of one, but not another (yet related) event. If this appears too abstract, it is. A plastic example might help.

A country that shall remain unnamed has suffered a recent spate of vicious and unprovoked terrorist attacks on civilians. You have been hired by the government of this country to advise on a rational response to this unacceptable barbarism. Specifically, the question is whether the government should implement profiling measures against the known characteristics of the perpetrators, and if so, to what degree.

Here are the known facts:

*All* terrorists have been bearded. This translates to $p$ (Bearded|Terrorist) = 1.
One in a million people is a terrorist. This translates to $p$ (Terrorist) = 0.000001.
One in 5 people is bearded. This translates to $p$ (Bearded) = 0.2.

And that's all she wrote. In reality, having solid numbers on this is probably a better position than most governments can manage, so you are in a strong position.

What you really would like to know is the probability that someone is a terrorist given that he is bearded, but you only have the probability that someone is bearded if you already know he is a terrorist. This doesn't help. If you already know someone is a terrorist, they might either already have committed their heinous act, or are in hiding. Language can be misleading here. Even though the probability that someone is bearded is 1, given that they are a terrorist, being bearded and being a terrorist are *not* synonymous in this case. Intuitively, if they were, there would have to be many more terrorists around (more than 1 in a million), given that 1 in 5 people is bearded. But how much is the risk of being a terrorist increased, given that someone is bearded?

Here, Bayes comes to the rescue.

$p$ (Terrorist|Bearded) = $p$ (Bearded|Terrorist) * $p$ (Terrorist)/$p$ (Bearded)
$p$ (Terrorist|Bearded) = 1 * 1e − 6/0.2 = 5e − 6.

In other words, the probability is 5 millionth. To put this in perspective, the absolute probability that someone is a terrorist if he is bearded is extremely low, while the relative probability is 5 times higher. This makes sense, as we use strong diagnostic information to link the two (as $p$ (Bearded|Terrorist) is 1) and the base rate of bearded people in the country is 1 in 5.

You can now either advise the government that each bearded person should be scrutinized 5 times as much as a non-bearded person (to match the increased risk), or forgo profiling altogether, as the absolute risk is still so negligibly small. Both responses are rational.

What if the probability of being bearded given that someone is a terrorist was only half as strong—0.5? In other words, there are other kinds of terrorists around, they are not all bearded.

Plugging in the numbers yields

$p$ (Terrorist|Bearded) $= 0.5 * 1e - 6/0.2 = 2.5e - 6$.

This makes sense. As the strength of the diagnostic information declined, the value of the criterion of beardedness to indicate terroristic tendencies declined in kind.

What if beardedness was much rarer; say only one in a hundred thousand people is bearded (and the strength of the link was back to 1)?

$p$ (Terrorist|Bearded) $= 1 * 1e - 6/1e - 5 = 0.1$

Now there is a 10% absolute chance that the person is a terrorist, just by virtue of being bearded, and the a priori chance is increased a hundred thousand-fold. Now, this seems like a rational case for strong profiling measures.

---

**EXERCISE 3.19**

Write a program that plots the probability of being a terrorist given that someone is bearded for a range of values of $p$ (Bearded|Terrorist) from 0 to 1 (in steps of 0.01), as well as variable base rates (from 1 in a thousand to 1 in a million people being terrorists) and from 1 in 2 to 1 in a million people being bearded. Should the amount of resources allocated to monitor bearded people follow this distribution, on a per unit basis?

---

This admittedly somewhat contrived example was used to provide an intuitive feel for Bayesian statistics, using a striking and emotional case study.

While it is unlikely that you will be hired by a government in this position, it is far from unlikely that you will encounter Bayesian reasoning in your study of the neural and cognitive sciences. As a matter of fact, it has been suggested that the entire sensory apparatus of the brain works as one giant Bayesian machine. It would make sense that it might, as Bayes' theorem allows making use of previous experience (establishing base rates) in a rational fashion. It has been shown that people do use prior experience to gauge what to expect from future interactions with members of a given class.

Moreover, the value of Bayesian reasoning is obvious in everyday life. It helps to know it. "Experts" might give misleading answers. For instance, if you go to the doctor for an

AIDS test or a mammogram, you are not interested in the probability that the test is positive if you have the disease. You are interested in the probability that you have the disease if the test is positive! It has been conclusively shown that doctors frequently confuse these two probabilities, and give you wildly inaccurate odds (Gigerenzer and Hoffrage, 1995; Hoffrage and Gigerenzer, 1998). This concludes our introduction to Bayesian Reasoning—as well as our Mathematics and Statistics Tutorial for the purposes of this book. If you are interested in specific applications of Bayesian Reasoning in neuroscience, see chapter 22. For a more extensive treatment of Bayes' theorem and Bayesian inference, see MacKay (2003).

### 3.3.9 Outlook

There is an almost infinite number of concepts we could add at this point. We could discuss other distributions, such as the Chi-squared distribution. We could take Bayes one step further and talk about likelihood modeling. However, this will suffice as a conceptual introduction to mathematical and statistical fundamentals. Some of these issues, e.g., maximum likelihood estimations, will be revisited in later chapters where appropriate. For now, we feel that enough groundwork has been laid, and that—if you have been working with us through this material—it is solid enough to carry you through the next chapters, which is really all anyone can ask for.

## MATLAB FUNCTIONS, COMMANDS, AND OPERATORS COVERED IN THIS CHAPTER

**eye**
**inv**
**det**
**eig**
**cov**
**nchoosek**
**mean**
**var**
**std**
**normrnd**
**normcdf**
**erf**
**scatter**
**erfinv**
**fcdf**
**regress**

# Programming Tutorial: Principles and Best Practices

## 4.1 GOALS OF THIS CHAPTER

Unlike most other sections in *MATLAB® for Neuroscientists*, the focus here is not on learning new techniques in MATLAB, but on how to use those techniques *better*. The sections that follow introduce guidelines for code organization in small and large projects, defect (bug) control, and testing strategies in an attempt to communicate strategies for managing the complexity that comes with larger programming efforts.

In order to benefit maximally, basic proficiency with MATLAB coding is necessary. Working through the MATLAB tutorial should be adequate preparation; however, progressing through a few sections beyond the tutorial is an even better preparation. The additional experience with MATLAB will provide a stronger foundation for understanding the rationale for the suggestions that follow.

## 4.2 ORGANIZING CODE

### 4.2.1 A Few Words about Maintenance

Code should be written with the expectation that the author will not be maintaining the code. This is especially true in many laboratories, where code is passed down from student to student, sometimes with very little direct communication. It can be very tempting to whip out a few lines of code to solve a simple problem, and think that the code might be a throwaway solution or only used by the code's author. As with other technical solutions to scientific challenges, solutions to computational problems are rarely entirely unique. A part of yesterday's throwaway code might be the kernel of someone else's thesis. *Maintainability should be as important as functionality in writing new code.* It is in this spirit that the following sections offer suggestions for writing more maintainable code.

### 4.2.2 Variables and How to Name Them

Simply put, a variable denotes a storage location. That location can hold a number, a function, a matrix, or even more complex entities, such as cell arrays or MATLAB objects. In the context of MATLAB code, a variable is referenced by name and scope. This section will discuss variable naming strategies. Variable scope is equally important, but it will be discussed in the next section.

Variable names can be any contiguous set of alphanumeric (i.e., 0–9 and a–z) characters plus the underline character, and they begin with a nonnumeric character. For example, this_is_a_variable and th1s_1s_4ls0_4_v4r14bl3. With such flexibility in name choice, it is surprising how often poor names are chosen.

Here is a simple function written twice to demonstrate the impact of good variable name choices. First, a version of the function with poorly chosen variable names.

```matlab
function out = align_waveforms(x)
    % Determines alignments for a set of waves relative
    % to the initial waveform using
    % cross correlation.
    % Input parameters
    % x:        MxN matrix of waveforms, where x(m,:) is the nth waveform
    % Output parameters
    % out:      vector of length M, where out(m) is the offset relative
    %                   to the first wave

    n = size(x);
    n = n(2);
    out = [];
    for w = 1:n
        c = xcorr(x(:, 1), x(:, w));
        s = find(c == max(c));
        d = s - length(c)/2;
        out = [out d];
    end

end
```

The same function with better variable names follows.

```matlab
function offsets = align_waveforms(waves)
    % Determines alignments for a set of waves relative
    % to the initial waveform using
    % cross correlation.
    % Input parameters
    % waves:    MxN matrix of waveforms, where x(m,:) is the nth waveform
    % Output parameters
```

```
% offsets:   vector of length M, where out(m) is the offset relative
%                to the first wave

wave_count = size(waves);
wave_count = wave_count(2);
offsets = [];
for wave = 1:wave_count
    c = xcorr(waves(:, 1), waves(:, wave));
    max_c_index = find(c == max(c));
    offset = max_c_index - length(c)/2;
    offsets = [offsets offset];
end

end
```

Clearly, variable name choice has impact on readability, even in short functions. Here are a few simple guidelines for naming variables that promote readability and maintainability.

**Avoid the names of global functions.** When MATLAB encounters a sequence of characters that forms a valid name, the variables in the current workspace are checked first for possible matches. MATLAB searches for functions, scripts, or classes only if an identifier fails to match any existing variable names. Choosing a name synonymous with a MATLAB function, or even a user-defined function, hides that function in the current scope. In the example code ahead, assigning the value 5 to a variable named "factorial" causes the subsequent attempt to call the factorial function to fail. Because MATLAB recognizes factorial as a variable, the interpreter attempts to resolve (4) an index into a vector. Since the variable factorial is a scalar, the index request fails and yields the error.

```
>> factorial(4)

ans =

        24

>> factorial = 5;
>> factorial(4)
??? Index exceeds matrix dimensions
```

Especially inappropriate choices can even cause difficult to identify errors. Another example ahead shows how setting gamma to a vector creates a situation in which an integer argument to the gamma function is misinterpreted as an index into the vector and yields the wrong value.

```
>> gamma(4) % the correct value for gamma(4) is 6

ans =

      6

>> gamma = [0  1  2  3  4  5];
```

```
>> gamma(4)

ans =
      3
```

While the MATLAB interpreter is able to evaluate the request for element 4 of the vector gamma without an explicit error, this is highly confusing to anyone familiar to the **gamma** function. If the intent of gamma(4) was actually to invoke the **gamma** function, the expression returns the wrong result silently. Such code needlessly complicates later maintenance and readability.

The **which** command is especially useful for determining if a variable might override an existing command. Typing **which** followed by a potential name for a variable displays information about the identity of that name. The **clear** command will remove a variable from the current workspace, which is quite useful when inadvertently overriding an important command. Note how **clear** alters how MATLAB resolves the identity of **gamma** in the following example.

```
>> gamma = [0  1  2  3  4  5];
>> gamma(4)

ans =

      3

>> which gamma
gamma is a variable.
>> clear gamma
>> gamma(4)

ans =

      6

>> which gamma
built-in (/sw/matlab-7.11/toolbox/matlab/specfun/@double/gamma) % double method
```

**Pick a mnemonic name**. A name that reflects the purpose of a variable improves readability significantly. Although it's quite tempting to choose short, one-character variable names such as n or x, variable names should reflect the variable's use or contents whenever possible. A common MATLAB task involves writing mathematical formulae as executable MATLAB code. When writing such code, the use of exceptionally short variable names is especially tempting, since variables used in mathematical notation are quite often single letters. In the simplest of functions, this is reasonable, especially if the function operates uniformly on all inputs, i.e., there is no specific meaning ascribable to the variable. This is fairly rare, however. Aside from simple mathematical functions or variables used as indices in **for** loops, single-letter variables should be avoided.

A mnemonic name is not an invitation to a stream of consciousness description of the code, however. An excessively long name might be a subtle clue of an overly broad or imprecise use. For example, a variable named indicates_yes_response_or_viable_value should probably be broken into two separate variables for simplicity. This guideline is particularly true for any variable expected to be used interactively. No one wants to type indicates_yes_response_or_viable_value over and over in an interactive session unless absolutely necessary.

**Retire variables after their specific purpose.** Repurposing variables can make code difficult to follow and maintain, particularly in a long or complicated sequence of code. Usually choosing mnemonic variable names automatically avoids this problem.

---

### EXERCISE 4.1

Review the following code and rename variables that could be better named. Use the comments as a guide to the intended functionality.

```
function psth,bins = bin_for_psth(rd, ...
                        sampling_rate_in_samples_per_second, ...
                        t, ...
                        q, ...
                        q2, ...
                        size_of_each_psth_bin_in_seconds)
% Locates events above threshold in raw data and generates PSTH from
% multi-trial recording. Trials should be contiguous.
%
% Input   parameters
%       rd:    raw input data
%       sampling_rate_in_samples_per_second:   sampling rate in Hz
%       t:   threshold for events, in same units as raw data
%       q:   number of contiguous trials
%       q2:   length of each trial, in seconds
%       size_of_each_psth_bin_in_seconds:   size of each PSTH bin, in
seconds
%
% Output parameters
%       psth:              count in each bin
%       bins:              center position of each bin, in seconds relative
to
%                          trial start

% first, threshold signal
vnts = rd > t;
% only positive threshold crossings (not sustained activity above
threshold)
vnts = diff(events) == 1;
vnts = [0 events];
% now, split into trials
```

```matlab
        vnts = reshape(vnts, q, ...
                            q2 * sampling_rate_in_samples_per_second);
        % vnts should be MxN, where M = trial and N = sample
        sum = sum(vnts);
        % sum should be the sum of events at each sample relative to
        % the start of the trial
        max_vnt_count = max(sum);
        for count = 0:max_event_count-1
                above_count = find(sum > count);
                vnt_offsets = [vnt_offsets above_count];
        end
        % vnt_offsets should be the offset in sample counts where events
    occur
        vnt_ts = vnt_offsets / sampling_rate_in_samples_per_second;
        [psth, bins] = hist(vnt_ts, q2/size_of_each_psth_bin_in_seconds);
    end
```

## 4.2.3 Understanding Scope

Scope refers to the extent of a variable within the code. During execution, variables move in and out of scope. For example, under normal circumstances, a variable created within a MATLAB function ceases to exist once the function ends. One of the most important aspects of scope is that scope together with name uniquely denotes a variable. Two or more variables with identical names can coexist separately in different scopes.

Related to the idea of scope is the MATLAB workspace, which acts as a container for variables within a specific scope. In a sense, workspaces implement scope. Unlike the abstract concept of scope, workspaces in MATLAB are entities that can be viewed and interacted with. The most visible workspace is the workspace associated with the command line, which is visible in the workspace window during interactive sessions, but other workspaces are created, suspended, and destroyed as necessary to implement other scopes during execution.

MATLAB recognizes three basic scopes: the command-line scope, global scope, and function-level scope. Each of these has one or more corresponding workspaces during execution. With a few exceptions, when the MATLAB interpreter encounters a legal variable name, the current workspace is checked for a match. Which workspace is current changes throughout execution. If a match is located, the identifier is treated as the corresponding variable. We will now discuss each of these types of scope (command line, global, and function level) and their corresponding workspaces.

The command-line scope consists of all variables created interactively at the command line or in a script file. Unlike functions, scripts operate under the command-line scope. Thus, scripts have access to all variables present in the command-line scope and their values. This makes scripts especially useful for executing sets of commands that one would normally type at the command line, such as commands to set up an environment for a specific type of analysis or visualization. It also means that scripts can easily overwrite variables in the command line scope.

Function-level scope occurs at the entry of a function (i.e., the beginning of execution, at a function call) and continues until the function ends, usually at the return. Variables within the function-level scope do not persist after the execution of a function call. Cases such as the code ahead may appear to counter this assertion, but a careful analysis demonstrates that this is not the case.

----- in square.m-----

```
function x = square(x)
       x = x * 2;
```

------at command line-----

```
>> x = 6;
>> x = square(x);
>> x

x =
    12
```

At the command line, it may appear that the value of $x$ persists beyond the call to square () or within the call to square(). However, this is not the case. Initially, there is a variable $x$ declared at the command line. This variable contains the value 6. Then, the assignment statement x = square(x) is executed, and the function square() is called, with the parameter $x$. During the execution of the call to square, all the expressions in the parameter list are evaluated prior to the call. In this case, the variable $x$ is evaluated, and it refers to the variable $x$ at command line scope. So the value is 6, and square() is called.

When square() is called with the parameter 6, the value 6 is bound to the function-level scope variable $x$ during the execution of square(). It is crucial to note that this $x$ at the function-level scope has no relationship with the variable of the same name at the command line scope aside from the confusing nature of their identical names. During the execution of square(), the function-level variable $x$ is set to 2 times itself, or 12, and the function returns. At the return of the function, the value of function-level $x$ is obtained as the return value of the function (this return value is 12) and bound to the command-line scope variable $x$ as the final part of the assignment statement. Thus, the command-line scope variable $x$ will then contain the value 12.

As demonstrated above, when examining functions and function calls, it is important to remember that all parameters in a function call are evaluated before the call and then bound to variables in the function-level scope. In other words, variable names in a function call have no direct relationship with variable names in the function's code. So, in the previous example, the command-line variables could have been named x2, y, or even not_the_x_in_square, and the example would have produced the same result. Formally, parameters in the function's code are termed *formal* parameters to distinguish them from parameters in the calling of a function, which are called *actual* parameters.

Function-level scope can become especially complex with recursive functions. Recursive functions invoke themselves, albeit (usually) with different parameters each time. Here is an example function that implements a factorial, N!. The function is called factorial2 to avoid conflict with the MATLAB built-in function factorial.

```
function f = factorial2(g)
if g == 1
     f = 1;
     return
end
f = g * factorial2(g-1);
end
```

So, calling **factorial2** with a value of 3 causes it to call **factorial2** with a value 2, which calls **factorial2** with a value of 1. The innermost **factorial2** call which was passed a value 1 terminates, returning 1. Then, the **factorial2(2)** call resumes, calculating 2*1 and returning the result 2. Finally, the original call **factorial2(3)** resumes, calculating 3*2 (2 being the result from **factorial2(2)**) and returning 6. This sequence of calls and the associated creation, suspension, and deletion of scopes is illustrated in Figure 4.1.

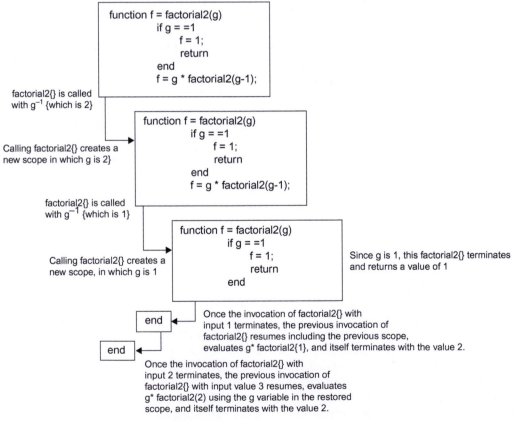

FIGURE 4.1   Sequence of scopes during a recursive call

It is important to note that each invocation of **factorial2** creates its own scope. Consequently, each variable *g* in an invocation is different from the variables named g in other invocations. As each execution of **factorial2** invokes **factorial2** with a slightly smaller input parameter, the existing scope is suspended for the execution of the inner call and resumed when the call returns.

In the context of execution, each scope created by the invocation of a function call is sometimes called a stack frame. The collection of all the frames existing at any point in the execution, suspended or live, is sometimes called the execution stack or call stack. **factorial2()** can be modified to make the stack frame a little more visible during execution by displaying input parameter g and the current stack using **dbstack**:

```
function f = factorial2(g)
g
dbstack
if g == 1
   f = 1;
   return
end
f = g * factorial2(g-1);
end
```

Invoking **factorial2()** on the input 3 at the command line involves 3 calls to **factorial2**.

```
>> factorial2(3)

g =

     3

In factorial2 at 3

g =

     2

In factorial2 at 3
In factorial2 at 8

g =

     1

In factorial2 at 3
In factorial2 at 8
In factorial2 at 8

ans =

     6
```

Global scope has characteristics of both command-line and function-level scopes. Like command-line scope, variables in global scope persist for the lifetime of the MATLAB interpreter. Like function-level scope, variables in global scope are accessible during function execution. One substantial difference between global scope and either of the two other scopes is the necessity of explicitly specifying global scope for those variables which need it. This is done with the **global** keyword. Here is an example showing two functions referencing the same variable in global scope.

```
function x = increment()
   global increment_value
   x = increment_value + x
end
function set_increment(y)
   global increment_value
   increment_value = y
end
```

Without the statement **global increment_value** in either one of the functions, subsequent statements would attempt to reference a variable of the same name (**increment_value**) with function-level scope. As one might expect from the scope of global variables (i.e., available wherever **global** appears), all global variables share a workspace separate from command-line and function-level workspaces.

### 4.2.4 Script or Function?

Script files and function files in MATLAB have distinct purposes, tightly related to both files' respective use of scope. Because script files operate in the workspace used at the command line, script files act much like commands typed into the interactive prompt. This means that all variables accessible from the command line are visible and can be modified. This workspace sharing can be spectacularly useful for tasks like automating a set of commands that one would normally type into the command line without modification.

Functions, on the other hand, create their own workspaces upon execution. Function parameters are the primary means of moving data from an external workspace into the function's workspace. As such, any local variables used in the function are isolated from variables of the same name outside the function.

When beginning to code in MATLAB, putting all code in script files may feel most natural, since script file execution operates so similarly to commands executed at the command line. However, over time, this becomes problematic. Using scripts as functional units requires designating a specific set of variables to be used to move data between scripts. If those variables are inadvertently used in scripts for other purposes, certain scripts may become unusable. By operating in isolated workspaces, functions avoid this problem. Thus, scripts are best for sequences of commands that require no parameterization.

### 4.2.5 The Art of Commenting

Nothing guarantees the obsolescence of code like the absence of comments. Even the clearest code will have areas whose details or larger goals are not self-evident. While too many obvious comments can obfuscate the code, it is better to err on the side of too much rather than too few.

In MATLAB, comments immediately following the declaration of functions or objects are particularly useful, since the help command will display these comments. Such usage comments should include the name, a brief summary of purpose, input and output parameters, and any data structures or nonobvious steps preliminary to usage required. Especially important to note in comments are nonobvious side effects (see side effects, ahead), such as modifying a global variable or altering a file.

## 4.3 ORGANIZING MORE CODE: BIGGER PROJECTS

### 4.3.1 Why Reuse Code?

As a project grows in scope, the amount of code becomes difficult to manage. Every line of code is a site for an unforeseen error. Minimizing code is an effective strategy for simplifying maintenance. One of the most significant ways of minimizing code is the reuse of existing code.

Very frequently, sections of code are very similar, save a few parameters. Placing these functional units within a separate function file has many advantages. In addition to reducing the amount of code, such an approach allows for testing the functionality independently. When a project is constructed from such independently testable, reusable parts there are fewer errors, and those errors that do occur are typically simpler to eradicate than in code which lacks this organization. Such projects also often require substantially less new code and are faster to implement.

This section focuses on guidelines which foster reuse and maintainability. Low coupling isolates unrelated areas from future changes in a logical unit. High cohesion and separation of concerns push related units together so that maintenance to a given logical unit requires modifying as little code as possible. Side effects often limit reuse. Finally, object oriented design provides one means of reducing coupling and improving cohesion.

### 4.3.2 Coupling and Cohesion

Coupling describes the flow of information to and, to some degree, the degree of dependency between two or more logical units. Functions requiring a greater quantity of structured information for use, such as parameters or global variables, are described as having high or strong coupling. Functions requiring minimal amounts of parameters or global variables are likewise described as having low or weak coupling.

```
function s = square(m)
    s = m .^2
end

function s = square_field(m)
  s = m.matrix_data.^2
  global global_sum
  global_sum = global_sum + sum(s)
end
```

In the function square, there is a small degree of coupling between square and a calling function. The function square expects a single input parameter and returns a single output parameter. The function square_field is far more coupled to its caller. Like square, square_field expects a single input parameter and returns a single output parameter. However, square_field expects the input parameter to be a MATLAB object with a field named matrix_data. Moreover, square_field uses a global variable to track the global sum of squares. These two aspects of square_field's functionality must be understood and managed by any function invoking square_field.

High coupling marks a more complex relationship between two or more functions. Usually, a more complex relationship is more difficult to manage in the event of modification. In the example above, if the name of the field matrix_data needed to be changed, every caller of square_field would need to be changed in addition to square_field itself.

Low coupling has a number of advantages for reuse. With fewer parameters to construct and provide, simpler interfaces are easier to integrate into existing code. The simpler relationships of weaker coupling also mean less to understand when reading or maintaining the code in the future.

Cohesion describes the degree to which a unit's functionality achieves a single purpose. Functions in which the functionality of each constituent part implements some necessary aspect of a clear purpose have high cohesion. The example functions above differ in cohesion as well as coupling. The function square has high cohesion. Its single statement implements the clear functionality of the unit, which is an element-wise squaring of the input. The function square_field, on the other hand, could be described as less cohesive. The rationale for tracking the global sum of elements is not clear from the function's name or other code. As such, one could argue that the statements dealing with the global variable global_sum are not aligned with the primary purpose of the unit, which is to calculate the square of the input variable's field matrix_data.

Functions with lower cohesion are often more difficult to reuse. Such functions often have additional input/output variables or even global variables that must be accounted for in the calling code. Accounting for unnecessary aspects of the function takes additional time and effort that, themselves, should not be necessary. Code focused on a single purpose is usually simpler, which is often easier to read, maintain, and debug.

### 4.3.3 Separation of Concerns

Separation of concerns is a useful guiding principle in organizing a larger project. Under separation of concerns, all code related to providing a distinct feature is grouped together in a separate logical unit. This logical unit could be a single-named function for simpler features or a set of functions or objects for more complex features. Thus, if systems have overlapping common features (i.e., concerns), then separation argues for separating out those common concerns into a new logical unit.

This is quite similar to, but distinct from, high cohesion. High cohesion demands conformity of purpose within a logical unit. Separation of concerns seeks to collect similar functionality across a system within a single logical unit. Often, maintaining high cohesion will result in a very natural separation of concerns among logical units.

Separation of concerns provides the substantial benefit that different areas of the project can be tested and modified independently of other functional units. Let's take a hypothetical example of an application that presents auditory stimuli and records EEG data during the stimulus presentation. In such an application, there are many concerns, including randomization of stimuli, playback of stimuli through the speakers, displaying EEG data in realtime, collecting EEG data through hardware, and storing the EEG data. This list is only an example. Different scenarios might lend themselves to a different set of concerns. Under a separation of concerns model, all the code dealing with one of these concerns, for example, realtime display, would be confined to one or more functions that would be limited as much as possible to realizing realtime display functionality.

### 4.3.4 Limiting Side Effects, or the Perils of Global State

Using nonlocal variables, either through script files or through variables declared as **global**, strongly limits reuse. Imagine a function that counts spikes in a recording and uses a global variable to track the total number of spikes over all recordings:

```
function interval_count = count_spikes(r, threshold)
      global global_count;
      % above_threshold will be for every sample above threshold
      above_threshold = r > threshold;
      % counting only points where diff(above_threshold) > 0
      % counts the number of contiguous blocks of samples above threshold
      interval_count = diff(above_threshold) > 0;
      global_count = global_count + interval_count;
end
```

This may be a convenient way of tracking the overall count, but this mechanism imposes significant constraints on how count_spikes could be used. Now imagine a set of extracellular recordings over time intervals for multiple sites, made simultaneously, stored in an interleaved fashion (i.e., site 1 for interval 1, site 2 for interval 1, site 3 for interval 1, site 1 for interval 2, site 2 for interval 2, site 3 for interval 2, etc.). If the intervals are processed in order, the global count will include spikes above thresholds at all three sites instead of counting the total spikes at each site separately.

Outside of limitations imposed by complex usage patterns, the use of global variable global_count could also limit how other functions are used in the same project. Since all global variables share a workspace, any other function that uses a global variable named global_count could disrupt the accumulation of results in global_count.

Modifying global variables in a function is a specific case of what is termed a side effect. A side effect is any change in the run time state outside the scope of a function. Sometimes, side effects are absolutely necessary. For example, printing text to the screen and writing to a file would both be considered side effects; such actions change the console (printing) or file system (writing to a file). Often, side effects are unnecessary, as in the previous example. Tracking all spikes at a given site is best left to the caller of count_spikes, as no information about the site context is provided to count_spikes.

## 4.3.5 Objects

Object-oriented programming (OOP) has been a popular programming paradigm for more than a decade. The fundamental kernel of object-oriented programming is the capability to package together cohesive units of code and data as "objects," entities that can be manipulated programmatically. Objects can refer to physical real world objects or can be highly abstract concepts. Most programming languages in common use support some mechanism for object-oriented programming, and MATLAB is no exception.

Object-oriented programming provides a mechanism for separating concerns, reducing coupling, and increasing cohesion. First, the object-oriented paradigm allows otherwise difficult to extract bits of code from multiple routines to be grouped together logically. Secondly, large, relatively inflexible function parameter sequences can be replaced by one or more flexible objects. In this case, the same data is passed between functions, but the semantics of objects allow many types of changes to objects without requiring alterations to those objects' users, which reduces coupling. Thirdly, by grouping related bits of code from throughout the system into logical units, the cohesion of the routines from where the bits of code originated improves.

Beyond the capacity to create and manipulate objects, there is no strict set of features that compose the object-oriented programming paradigm. Programming languages differ significantly on the functionality that their respective object models provide. Even MATLAB supports two different object models, with varying functionality. Within that variety of models, the following features are strongly associated with the object-oriented programming paradigm, and many object models support a majority if not all of them:

**Encapsulation**: the grouping together of data and relevant code in cohesive units
**Data hiding**: limiting access to data or executable routines related to functionality internal to the object
**Inheritance**: allowing "descent of objects"; objects can be defined as descendants of other objects, gaining their data and executable code
**Subtype polymorphism**: functioning as objects of a parent type in places where an object of the parent type are expected
**Dynamic dispatch**: the capability to differentiate among multiple implementations of a routine at runtime depending on the identity of an object

As mentioned earlier, MATLAB has supported two separate object models. The more recently introduced object model, available in MATLAB R2008a and later, provides for all the features listed above. Only this object model will be discussed here.

### 4.3.5.1  Creating Objects

Under the MATLAB object model, one specifies the data held by an object and associated routines in a class definition. These data are called member variables or properties, and the executable routines are called methods. This terminology is fairly standard among object models. Once a class is defined, any number of objects (limited by memory, of course) can be created from that class through a process called instantiation. Each object can hold its own copy of the member variables and operate on them independently of objects of the same class.

To illustrate the benefits of object-oriented programming, we will create a set of objects that will present a unified model for locating auditory recordings, regardless of the underlying format representation. At this time, we anticipate that other code that analyzes the auditory recordings will require a sampling rate and a time stamp for the start of the file in addition to the raw data of the recording.

The initial class we define will provide a basic interface for all recordings for obtaining raw data, sampling rate, and time of recording. Here is code for this initial class.

```
classdef recording
    properties
        filename
    end
    methods
        function obj = recording(filename)
            obj.filename = filename;
        end
        function t = timestamp(obj)
            % get the time stamp by getting the date from dir
            d = dir(obj.filename);
            t = d.date;
        end
        function r = sample_rate(obj)
            r = -1;
        end
        function d = raw_data(obj)
            d = [];
        end
    end
end
```

The class definition begins with the reserved word **classdef**. Like most statements that introduce blocks in MATLAB, **classdef** has a matching end. Within **classdef**, there are properties and methods sections. We'll discuss the methods section in a moment. Names for data managed by the class are specified in the properties section. In this case, a property called "filename" is defined.

The methods section contains the executable routines specific to the class. Object-oriented programming has a number of terms to describe the subtly different invocation of functions in the context of objects. Functions bound to objects and operating on them are called methods. Outside of MATLAB, data held by objects are often called members or member fields (MATLAB calls them properties).

Examining the methods, one will quickly discover that all but the first method have an initial parameter, **obj**. This initial parameter is the object being referenced. This should be fairly clear in the implementation of **timestamp()**. The code in **timestamp()** obtains the name of the file through the filename property of the referenced object, which is then used to locate the date through the **dir()** function. With the exception of this initial parameter, methods operate similarly to normal functions.

Now we look at the first method. This method has the same name as the class as specified after the **classdef** statement. That marks this method as a special method, called a constructor. A constructor is eponymous with the class and includes initialization code to be executed when the object is *constructed*. The parameter list of the constructor lists all the parameters required to create an instance object of the class. Unlike the other methods, there no is referenced object as a first parameter, since the object has not been created yet. Instead, the constructor has the output parameter that appears to go unassigned. It is this variable that holds the newly created object during the invocation of the constructor. To initialize our recording object, the value of the filename variable passed into the constructor must be copied into the property of the same name in **obj**. This may seem unnecessary, but the two variables named filename are entirely distinct and live in entirely different scopes, one within the class recording, and one as a local variable in the constructor for recording.

To create a recording object, type the following:

```
>> r = recording('test.wav');
```

Look at the time stamp:

```
>> r.timestamp()
```

Note that the methods to load the data and return the sample rate are unimplemented:

```
>> r.sample_rate()
>> r.raw_data()
```

Creating implementations for these methods are the focus of the next section.


### 4.3.5.2 Inheritance

At this stage, it would be helpful to be able to load a sound file. The example below shows code for a **wav_recording** class, which loads WAV files. The code for the **wav_recording** class is fairly similar to the recording class, with a few differences.

```
classdef wav_recording < recording
    methods
        function obj = wav_recording(filename)
            obj = obj@recording(filename);
        end
        function r = sample_rate(obj)
            [data, r] = wavread(obj.filename);
        end
        function data = raw_data(obj)
            [data, r] = wavread(obj.filename);
        end
    end
end
```

The most noticeable difference is the less than sign and "recording" at the beginning of the class definition, immediately after the class name. These denote that the **wav_recording** class should inherit from recording. This inheritance means that objects of the **wav_recording** class have their own copies of the properties and methods in recording. Through inheritance, all objects of **wav_recording** are also objects of recording. Note the constructor. The unusual function call in the constructor references the constructor of the parent class, recording. Calling the constructor in recording ensures that the filename input parameter in the **wav_recording** constructor will be copied to the filename property during the execution of the constructor of recording. Also, note the absence of the timestamp method here in **wav_recording**. Since the functionality provided in the parent class is sufficient, there is no need to override it here.

```
>> r = wav_recording('test.wav')
>> r.timestamp()
```

The inheritance is also clear in the implementations of **sample_rate()** and **raw_data()**. In these methods, the filename property of the object is referenced, and this relies upon the definition of the parent class recording.

Try obtaining the sample rate:

```
>> r.sample_rate()
>> plot(r.raw_data())
```

The previous functionality is still available, simply by instantiating a recording object:

```
>> r = recording('test.wav')
>> r.sample_rate()
>> plot(r.raw_data())
```

The capability of the MATLAB interpreter to choose the proper method based on the class identity of the object is called *dynamic dispatch*. For dynamic dispatch to work properly, the method name and input parameter lists must be the same throughout the class hierarchy.

Now, we will add support for PCM audio files. PCM (pulse code modulation) is a simple file format that stores digitized samples as 16 bit integers. Unlike WAV files, PCM files include only data, and the sample rate must be stored elsewhere (e.g., in experimental notes or in a separate file). Because of this, we will include the sample rate as a parameter on the constructor. Our PCM reading–recording class will also require PCM-specific implementations of **sample_rate()** and **raw_data()**, as did the WAV reading class. Here is code for a PCM reading class:

```
classdef pcm_recording < recording
    methods
        function obj = pcm_recording(filename, sample_rate)
            obj = obj@recording(filename);
            obj.sample_rate = sample_rate;
        end
        function r = sample_rate(obj)
            return obj.sample_rate;
```

```
        end
        function data = raw_data(obj)
                fid = fopen(obj.filename, 'r');
                if fid  == -1
                        error('Unable to open file' + obj.filename);
                        data = [];
                        return
                else
                        data = fread(fid, inf, 'uint16 = >double', 0, 'l');
                        fclose(fid);
                end
        end
    end
end
```

Note that the **sample_rate()** method does nothing with the file, but it returns the sample rate specified through the constructor.

Try the following:

```
>> r = pcm_recording('test.pcm', 20000)
>> r.sample_rate()
>> r = pcm_recording('test.pcm', 40000)
>> r.sample_rate() % note 40 kHz rate now
```

One of the major benefits of working under an object-oriented paradigm is the ability to write code that works under a variety of cases and that, at the same time, isolates those cases in separate pieces of code. This strongly promotes both high cohesion and lower coupling. The example ahead shows a set of functions that scan for events above a threshold and report on their threshold crossing times.

```
function dates = threshold_crossings(wav_filename, start_time)
    % Input parameters
    % wav_filename : filename for WAV file
    % start_time : start time of the WAV recording
    raw_data, sampling_rate = wavread(wav_filename);
    above = raw_data > threshold;
    threshold_crossings = diff(above)  == 1;
    threshold_times = find(threshold_crossings);
    % threshold_times is the sample count since start_time
    % for each threshold crossing
    threshold_sec = threshold_times / sampling_rate;
    dates = zeros((length(threshold_sec), 1));
    for ii = 1:length(threshold_sec)
        dates[ii] = addtodate(datenum(start_time, threshold_sec, 'second'));
    end
end
```

At the moment, this function works only for WAV files. We can change **threshold_crossings()** to work with our **wav_recording** objects with a small amount of work. The next example shows **threshold_crossings()** modified to use recording objects. In making the change, we reduce coupling somewhat, as **threshold_crossings** now only requires a single input parameter, even though new constraints are placed on that parameter (it must support **raw_data**, **sampling_rate**, and **start_time** methods). Cohesion is improved as well, as the WAV loading code is no longer in **threshold_crossings()**.

```
function dates = threshold_crossings(rec)
    % Input parameters
    % rec : audio recording object

    raw_data = rec.raw_data();
    sampling_rate = rec.sampling_rate();
    start_time = rec.start_time();
    above = raw_data > threshold;
    threshold_crossings = diff(above) == 1;
    threshold_times = find(threshold_crossings);

    % threshold_times is the sample count since start_time
    % for each threshold crossing
    threshold_sec = threshold_times / sampling_rate;
    dates = zeros((length(threshold_sec), 1));
    for ii = 1:length(threshold_sec)
        dates[ii] = addtodate(datenum(start_time, threshold_sec, 'second'));
    end
end
```

One substantial benefit in making the change is near effortless support of PCM files obtained through data hiding and encapsulation. Since all the code specific to PCM loading is isolated in the PCM class, we can safely make changes in **threshold_crossings()** to support generic recordings without worrying about PCM support within **threshold_crossings()**.

---

### EXERCISE 4.2

Write a simple function **play_recording** that uses sampling rate appropriately to play an audio recording. Use the sound() function with raw data and sampling rate to produce correctly timed sound.

---

### EXERCISE 4.3

Write **play_recording** from Exercise 4.2 as a method on the recording class.

---

---

### EXERCISE 4.4

File test_audio.hdf5 contains an audio recording within an HDF5 file. HDF5 is a high performance hierarchical data format used to create structured files (i.e., files containing many types of data in an organized, labeled manner). MATLAB has substantial support for HDF5 files. The function **h5disp()** displays the content of an HDF5 file:

    h5disp('test_audio.hdf5')

There are also functions to read datasets (h5read) and attributes on datasets (h5readatt). Within the file test_audio.hdf5 at location "/audio/recording1" is audio data. Attached to that dataset as an attribute "sampling_rate" should be a sampling rate, in Hz.

Write another child class of **recording** that provides an implementation of this structure within an HDF5 file. You can test your code on test_audio2.hdf5, which should conform to the same format.

---

### EXERCISE 4.5

Generalize the class written for Exercise 4.4 to work with any HDF5 file in which a dataset containing a vector of values and an associated attribute with sampling rate will work, regardless of the dataset location within the HDF5 file or the name of the sampling rate attribute. (*Hint:* The names of these two keys will need to be specified at object instantiation, in the constructor!) Try your solution on test_audio3.hdf5, which has audio data at /audio/recording2 and sampling information at "samples_per_sec" on /audio/recording2.

---

#### 4.3.5.3 *Passing Objects Around: The Handle Class*

Much like other types of variables, MATLAB objects are copied in and copied out during function calls. Here is an example that demonstrates this phenomenon for a vector:

```
function x = no_change(in_vector, index, new_value)
      in_vector(index) = new_value;
end
```

Type the following:

```
>> x = 1:5;
>> no_change(x, 4, 2)
>> x

x =

      1   2   3   4   5
```

For the change to be permanent, the input parameter must be moved to the output parameter, as in the following:

```
function x = change(in_vector, index, new_value)
        in_vector(index) = new_value;
        x = in_vector;
end

>> x = 1:5;
>> y = change(x, 4, 2);
>> x

x =

     1   2   3   4   5
>> y

y =

     1   2   3   2   5
```

While the original input $x$ is still unchanged, the modification does leave the function as an output, which is copied into the variable $y$. Objects operate much the same. This is even the case for methods that modify properties. To preserve the change, the modified object must be copied out:

```
classdef example
    properties
            name
    end
    methods
            function change_name1(obj, new_name)
                    obj.name = new_name
            end
            function change_name2(obj, new_name)
                    obj.name = new_name;
            end
    end
end

>> r = example;
>> r.change_name1('new name');
>> r

r =

  example

  Properties:
    name: []

  Methods

>> r.change_name2('new name');
```

```
>> r

r =

    example

    Properties:
        name: []

    Methods

>> r = r.change_name2('new name');
>> r

r =

    example

    Properties:
        name: 'new name'

    Methods
>> r2 = r

r2 =

    example

    Properties:
        name: 'new name'

    Methods

>> r2.name = 'testing';
>> r2

r2 =

    example

    Properties:
        name: 'testing'

    Methods
>> r

r =

    example

    Properties:
        name: 'new name'

    Methods
```

Only passing the modified object out and storing the result in allows for the method to change the property value. Also note that assigning $r$ to a new variable $r2$ causes the object held by $r$ to be copied. The result is a new object in $r2$ distinct from that held by $r$.

The situation may arise where this copying of objects is undesirable. This often is the case when objects are used to manipulate the state of a non-duplicating resource, such as a file reference or a graphical object. Another case in which these semantics are undesirable occurs when creating objects where the internal state of the object could change without the awareness of the invoking code. The WAV reading class written earlier illustrates an example of this later use.

Because the MATLAB function **wavread()** always reads the whole WAV file, even if only the sampling rate is needed, invoking **wav_recording.sample_rate()** will still read the entire WAV file into memory. Moreover, invoking **wav_recording.raw_data()** immediately afterwards reads the WAV file into memory again. Ideally, the contents of a file could be stored away as a property when read. Whenever the data was requested through the **raw_data()** method, the previously read contents of the file could be returned if available, saving the overhead of an extra read. Unfortunately, this would require the parent class' **raw_data()** to return the current object as a parameter in addition to the read data. This also makes for a messier call, since the call would be something like

**[data, r] = r.raw_data()**

Fortunately, MATLAB provides an alternate method of passing objects. Under this alternate object passing mechanism, variables refer not to objects directly, but to handles of objects. With this paradigm, multiple variables can "point" to the same object, and the normal copying of variables during function calling only copies a handle. To specify that a class use this alternate passing mechanism, classes must inherit from **handle**. Open up the example class above, and add "< handle" to the end of the first line. Change the class name to example2, and save as example2. The first two lines of example2 should be:

```
classdef example2 < handle
        properties
        .
        .
        .

```

Then, type

```
>> r = example2;
>> r.change_name1('new name');
>> r

r =

   example2 handle

   Properties:
      name: 'new name'

   Methods, Events, Superclasses
```

```
>> r2 = r

r2 =

    example2 handle

    Properties:
      name: 'new name'

    Methods, Events, Superclasses

>> r2.name = 'testing'

r2 =

    example2 handle

    Properties:
      name: 'testing'

    Methods, Events, Superclasses

>> r

r =

    example2 handle

    Properties:
        name: 'testing'

    Methods, Events, Superclasses
```

Invoking **change_name1** on *r* modified *r* itself. Likewise, *r* and *r2* refer to the same object, so changes made using *r* are visible when examining the object through *r2*.

For completeness, here is a class **wav_recording2** that implements the caching discussed earlier. A given file is only read once, regardless of how many times **sample_rate()** or **raw_data()** is invoked. The parent class is **recording2**, which is identical to recording except that it inherits from handle. A listing of the first few lines follows the listing for **wav_recording2**. If a class inherits from handle, then all parent classes must as well.

```
classdef wav_recording2 < recording2
    properties
            stored_data = [];
            stored_rate = -1;
    end
    methods
            function obj = wav_recording2(filename)
                    obj = obj@recording2(filename);
        end

        function r = sample_rate(obj)
```

```
                if obj.stored_rate > = 0
                     r = obj.stored_rate;
                else
                     [data, r] = wavread(obj.filename);
                     obj.stored_data = data;
                     obj.stored_rate = r;
                end
        end

        function data = raw_data(obj)
                if ~isempty(obj.stored_data)
                     data = obj.stored_data;
                else
                     [data, r] = wavread(obj.filename);
                     obj.stored_data = data;
                     obj.stored_rate = r;
                end
        end
    end
end
classdef recording2 < handle
    properties
            filename
    end
    .
    .
    .
```

One last note about objects descending from **handle**: Creating objects derived from **handle** causes allocation of memory that is not automatically cleaned up when the variable is cleared or goes out of scope. Clearing a variable holding an object derived from handle only removes the reference to the object. To remove the object itself, **delete** must be used. Since **delete** removes the object itself from memory, other references to the same object automatically become invalid after deletion.

```
>> r = recording2('test.wav');
>> r2 = r;
>> delete(r)
>> r.filename
??? Invalid or deleted object.

>> r2.filename
??? Invalid or deleted object.
>> r2

r2 =

deleted recording2 handle
```

**Methods, Events, Superclasses**

Unfortunately, in addition to deleting handle-derived objects, **delete** can also be used to remove files when used as a command. To avoid inadvertently deleting files, always be sure to use the functional form of **delete** (i.e., with parentheses).

#### 4.3.5.4 Summary

MATLAB's object model supports much, much more than what is touched on here, including access control, events, and complex inheritance patterns. For simple data analysis, simple representation of data as matrices is suitable. Object-oriented programming provides a tool for organizing larger efforts that promotes maintainability and minimizes code duplication. Object-oriented programming is particularly suited to GUI programming, where the programmatic objects are a natural analog of the controls and other visual entities on the screen such as the mouse cursor or menus.

## 4.4  TAMING ERRORS

### 4.4.1  An Introduction to the Debugger

At some point, despite great care in design and implementation, errors will rear their ugly heads. One of the greatest tools for eradicating errors is a debugger.

A debugger allows for running MATLAB code in an environment where the program state (e.g., variable value, interpreter location, etc.) can be explicitly controlled. The following code shows a naïve implementation of a factorial function.

```
function f = factorial2(g)
if g == 1
    f = 1;
    return
end
f = g * factorial2(g-1);
end
```

Typing **factorial2**(5) yields the expected 120. Try typing **factorial2**(5.1). Clearly, this behavior is not desirable. In this simple example, finding the error by inspecting the code alone is entirely plausible. That same simplicity also argues for this as a good example for demonstrating the debugger.

The easiest way to invoke the debugger on a function like **factorial2** is to open the function in the editor and add a breakpoint in the editor. A breakpoint denotes a location in the code where the MATLAB interpreter will always stop. Not all lines can support a breakpoint. In the editor, lines where a breakpoint can be placed will have a horizontal line to the left of the code. Clicking on that horizontal line to the left of the text places a breakpoint at that line. Clicking again removes the breakpoint. For this example, place a breakpoint on the line "if g == 1." Figure 4.2 illustrates the editor/debugger window with a breakpoint set.

With breakpoint in place, type **factorial2**(5.1) again. MATLAB should reposition the editor/debugger window to the front and place an arrow at the breakpoint. The command line prompt should also change. The MATLAB interpreter is now inside the function.

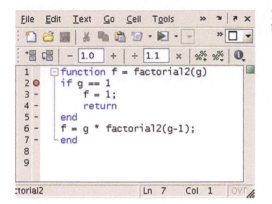

FIGURE 4.2    Editor/debugger window showing a set breakpoint

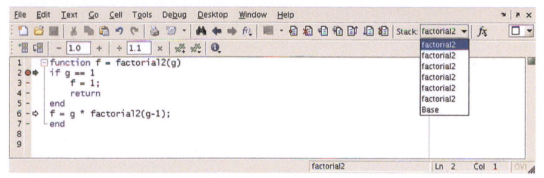

FIGURE 4.3    Editor/debugger window showing the stack list and an active debugger (green arrow)

The current workspace is not the command line (Base), but the current factorial scope. Because the interpreter is in the scope of the current invocation of the function, variable values can be inspected. Typing "g" shows the value of $g$ in the current scope (5.1). One can also observe the value of $g$ in the workspace window. In addition to the $g$ in the current scope, the workspace window allows viewing variables in other scopes as well. Selecting a different scope in the scope dropdown will cause all the variables in that scope to be displayed.

Selecting "Continue" from the Debug menu in the editor/debugger window will resume execution until encountering the next breakpoint. Note the contents of the stack list box after continuing. The stack dropdown allows selecting a specific scope (i.e., stack frame) in which to operate (see Figure 4.3). In the command window, type

>> g

After observing the value of $g$, select a different frame from the dropdown and inspect the value of that frame's $g$ value by typing

>> g

Are they the same?

After continuing four more times (a total of five), inspect the value of "g."

The variable "g" should be 0.1. Continuing again and inspecting $g$ will reveal the error; $g$ becomes negative. From this, it should be clear that the termination condition for the **factorial2** function in the if statement is not specific enough. Checking whether the input is exactly 1 will miss any non-integer argument. At this point, the error could be addressed in a number of fashions, depending on the desired functionality when non-integers are specified as the initial argument.

Change the value of "g" to 1 from $-0.9$. This can be done by typing g = 1 at the command line or by clicking on "g" in the workspace window. After setting $g$ to 1, continue executing the function by selecting "Continue" from the debug menu in the editor/debugger window. From this point forward in the execution, the value of $g$ at that scope will be 1. Since the next line after the breakpoint is the if statement comparing the value of $g$ to 1, the invocation of **factorial2** at the level where $g$ was modified quickly terminates and returns a value of 1 to the previous calling level.

Again, type **factorial2**(5.1), and continue until $g$ is negative. Before continuing again, set the value of $g$ to 1. Before continuing, place a breakpoint on the last end of the function. Now, the MATLAB interpreter will stop after calculating each part of the factorial. After continuing, examine the value of $g$. Is it what you expect? It is important to remember that the variable "g" at each scope is a distinct and separate variable.

In addition to continuing after a breakpoint, the debugger allows stepping through code line by line. Type **factorial2**(3). When the interpreter reaches the first breakpoint, step one line by selecting "step" from the debug menu. Continue stepping through the function until the final answer is calculated. Stepping through line by line demonstrates how the calculation invokes a series of calls that only end once termination condition is reached. Normally, step does not enter called functions. To enter a called function, step in can be used instead of step. Likewise, when inside a called function, step out will return to the calling function.

Finally, when finished, clear the breakpoints, either by clicking on the breakpoints in the editor or by selecting "clear all" from the debug menu in the editor window.

## 4.4.2 Logging

In larger programs, running under the debugger may not be feasible. Larger programs have more complex states, and it may not be possible to duplicate the bug within the debugger. For such cases, logging may be an appropriate methodology for tracking down bugs. Logging is simply printing out the internal status of the program. Usually, the log will be written to an external file to preserve the record for later debugging.

```
function f = factorial2(g, log_file)
fprintf(log_file, 'Entering factorial2(), g = %d', g)
if g == 1
    f = 1;
    return
end
f = g * factorial2(g-1);
end
```

This example has been simplified to demonstrate logging. This example requires the log file to remain open for the running of the program. Ideally, one would want the file only open when the log was being updated. To realize this, the file name could be passed instead of an open file, but that would require error-handling code in every function using the log. A better approach would be to create a logging object, capable of maintaining a file name and encapsulating all the file handling code. This approach improves the cohesion of the **factorial2()** function, as only the portion of the logging relevant to supplying **factorial2**'s behavior would remain in **factorial2()**. The code ahead demonstrates such a class and its usage (post-R2008a semantics).

```
classdef logger
    % Provides simple logging functionality.
    % To create, use the constructor with a filename.
    % Log entries are appended to the end of the file, with each
    % entry added on its own line.
    properties
        filename;
    end

    methods
        function obj = logger(filename)
            obj.filename = filename;
        end

        function message(obj, msg_str)
            % Outputs a message to the logging file.
            % Messages should be a string.
            % Usage:
            % log.message('Within function message()')
            fid = fopen(obj.filename, 'a + ');
            if fid  == -1
                warning('Unable to open log file')
                return
            end
            fprintf(fid, '%s\n', msg_str);
            fclose(fid);
        end
    end
end

function f = factorial2(g, log)
    log.message(sprintf('Entering factorial2(), g = %d', g));
    if g  == 1
        f = 1;
        return
    end
    f = g * factorial2(g-1);
```

```
end
>> log = logger('factorial-log.log');
>> f = factorial2(g, log);
```

### 4.4.3 Edge Cases and Unit Testing

Greater modularity and tighter cohesion lend themselves to simpler testing. With modular code, small portions of a larger program can be isolated and tested in a rigorous fashion. Such automated testing of smaller logical components through isolating them from the other parts of the program is called unit testing.

When unit testing, one wants to verify that all possible inputs have an expected result. Since testing all possible inputs is not feasible (for a single two element vector alone, this is $2^{128}$ different possibilities!), unit testing focuses on what are termed edge cases: those states or sets of parameters on the edges of different parameter spaces. Because these edge cases are usually the boundary between qualitatively different types of functional behavior, these types of inputs are often likely to evoke erroneous behavior because these types of values are often unplanned for. To illustrate the selection of edge cases, we will use the **factorial2**() function from previous sections:

```
function f = factorial2(g)
if g < = 1
    f = 1;
    return
end
f = g * factorial2(g-1);
end
```

If the input is limited to scalars, the behavior is likely to differ for integers (positive and negative), reals (positive and negative), 0, and 1. Thus, good edge cases would include $-1$, 0, 1, positive values and negative distant from 0, and small real values. Each test case should test a single edge case. Test cases should invoke the function with known inputs and compare the result to the expected output. Errors are acceptable as long as the error is the expected output for the function, given the input.

Here is a script with test cases for a $-1$, 0, 1, and a sequence of positive integer inputs.

```
% test 1
if factorial2(-1) ! = -1
    error('incorrect output for test 1: negative numbers');
end
% test 2
if factorial2(0) ! = 1
    error('incorrect output for test 2: 0');
end
% test 3
if factorial2(1) ! = 1
    error('incorrect output for test 3: 1');
end
% test 4
a = [2  3  4  5];
```

```
a = factorial(a);
a2 = [2  3  4  5];
for ii = 1:length(a)
    a2(ii) = factorial2(a2(ii));
end

if a2 ~ = a
    error('incorrect output for test 4: positive value check');
end
```

---

### EXERCISE 4.6

As coded previously, **factorial2**() does not pass the test cases. Modify **factorial2**() to pass the test cases.

---

### EXERCISE 4.7

Add test cases to the test script for other edge cases (reals, negative integers).

---

Even though the test script demonstrated clear deficiencies in the existing **factorial2**() function, one of the benefits of a set of unit tests is quickly capturing bugs observed after the initial design or implementation. Once an aspect of a unit's functionality is captured in a test script, regressions in that aspect can be spotted quickly. For example, if someone later modifying the code changed the termination condition from $g <= 1$ to $g == 1$, executing the unit test script would identify the error immediately. Unit tests capture the expected behavior of a function and allow divergences to be identified much more quickly than embedded in a large program.

---

### EXERCISE 4.8

The following code thresholds a raw recording, separates the event set into trials, and sorts the events to generate data for a PSTH (peri-stimulus time histogram).

```
function psth,bins = bin_for_psth(raw_data, sampling_rate, threshold,
trial_count, trial_length, bin_size)
    % Locates events above threshold in raw data and generates PSTH from
    % multi-trial recording. Trials should be contiguous.
    %
    % Input parameters
```

```
%      raw_data:        raw input data
%      sampling_rate:   sampling rate in Hz
%      threshold:       threshold for events, in same units as raw data
%      trial_count:     number of contiguous trials
%      trial_length:    length of each trial, in seconds
%      bin_size:        size of each PSTH bin, in seconds
%
% Output parameters
%      psth:            count in each bin
%      bins:            center position of each bin, in seconds relative to
%          trial start
% first, threshold signal
events = raw_data > threshold;
% only positive threshold crossings (not sustained activity above
threshold)
events = diff(events) == 1;
events = [0 events];
% now, split into trials
events = reshape(events, trial_count, trial_length*sampling_rate);
% events should be MxN, where M = trial and N = sample
summed_events = sum(events);
% summed_events should be the sum of events at each sample relative
to
% the start of the trial
max_event_count = max(summed_events);
for count = 0:max_event_count-1
    above_count = find(summed_events > count);
    event_offsets = [event_offsets above_count];
end
% event_offsets should be the offset in sample counts where events
occur
event_times = event_offsets / sampling_rate;
[psth, bins] = hist(event_times, trial_length/bin_size);
End
```

Determine edge cases for testing the input parameters of **bin_for_psth**, above.

---

### EXERCISE 4.9

Write a unit test script for the edge cases identified in Exercise 4.8.

### 4.4.4 A Few Words about Precision

Like most quantitative software, MATLAB does not represent most real numbers exactly. Where this representation fails to capture values exactly is often a source of bugs in quantitative code. This section discusses how MATLAB represents real numbers so that such problems can be diagnosed and avoided.

MATLAB labels the specific representation of every variable, and this is visible in the workspace. You may have noticed that nearly all variables have the label "double." This representation is the default representation type for values in MATLAB. Double here is in deference to single precision floating point, a lesser used floating point representation that consumes half the memory. Floating point representations are so called because the representation does not fix the number of digits on either side of the decimal point. Since the standards body IEEE (Institute of Electrical and Electronics Engineers) oversees the specification of this format, it is commonly known as IEEE 754 or 64 bit IEEE floating point. Similarly to MATLAB, most quantitative software uses this format for representing real numbers (Figure 4.4).

The sign bit denotes whether the number as a whole is positive or negative. The exponent is a base 2 number biased by $2^{10} - 1$, or 1023. The representation of exponents are 1023 plus the exponent's value. This system allows for exponents in the range $-1023$ to 1023.

The representation of the mantissa is the most complex portion of this standard. The digits in the mantissa represent a binary fraction, where each successive digit represents a successive fractional power of 2. Additionally, the mantissa is the fractional part of the number; there is a 1 implicit in the number not represented in the format.

Here is an example to illustrate how a decimal floating point number is represented internally by MATLAB.

Take 15.1875.

In base 2, 15 is $1111_2$. As a binary fraction, 0.1875 is $0.0011_2$. (0.1875 is 3/16, or 1/8 + 1/16, or $0*1/2 + 0*1/4 + 1*1/8 + 1*1/16$.) In total, 15.1875 is $1111.0011_2$. This must be converted to binary exponential notation: $1.1110011_2 \times 2^3$. To save space, the IEEE format assumes a leading one, so we must do the same. To store this as in double precision format, we need to discard the initial 1 from the mantissa and bias the exponent (Figure 4.5).

Why is this important? MATLAB can only represent a small subset of real numbers with absolute precision. These numbers are those whose fractional part is a sum of fractional powers of 2. For example, 7/16 can be perfectly represented (1/4 + 1/8 + 1/16), but 7/17

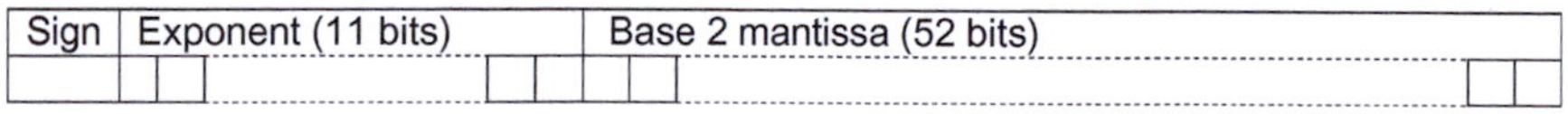

| Sign | Exponent (11 bits) | Base 2 mantissa (52 bits) | |
|---|---|---|---|
| | | | |

**FIGURE 4.4** Representation of floating-point numbers.

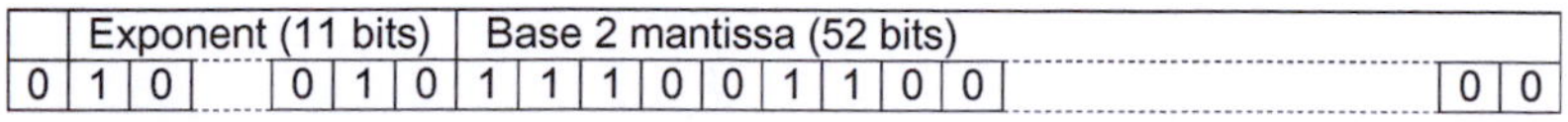

| | Exponent (11 bits) | | | | Base 2 mantissa (52 bits) | | | | | | | | | | | | |
|---|---|---|---|---|---|---|---|---|---|---|---|---|---|---|---|---|---|
| 0 | 1 | 0 | 0 | 1 | 0 | 1 | 1 | 1 | 0 | 0 | 1 | 1 | 0 | 0 | | 0 | 0 |

**FIGURE 4.5** Representation of the number 15.1875.

cannot. Understanding the limitations of floating point representation can also help to diagnose difficult to see errors.

One such error is testing for equality with floating point values. This often occurs when testing that a variable is zero or one valued. For example, the test below is attempting to verify that the variable $x$ is 0:

**if x == 0**

However, this will often not work if $x$ is the result of extensive calculations. When testing for zero, it is usually best to check a range around zero because the value is likely some extremely small floating point number rather than exactly zero:

**if abs(x) < 1.0e-6**

This scenario most often happens when checking for equality with zero, but any comparison involving floating point values and an arbitrary value, such as 0 or 1, should use a small interval instead.

Another such error occurs with operations on two values of extremely different magnitudes. The mantissa portion of the IEEE format has only 52 bits of precision. Thus, values differing by more than $2^{53}$ cannot be reliably added. This example demonstrates the problems inherent with sums of large and small magnitude numbers.

```
>> format compact
>> format long
>> 2^52
ans =
   4.503599627370496e + 15
```

Note that this is exactly $2^{52}$; there are no values hidden from view. To demonstrate, we can add one:

```
>> 2^52 + 1
ans =
   4.503599627370497e + 15
```

This cannot be done for $2^{53}$:

```
>> 2^53
ans =
   9.007199254740992e + 15
>> 2^53 + 1
ans =
   9.007199254740992e + 15
```

The result is identical to the original value, $2^{53}$. It is important to note that this is not the result of the unit's place being hidden from view. The following demonstrates that, under MATLAB, $1 + 2^{53}$ is equal to $2^{53}$. The equivalent example with $2^{52}$ is shown for comparison.

```
>> 2^53 == 2^53 + 1
ans =
```

```
      1
>> 2^52 == 2^52 + 1
ans =
      0
```

Similar problems occur when multiplying or dividing exceedingly small numbers. For example:

```
>> 10^( − 200) * 10^( − 100) * 10^( − 100)
ans =
      0
>> 10^( − 200) * 10^( − 100) * 10^( − 100) * 10^(300) * 10^(300)
ans =
      0
```

Rearranging the terms yields the correct answer:

```
>> 10^( − 200) * 10^(300) * 10^( − 100) * 10^( − 100) * 10^(300)
ans =
      1.000000000000000e + 200
```

Such problems can be avoided by carefully considering the magnitudes of the values in the calculation. When adding or subtracting terms of varying magnitudes, keeping the large and small values separated as distinct terms as long as possible often avoids this problem. Moving the calculation to logarithms is particularly effective for problematic multiplication or division.

### 4.4.5 Suggestions for Optimization

Occasionally, the situation arises where code does produce the expected result, but the code does so too slowly. In such cases, the code can be optimized. Optimization here means the rewriting of a portion of the code to improve the performance of the overall program. Since optimization involves scrutinizing working code, it is best to limit substantial optimization efforts to known hot paths—places in a program where the MATLAB interpreter spends a substantial proportion of the execution time.

Identifying hot paths from source code is difficult and quite error prone for larger pieces of code. Before engaging in a substantial optimization effort at a poorly performant site in the code, it is best to verify that the site is actually the cause of the perceived performance problem. This can be done by timing experiments with tic/toc or using the MATLAB profiler. Once identified, addressing efficiencies in the hot paths of a program can yield substantial returns.

#### 4.4.5.1 Vectorizing Matrix Operations

MATLAB is particularly efficient in executing matrix operations relative to the same operations. Taking full advantage of matrix operations in code often doesn't occur when first learning MATLAB, as the syntax is not as straightforward. This and the following sections offer suggestions for moving common non-matrix operations to matrix form. Code transformations of this type are called **vectorization**, as the type of matrix operations MATLAB offers are termed **vector** operations. (The use of **vector** to characterize MATLAB matrix operations indicates multi-valued operations, in deference to scalar (single-valued) operations, and does not refer to the mathematical objects operated on.)

A primary benefit of vectorizing code is a potential speed up in execution with a substantial change to the larger algorithm in the code. Here is an example contrasting two different approaches to adding matrices.

```
A = ones(4, 4); * 3; % matrix of threes
B = ones(4, 4); * 6; % matrix of sixes
C = zeros(4, 4);
for ii = 1:4
      for jj = 1:4
            C(ii, jj) = A(ii, jj) + B(ii, jj);
      end
end
```

or

```
A = ones(4, 4); * 3; % matrix of threes
B = ones(4, 4); * 6; % matrix of sixes
C = zeros(4, 4);
C = A + B;
```

While both pieces of code accomplish the same task, the second executes measurably faster. Note that the second snippet avoids the nested for loops.

Understanding why these two bits of code execute so differently requires a brief explanation of how MATLAB evaluates code. Individual operations in MATLAB execute as compiled machine code, at high speed. For example, the matrix addition in the second code section executes in this manner.

However, in the case of the first example, evaluation of the inner statement alone requires evaluating each of the two index variables, three matrix lookups, a scalar addition, and storing the scalar result. In between operations, the interpreter must be constantly consulted to determine the next step.

#### *4.4.5.2 Conditional Expressions*

Using relational operations can often function as an alternative to an if statement nested within a for loop. A relational operator acting on a matrix returns a matrix of the same shape with values of 1 for true and 0 for false.

```
A = ones(4, 4);
B = rand(4, 4);
for ii = 1:4
      for jj = 1:4
            if (B(ii, jj) > 0.5)
         A(ii, jj) = A(ii, jj) + B(ii, jj);
            end
```

    **end**
**end**

Compare the above with the following.

**A = ones(4, 4);**
**B = rand(4, 4);**
**A = A + (B .\* (B > 0.5));**

In the latter example, the single expression takes the place of the nested for loops and if statement. The inner relational expression evaluates to a $4 \times 4$ matrix whose elements are 1 if the corresponding element of B is greater than 0.5. Thus, the element-wise multiplication of this matrix with B generates a matrix whose elements are either the corresponding element of B, if B is greater than 0.5, or 0, if that element of B is less than or equal to 0.5.

### 4.4.5.3 *Extracting Subsets from Arrays*

Many times, an if statement nested within a for loop is used to extract some subset of values from a matrix. The use of matrix relational operations and **find** can eliminate the need for the iteration. The function **find** returns all the indices of the input for which the input is non-zero. For example,

```
>> A = [1 2 3 4];
>> find(A < 3)
ans =
   1   2
```

Specifying a set of values for the index of a matrix will return a subset of the matrix values. This can apply to the results of **find**

```
>> A = [8 9 10 11];
>> find(mod(A,2) == 0)
ans =
   1   3
>> A(find(mod(A,2) == 0))
ans =
   8   10
```

# MATLAB FUNCTIONS, COMMANDS, AND OPERATORS COVERED IN THIS CHAPTER

**which**
**global**
**classdef**
**delete**
**clear**
**dbstack**

# 5

# Visualization and Documentation Tutorial

## 5.1 GOALS OF THIS CHAPTER

This chapter represents the last chapter of the fundamentals before moving on to the later parts of the book that contain more specific and more modular material. The content of this chapter, Visualization and Documentation, will be revisited in each subsequent one and quite likely as long as you use MATLAB®. Therefore, it is worthwhile to devote a chapter to it at this point, getting a firmer grip on these elementary issues, allowing you to focus on the specific new content that is introduced later on.

## 5.2 VISUALIZATION

The ability to rapidly and effectively visualize data that is afforded by MATLAB is one of the key reasons why MATLAB is so popular in the first place, perhaps only second to its efficient computation of matrix operations. In the previous chapters, we have already seen how easy it is to create figures from data in MATLAB. It is so easy that anyone can do it. However, this low threshold can be treacherous. While it is easy to make the figure, comparatively few people know how to make the figure so that it looks just how they want it to look. This causes much frustration and often drawn-out and lengthy modification of figures with other image processing software. It is better to avoid this altogether by taking complete control of the figure and its appearance from the start.

We already encountered the function **set** in Chapter 2 when we manipulated the color of individual subplots. **set** is a key function in this context. It allows you to set the value of any figure attribute you want.

*MATLAB® for Neuroscientists.*
DOI: http://dx.doi.org/10.1016/B978-0-12-383836-0.00005-9

We will start with creating a figure in itself. By now, you have probably noticed that MATLAB creates figures by default with a certain size and in a certain position. For many purposes, this default is too small, which means you have to resize the figure manually each time the program is executed. A better way to do this is to create the figure with the right size from the get-go.

Let's try it:

```
>>figure
>>set(gcf,'Position',[100   200   400   300])
```

This code creates a figure at a position on the screen that is at a distance of 100 pixels from the left edge of the screen, is 200 pixels from the bottom, extends 400 pixels to the right (width), and extends 300 pixels upwards (height). The function that accomplished this is **set**, which allows us to set the value of object attributes (the object in this case is a figure, and attributes are called "properties" within MATLAB).

**set** expects three values in the parentheses: the handle of the object we refer to, the property we want to change, and the new property value, in this order and separated by commas.

In this case, we told the function **set** that we want to refer to a figure not by giving it the object handle, but by using the function **gcf**, which stands for "get handle for current figure." This is adequate, as the current figure is the one we just created. If there are multiple active figures, it is better to create the figures with a handle, and later specify the handle of the figure object we want to refer to. The second value here was 'Position'; it is important to put it into quotes, so that MATLAB recognizes it as a property. The third and last value is a vector of the form

[Left_edge Bottom_edge Width Height]

You can think of the figure as a rectangular window that starts at the point defined by the **Left_edge** and **Bottom_edge** values as their x and y and extends from there, as specified by width and height.

But how are you supposed to know all of this? How do you know which properties *can* be set and what values they expect?

That is an excellent question. Luckily, the solution is relatively straightforward. Try

```
>>get(gcf)
```

This command displays a long list of figure properties that, as we just created the figure, are set to default values, except for position, which should be what we set it to be. You can use the function **get** to either get *all* of the figure properties and their values as with the previous command, or a specific one, as with the command ahead, for the case of position.

```
>>get(gcf,'Position')
ans =
   100   200   400   300
```

Note that you can either use this for your edification as in the previous command, in which MATLAB is telling you what the position vector of the figure is, or you can store it for further use, as with any other vector in MATLAB, and as in this command:

>>**tempfigpos = get(gcf,'Position')**

You can now use the variable **tempfigpos** to do calculations, like so

>>**figpos = 2 .\* tempfigpos %Doubling everything**
>>**set(gcf,'Position',figpos)**
**figpos =**
  **200   400   800   600**

Does it still fit on your screen? (I do realize that—given the rapid advance of technology—the answer to this question will strongly depend on when you are reading this).

Speaking of figures fitting on the screen: For most applications, it is opportune to set the figure size to the screen size. This will allow you to take advantage of the entire screen real estate, which is beneficial for complicated figures, particularly if you have a second monitor—which allows you to code in one window and look at the figures in another. If you don't know the number of pixels on your screen or you change monitors often, it is better to ask MATLAB than to hardcode this, like so:

>>**temp = get(0,'Screensize')**
>>**set(gcf,'Position',temp)**

"Screensize" is a property of root, which can be accessed by giving the function set "0" as the object handle. But how were you supposed to know that 0 would work as a handle?

This is a good time to introduce the hierarchy of graphics objects within MATLAB. At the top of the hierarchy is root, the screen itself (handle is 0). The screen (or root) can contain any number (limited by memory) of figures. Each figure has a handle which allows you to access its properties. If you don't specify a handle, the figure handles are simply consecutive integers, in order of figure creation, starting with 1. So passing an object handle of "0" accesses root, whereas "1" accesses the first figure, "2" the second figure, and so on. Each figure in turn contains any number (within reason) of individual axes, which can also have their own handles. You can change the property values of any property of any graphics objects in MATLAB by specifying its corresponding object handle. Understanding this will allow you, as promised, to take complete charge of the appearance of the figures you create. As you get more practice, this will substantially cut down on the amount of post-processing you will have to do.

---

### EXERCISE 5.1

What other properties does the screen have? What are their values?

---

*Note*: Some of the root properties will have byzantine names and a possible use that will likely escape you at this point. Do not be discouraged by this. It will come, in time.

Back to the figure. Let's say that for this particular project, you want a figure that doesn't feature the usual MATLAB figure gray as a background (represented by the RGB color vector [0.8 0.8 0.8]), but uses red instead. There are many ways to do this. The most straightforward one involves passing the current figure handle and property a new value, in this case "r" (for red).

Here you go:

```
>> set(gcf,'Color', 'r')
```

Your figure background should look red now.

MATLAB knows eight different character codes for all integer combinations of the three-element RGB vector, such as "r" for red [1 0 0], "g" for green [0 1 0], "b" for blue [0 0 1], and so on. If you need more nuanced coloring, you can pass the vector directly, using non-integral values. For instance, if you want a darker shade of red with a touch of purple, this would do it:

```
>> set(gcf, 'Color', [0.5   0   0.2])
```

---

**EXERCISE 5.2**

Explore 50 different shades of red. Which one do you like best?

---

Time to plot something. To make matters easy, we'll just plot a sine wave. This will do it, creating a nice sine wave:

```
>> x = 0:0.01:20;
>> y = sin(x);
>> plot(x,y)
```

Your figure should now look something like Figure 5.1.

This command did two things; first, it created an axis in our figure, and then it created an object within that axis. Both can be addressed.

One issue that is immediately obvious is that the axes are now hard to make out, as they are plotted in the default black. Setting them (both x and y) to a brighter shade of gray should solve the problem:

```
>> set(gca,'xcolor', [0.7   0.7   0.7]) %Changing the color of the current x-axis.
>> set(gca,'ycolor', [0.7   0.7   0.7]) %Changing the color of the current y-axis.
```

Voila. Now you can read the axis values again. However, the actual sine wave still looks positively anemic. It might be prudent to increase the line width. Alas, you plotted it without giving it an explicit object handle. It still has one, but you don't know it. What to do?

There are several ways to retrieve the object handle.

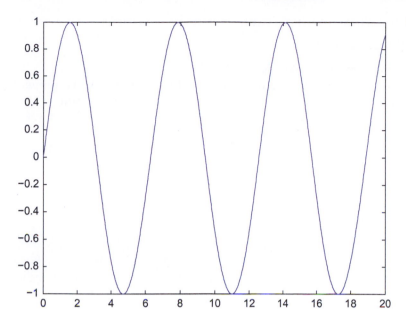

FIGURE 5.1  A sine wave.

One way is to use the **findall** command. **findall** returns a list of all objects for a given handle. In this case, we'll use the current axis:

>> temp = **findall(gca)**
**temp =**
   686.1190
   687.1194

This yields two object handles, one for the axis itself and one for the line (represented by numbers; your numbers might vary, as they are assigned by MATLAB). But how to find out which one is the line and one which one is the axis?

Type

>> **get(temp(1))**

and

>> **get(temp(2))**

Note that in the output, one will list "axes" as "type," the other "line." We want to modify the line.

We can now either access this object handle, or—for future use—redefine temp as only the object in the current axis out of all the objects that is a line, like this:

>> temp = **findall(gca,'type','line')**

**temp =**
   687.1194

Now set the line width to a thicker strength, but hash it at the same time:

```
>> set(temp,'linewidth',2,'linestyle',':')
```

The figure should now look something like Figure 5.2.

*Note*: If you were to print this figure, it would use up a lot of colored ink. To prevent this, MATLAB prints the background as white and the axes in black by default. If you want MATLAB to print things exactly as they look on the screen, you need to set a figure property, like this: **set(gcf, 'InvertHardCopy', 'off')**.

Of course, things really get interesting as one introduces multiple objects in the same axes. Let's do it.

In order to do this without erasing the other object, we need to put the hold on, so type

```
>> hold on
>> z = cos(x);
>> h = plot(x,z);
```

This adds a cosine to the mix. Now we have two line objects. The sine wave from before, and the new one. This time, we labeled our object with an explicit object handle, "h." We can now access it.

---

**EXERCISE 5.3**

Take a look at the object properties of the second line object, both ways. Once by using the explicit handle *h*, and once by using the MATLAB-internal handle.

---

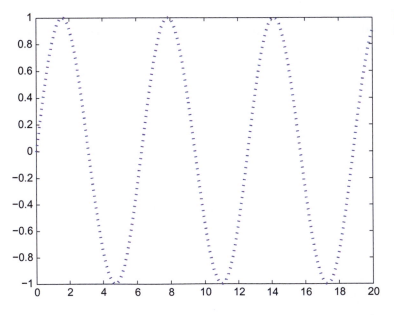

**FIGURE 5.2** The modified sine wave figure.

Say that in a not so distant future you have a busy figure with a large number of lines. For some reason, you want to take all lines of a particular color and change a particular property, e.g., their line width. To do this, you can use the function **findobj**. It returns the handles of all objects with a particular property. For example,

>> temp = findobj(gca,'color','b')

temp =
   692.1171
   687.1194

finds the handles of our two lines again (as they are plotted in blue, per MATLAB default). We can now set their line width to a uniform 3 and, while we are at it, change their color to green (see Figure 5.3).

>> set(temp,'color','g','linewidth',3)

If this is what your figure was supposed to look like, you can declare victory at this point.

Of course, there is a lot left to be done. One important concept to be understood is that of children in MATLAB. Another one is that of multiple axes in the same figure. We'll tackle both at the same time.

If you type >> get(0,'children'), you should get "1" as an output. You are asking MATLAB how many children (figures) the screen has. At this point, it has 1. If you open another figure (please do so, by typing "figure"), the output will now be a vector with two elements: 1 and 2.

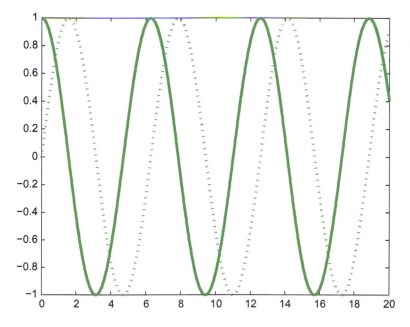

FIGURE 5.3 The sine and cosine waves plotted, in green, with line width set to 3.

The second figure seems to be empty at this point, as confirmed by

>> get(gcf,'Children')
ans =
Empty matrix: 0-by-1

but that is an illusion. MATLAB sometimes uses hidden handles, particularly if it doesn't want the user to accidentally do something foolish. However, even these handles can be displayed and accessed by using the **allchild** function. For example,

>> temp = allchild(gcf)

returns plenty of handles.

---

### EXERCISE 5.4

Find the handle that corresponds to the **uitoolbar** (which allows you to save, print the figure, etc.), and turn it off by making it invisible.

---

There are, of course, a lot of figure, axes, and line properties to modify. It will take time to become familiar with all of them. With some simple examples, this tutorial demonstrated how to access these properties in principle. For more information on these properties and their potential values, search the help (or the function browser) for **figure_props** (for figure properties), **axes_props** (for axes properties), and **line_props** or **linespec** (for line properties). There are a lot of them.

Finally, what is left to do is to explore the syntax to add multiple axes to a given figure.

In Chapter 2, we already discussed how to add a tiled (and ordered) number of axes to a figure by using the subplot command. While this is sufficient for many purposes, it is good to know how to add axes at arbitrary positions in the figure.

This is done in relative figure coordinates, where 0,0 corresponds to the lower left and 1,1 to the upper right.

For instance, if we want to place three unequally shaped plots on a figure, we could type

```
>> h1 = axes('position',[0     0.8   1     0.2])
>> h2 = axes('position',[0.8   0     0.2   0.8])
>> h3 = axes('position',[0     0     0.8   0.8])
```

to create the 3 axes and their handles, then type

```
>> set(gcf,'CurrentAxes',h1)
>> plot(x,y)
>> set(gcf,'CurrentAxes',h2)
>> plot(x,y)
>> set(gcf,'CurrentAxes',h3)
>> plot(x,y)
```

to get a figure that looks something like Figure 5.4.

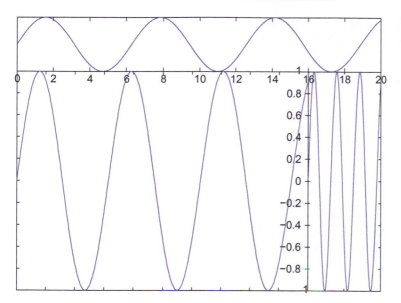

**FIGURE 5.4** Three unequally shaped plots.

## 5.3 DOCUMENTATION

There are several interpretations of the notion of "documentation." One of these, in the sense of "commenting your code," was already covered in the previous chapter. If you skipped it, you might want to revisit that. All I want to say about it here is that you do want to comment your code as neatly as possible. I can guarantee you that the very code you just wrote, sans comments, will make perfect sense to you now, but it will be so opaque to you in about 6 months that it might as well have been written by someone else. This is a problem. It happens surprisingly often; for instance, it happens when you need to revisit your code because reviewer 2 suddenly asks for a different analysis in the second round of reviews of your paper.

Another sense of "documentation" is the sense of a protocol. This can come in handy when you are in need of documenting your work, e.g., for a class. Copying and pasting individual inputs and outputs from the command window to a word processor can get tedious quickly. It might be easier just to copy the entire command history and paste that, but it lacks the outputs. A simple solution is to type **diary**, which toggles the function diary on (it is off by default). If it is invoked in the absence of a filename, "diary" will be the filename, in the "current folder" directory. Of course you can specify a filename, e.g., by typing

>> **diary('report.txt')**

Yet another sense of "documentation" is the documentation of MATLAB itself. We already covered the generic help function in Chapter 2; however, as you might appreciate by now, MATLAB uses a great deal of punctuation, all of which has distinct meaning. This can seem overwhelming to the beginner. But don't despair. You don't have to memorize this all. Proficiency will come with use. In the meantime, you can rely on specific MATLAB help functions for reminders, such as:

```
>> help punct %Details on punctuation
>> help relop %Details on logical and relational operators
>> help paren %Explains parentheses, braces, brackets and their use
>> help colon %Explains the use of the colon operator
>> help lists %Comma separated lists
```

Executing these functions by themselves (without help) doesn't do anything. They are specific help functions to remind you of the syntax.

## MATLAB FUNCTIONS, COMMANDS, AND OPERATORS COVERED IN THIS CHAPTER

**Get**
**Set**
**Gca**
**Gcf**
**Findall**
**Allchild**
**Findobj**
**Axes**
**Diary**
**Punct**
**Relop**
**Paren**
**Colon**
**Lists**

# PART II

# DATA COLLECTION
# WITH MATLAB

# 6

# Collecting Reaction Times I: Visual Search and Pop Out

## 6.1 GOALS OF THIS CHAPTER

The primary goal of this chapter is to collect and analyze reaction time data using MATLAB®. Reaction time measures to probe the mind have been the backbone of experimental psychology at least since the time of Donders (1868). The basic premise underlying the use of reaction times in cognitive psychology is the assumption that cognitive operations take a certain and measurable amount of time. In addition, it is assumed that additional mental processes add (more or less) linearly. If this is the case, increased reaction times reflect additional mental processes. Let us assume for the time being that this is a reasonable framework. Then, it is highly useful to have a program that allows you to quickly collect reaction time data.

## 6.2 BACKGROUND

Understanding how the mind/brain decomposes a sensory scene into features is one of the fundamental problems in experimental psychology and systems neuroscience. We take it for granted that the visual system, for example, appears to decompose objects into different edges, colors, textures, shapes, and motion features. However, it is not obvious a priori which features actually represent primitives that are encoded in the visual system. Many neurophysiological experiments have searched for neurons that are tuned to features that were chosen somewhat arbitrarily based on the intuitions of the experimentalists.

Psychologists, however, have developed behavioral experiments by which feature primitives can be revealed. For instance, a study by Treisman and Gelade (1980) has been particularly influential. This is probably due to the fact that it is extremely simple to grasp, yet the pattern of results suggests provocative hypotheses about the nature of perception (e.g., feature primitives, serial search, etc.).

So what is the visual search and pop-out paradigm that was used in the Treisman study?

*MATLAB® for Neuroscientists.*
DOI: http://dx.doi.org/10.1016/B978-0-12-383836-0.00006-0

Research participants were asked to report the presence or absence of a target stimulus (in this case, a colored lowercase letter "o") among various numbers of distracter stimuli. If the distracter stimuli are just of a different color—that is, if they differ by a single feature—you usually find the "pop-out" effect: the reaction time to detect the target is independent of the number of distracters. Conversely, if more than one stimulus dimension has to be considered to distinguish targets and distracters (conjunction search), you typically find a linear relationship between reaction time and the number of distracters. See Figure 6.1.

As pointed out previously, this pattern of results immediately suggests the existence of "feature primitives", fundamental dimensions that organize and govern human perception as well as a serial scanner in the case of conjunction search, where one element of the stimulus set after the other is considered as a target (and confirmed or discarded). Often, the ratio of the slopes between conditions where the target is present versus where the target is absent suggests a search process that self-terminates once the target is found.

There are many, many potential confounds in this study (luminance, eye movements, spatial frequency, orientation, etc.). However, the results are extremely robust. Moreover, the study was rather influential. Hence, we will briefly replicate it here.

## 6.3 EXERCISES

In this section, we introduce and review some code that will help you to complete the project in Section 6.4. The first thing you need to be able to gather reaction time measures is a way to measure time. There are different ways to measure time in MATLAB. One of the most convenient (and, for our purposes, sufficient ones) comes in the form of the functions **tic** and **toc**. They work in conjunction and effectively implement a stopwatch. Try the following on the command line:

```
>> tic
>> toc
```

What is your elapsed time?

The time reported by MATLAB is the time between pressing Enter after the first statement and pressing Enter after the second statement.

Of course, operating in the real physical world, MATLAB also takes some time to execute the code itself. In most cases, this delay will be negligible. However, you should not

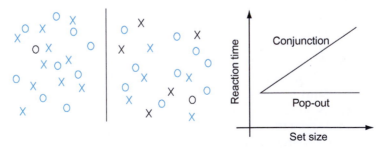

FIGURE 6.1   The pop-out task.

take this delay for granted. Try it. In other words, write a program (M-file) that contains only the following lines and execute it:

**tic**
**toc**

In my case, **toc** reported 0.000004 seconds.

You can now create some code to test if MATLAB takes equal amounts of time to increment an index or if it depends on the magnitude of the index. So open a new M-file and enter the following:

```
format long; %We want to be able to see short differences in time
ii = 1; %Initializing the index, ii
t = [ ]; %Initializing the matrix in which we will store the times
while ii < 11 %Starting loop
tic%Starting stop-watch
ii = ii + 1; %Incrementing index
t(ii,1) = toc ; %Ending stop-watch and putting respective time into the matrix
end % End the loop
```

Try to run the program. It should execute rather quickly.

Now create a little plot by typing the following on the command line:

```
>>figure
>>plot (t)
```

The result should be fairly reproducible but the exact shape of the curve as well as the absolute magnitude of the values depends on the computer and its speed.

The result should look something like that shown in Figure 6.2.

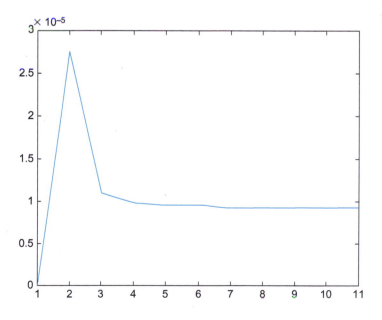

FIGURE 6.2    Task timing.

As you can see, after an initial transient, the time does not depend on the actual value of the index, unlike most human mental processes (e.g., Shepard and Metzler, 1971). This could be taken as evidence for a different kind of information processing in man versus machine.

If you want to know the average time it took for the index to increment, type

>> **mean**(*t*)

If you want to know the maximum and minimum times, you can type **max**(*t*) or **min**(*t*), respectively.

This example also illustrates several important points. First, when making an inductive claim about all cases all the time, you should sample a substantial range of the problem space (complete would be best). In this case, incrementing an index 10 times is not very impressive. What about incrementing it 100,000 times?

---

### EXERCISE 6.1

Increment your index 100,000 times. What does the resulting graph look like?
It should look something like the result shown in Figure 6.3.

---

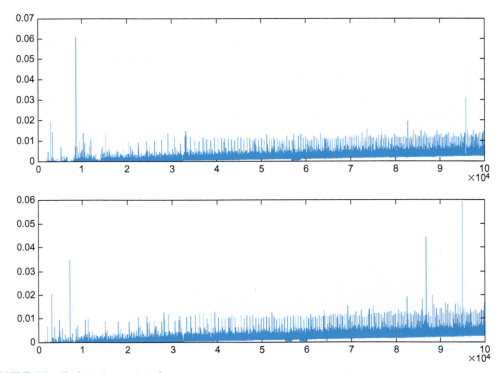

FIGURE 6.3    Task timing revisited.

Hence, we discourage premature conclusions on the basis of scant data. Second, it shows that while MATLAB inherently takes care of "plumbing issues" such as memory management or the representation of variables, it cannot avoid the consequences of physical processes. In other words, they might very well impact the execution of your program. Therefore, you should always check that the program is doing what you think it is doing. The structure of the peaks and their robust nature (the subplots show different runs of the program) indicate that they are reliable and not just random fluctuations, probably induced by MATLAB having to change the internal representation of the variable $t$, which takes longer and longer as it gets larger. This makes sense because MATLAB is shuffling an increasingly large array around in memory, looking for larger and larger chunks thereof. Some of the observed spikes in time taken appear to be distributed largely at random, mostly due to other things going on with the operating system.

Finally, it is a lesson on how to avoid problems like this—namely by preallocating the size and representation of $t$ in memory, if the final size is known in advance.

---

**EXERCISE 6.2**

Replace the **line t = [];** with **t = zeros (100000,1);** to preallocate the size of the variable in memory. Then run it. The result should look something like that shown in Figure 6.4.

---

There are still some issues left, but nowhere near as many as there were before. As you can see, the problem largely goes away (you should also close all programs other than MATLAB when running time-sensitive code). Note also the dramatic difference in the time needed to execute the program. The reason for this is the same—namely memory management. Therefore, if you can, always preallocate memory for your variables, particularly when you know their size in advance and if their size is substantial.

If you'd like to, you can save this data (stored in the $t$ variable) by clicking on the File menu from the MATLAB command window and then clicking on the Save workspace as entry. You can give your file any name you like. Later, you can import the data by clicking on Open in the File menu from the MATLAB command window (not the editor). Try this now. Save your workspace, clear it with **clear all**, and then open it again.

You can also use the **tic-toc** stopwatch to check on MATLAB. For example, the following code is supposed to check (use another M-file) whether MATLAB really takes a 0.5 second break:

```
tic %Start stopwatch
pause (0.5) %Take a 0.5(?) second break
toc %End stop-watch
```

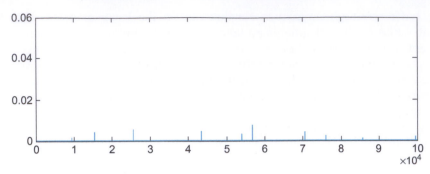

**FIGURE 6.4**   Problem solved—mostly.

By running this program several times, you can get a sense of accuracy and variance (precision) within internal timers in MATLAB. So much for time and timing.

What is still missing at this point is a way to handle random events. In the design of experiments, randomness is your friend. Ideally, you want everything that you don't vary systematically to behave randomly (effectively controlling all other variables, including unknown variables and unknown relations between them).

You encountered the random number generator earlier. Now you can utilize it more systematically. This time, it will be enough to generate random numbers with a uniform distribution. This is achieved by using the function **rand( )**. Remember that the function **randn( )** will generate random numbers drawn from a normal distribution. Conversely, **rand( )** draws from a uniform distribution between 0 and 1. This distinction is important to know, since you can use this knowledge to create two events that are equally likely. Now start a new M-file and add the following code:

```
a = rand(1,1) %Creates a random number and assigns it to the variable a
if a > 0.5 %Check if a is larger than "0.5"
b = 1 %We assign the value "1" to the variable b
else %If not,
b = 0 % We assign the value "0" to the variable b
end %End the condition check
```

Run this code a couple of times and see whether different random values are created every time. Note that this is a rather awkward—but viable—way to create integral random numbers. Recent versions of MATLAB include a new function **randi**, which draws from a uniform discrete distribution. For example, the command randi(2,30,1) yields a single column vector with 30 elements randomly drawn to be 1 or 2. This new function is a good example of how innovation in MATLAB versions makes previously accepted ways of doing things (such as generating discrete random numbers) obsolete. The old way still works, but the new one is much more elegant.

Next, we will introduce several functions and concepts that will come in handy when you are creating your program for the project in Section 6.4. The first is the concept of *handles*. A handle typically pertains to an object or a figure, for our purposes. For more detailed treatment of handles, see the Visualization and Documentation tutorial in Chapter 5.

It is simply declared as follows:

**h = figure**

This creates a figure with the handle $h$. Of course, the name can be anything. Just be sure to remember which handle refers to which figure:

**thiscanbeanything = figure**

This creates the handles as variables in the workspace. You can check this by typing **whos** or simply by looking in the workspace window.

Handles are extremely useful. They literally give you a handle on things. They are the embodiment of empowerment.

Of course, you probably don't see that yet because handles are relatively useless without two functions that go hand in hand with the use handles. These functions are **get** and **set**.

The **get** function gives you information about the properties that the handle currently controls as well as the values of these properties. Try it. Type **get(h).**

You should get a long list of figure properties because you linked the handle $h$ to a particular figure earlier. These are the properties of the figure that you can control. This capability is extremely helpful and implemented via the **set** function.

Let's say you don't like the fact that the pointer in your figure is an arrow. For whatever reasons you have, you would like it to be a cross-hair. Can you guess which figure property [revealed by **get(h)**] controls this? Try this:

**set(h, 'Pointer', 'crosshair')**

How do you know which property takes which values? This is something you can find in the MATLAB help, under Figure properties. Don't be discouraged about this. It is always better to check. For example, MathWorks eliminated the former figure property "fullcrosshairs" and renamed it "fullcross."

Of course, some of the values can be guessed, such as the values taken by the property **visible**—namely **on** and **off**. This also illustrates that the control over a figure with handles is tremendous.

Try **set (h,'visible','off')** and see what happens. Make sure to put character but not number values between'and'.

Of course, handles don't just pertain to figures; they also pertain to objects. Make the figure visible again and put some objects into it.

An object that you will need later is **text**. So try this:

**g = text (0.5, 0.5, 'This is pretty cool')**

If you did everything correct, text should have appeared in the middle of your figure. Text takes at least three properties: x-position, y-position, and the actual text.

But those are not all the properties of text. Try **get(g)** to figure out what you can do.

It turns out, you can do a lot. For example, you can change color and size of the text:

```
set(g,'color','r', 'fontsize', 20)
```

Also, note that this object now appears as a "child" of the figure **h**, which you can check with the usual method.

Now you should have enough control over your figures and objects to complete the project in Section 6.4.

Finally, you need a way for the user (i.e., the participant of the experiment) to interact with the program. You can use the **pause** function, which waits for the user to press a key before continuing execution of the program. In addition, the program needs to identify the key press.

So type this:

```
pause %Waiting for single key press
h725 = get(h,'CurrentCharacter')
```

The variable *h725* should contain the character with which you overcame **pause**. Interestingly enough, it is a figure property that allows you to retrieve the typed character in this case, but that is one of the idiosyncrasies of MATLAB. Be sure to do this within an M-file.

Another function you will need (to be able to analyze the collected data) is **corrcoef**. It returns the Pearson correlation between two variables, e.g.,

```
a = rand(100,2); %Creates 2 columns of 100 random values each, puts it in variable a
b = corrcoef(a(:,1),a(:,2)); %Calculates the Pearson correlation between the two columns
```

In my case, MATLAB returns a value close to 0, which is good because it shows that the random number calculator is doing a reasonable job.

```
b =
   1.0000   0.0061
   0.0061   1.0000
```

**Corrcoef** as a function can also take several parameters:

```
[magnitude, p] = corrcoef(a(:,1),a(:,2)) %Same as before, but asking for significance
magnitude =
   1.0000   0.0061
   0.0061   1.0000
p =
   1.0000   0.9522
   0.9522   1.0000
```

According to MATLAB, there is a probability of 0.95 that the observed values were obtained by chance alone (which is, of course, the case). Hence, you can conclude that the correlation is not significant.

By convention, correlations with $p$ values below 0.05 are called "significant."

The final function concerns checking of the equality of variables. You can check the equality of numbers simply by typing == (two equal signs in a row):

```
>>5 == 6
ans =
    0
>>5 == 5
ans =
    1
```

Since MATLAB 7, this technique is also valid for checking the equality of characters; Try this:

```
>> var1 = 'a'; var2 = 'b'; var1 == var2
>> var1 == var1
```

The more conventional way to check the equality of characters is to use the **strcmp** function:

```
>> strcmp(var1,var2)
>> strcmp(var2,var2)
```

Together with your knowledge from Chapter 2, "MATLAB Tutorial," you now have all the necessary tools to create a useful experimental program for data collection.

## 6.4 PROJECT

Your project is to implement the visual search paradigm described in the preceding sections in MATLAB. Specifically, you should perform the following:

- Show two conditions (pop-out search versus conjunction search) with four levels each (set size = 4, 8, 12, 16). These conditions can be blocked (first all pop-out searches, then all conjunction searches or something like that).
- It is imperative to randomly interleave trials with and without target. There should be an equal number of trials with and without targets.
- Make sure that the number of green and red stimuli (if you are red/green blind, use blue and red) is balanced in the conjunction search (it should be 50%/50%). Also, make sure that there is an equal number of $x$ and $o$ elements, if possible.
- Use only correct trials (user indicated no target present when no target was presented or indicated target present when it was present) for the analysis.
- Try to be as quick as possible while making sure to be right. It would be suboptimal if you had a speed/accuracy trade-off in your data.
- The analysis should contain at least 20 correct trials per level and condition for a total of 160 trials. They should go quickly (about 1 second each).
- Pick two keys on the keyboard to indicate responses (one for target present, one for target absent).
- Report and graph the mean reaction times for correct trials as a function of pop-out search versus conjunction search and for trials where the target is present versus where

the target is absent. Hence, you need between two and four figures. You can combine graphs for comparison (see below).

- Report the Pearson correlation coefficients between reaction time and set size and indicate whether it is significant or not (for each condition).
- Make a qualitative assessment of the slopes in the different conditions (we will talk about curve fitting in a later chapter).

*Hints:*

- Start writing one trial and make sure it works properly.
- Be aware that you effectively have an experimental design with three factors [Set size: 4 levels (4, 8, 12, 16), conjunction versus feature search: 2 levels, target present versus absent: 2 levels]. It might make sense to block the first two factors and randomize the last one.
- Be sure to place the targets and distracters randomly.
- Start by creating a figure.
- Each trial will essentially consist of newly presented, randomly placed text.
- Be sure to make the figure big enough to see it clearly.
- Make sure to make the text vanish before the beginning of the next trial.
- Your display should look something like Figure 6.5.
- Determine reaction time by measuring time from appearance of target to user reaction.
- Elicit the key press and compare with the expected (correct) press to obtain a value for right and wrong answers.
- Put it into a matrix, depending on condition. It's probably best to have as many matrices as conditions.
- Plot it.

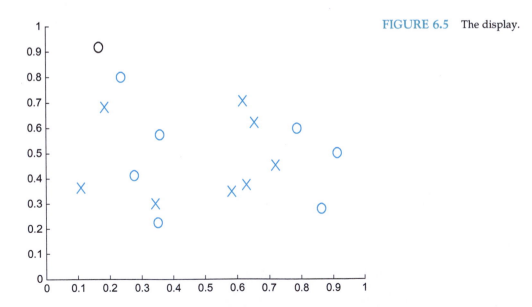

FIGURE 6.5   The display.

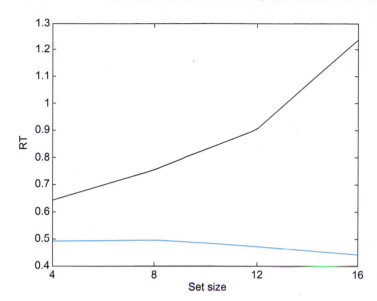

FIGURE 6.6    Typical results.

- Write a big loop that goes through trials. Do this at the very end, if individual trials work.
- You might want to have a start screen before the first trial, so as not to bias the times of the first few trials.
- If you can't do everything at once, focus on subgoals. Implement one function after the other. Start with two conditions.
- Figure 6.6 shows one of the exemplary result plots from a participant. Depicted is the relationship between mean reaction time and set size for trials where a target is present (only correct trials). Red: Conjunction search. Blue: Pop-out search. Pearson $r$ for conjunction search in the data above: 0.97.

## MATLAB FUNCTIONS, COMMANDS, AND OPERATORS COVERED IN THIS CHAPTER

**clear all**
**tic**
**toc**
**mean**
**min**
**max**
**pause**
**rand**
**randi**
**==**

**strcmp**
**whos**
**get**
**set**
**text**
**corrcoef**
**pointer**
**fullcrosshair**
**visible**
**fontsize**
**CurrentCharacter**

# Collecting Reaction Times II: Attention

## 7.1 GOALS OF THIS CHAPTER

The primary goals of this chapter are to consolidate and generalize what you learned in Chapter 6 about data collection in MATLAB®. Moreover, we will elaborate on data analysis in MATLAB beyond simple correlations. You will also learn how reaction time data can be used to infer the mental process of spatial attention.

## 7.2 BACKGROUND

As the pioneering American psychologist William James pointed out well over 100 year ago, we all have a strong intuition what attention is:

> Everyone knows what attention is. It is the taking possession by the mind, in clear and vivid form, of one out of what seem several simultaneously possible objects or trains of thought. Focalization, concentration of consciousness are of its essence. It implies withdrawal from some things in order to deal effectively with others, and is a condition which has a real opposite in the confused, dazed, scatterbrained state. ...
>
> *(James, 1890, p. 403)*

The idea of attention as a process by which mental resources can be concentrated or focused continues to pervade thinking in the scientific study of attention. Psychologists and neuroscientists have divided the concept into three different forms: space-based, object-based, and feature-based attention. In this chapter, we will focus on spatial attention. Helmholtz (1867) was one of the first experimentalists to demonstrate that one could covertly (i.e., by holding the eyes fixed) shift one's attention to one part of space prior to presentation of a long list of characters. He found that one could more effectively recollect the characters within the region of space to which the "attentional search light" was shifted.

In the modern study of attention, the Posner paradigm (Posner, 1980) has been particularly influential. This is likely owed to the fact that it is extremely simple to grasp, yet the pattern of results has potentially far-reaching implications for our understanding of spatial attention in mind and brain. In particular, this paradigm has been used to quantify the attentional deficits in patients with parietal-lobe damage (i.e., parietal hemi-neglect syndrome), leading to the theory that spatial attentional mechanisms may be localized in the parietal cortex.

### 7.2.1  So What is the Posner Paradigm?

In the Posner paradigm, research participants are asked to fixate in the center of the screen and not to break fixation for the duration of the trial. Then, a location on the screen is cued in some way (usually by highlighting or flashing something). After the cue, a target appears in either the cued location or in another location. Research participants are instructed to press a key as soon as they see the target. Figure 7.1 provides a schematic illustration of the paradigm.

Posner (1984) found that if the cue is valid, reaction time was substantially lower than if it was invalid. He interpreted this in terms of an "attentional spotlight" that is focused on a certain region in space and permanently shifting at a finite and measurable speed.

## 7.3  EXERCISES

Most of the functions needed to write software that allows you to gather reaction time data were already introduced in Chapter 6, "Visual Search and Pop Out." This time, we will introduce some functions that allow you to generalize the kinds of conditions in which such data are collected. To this end, we introduce another drawing function, **rectangle**, that will come in handy when creating your program in Section 7.4.

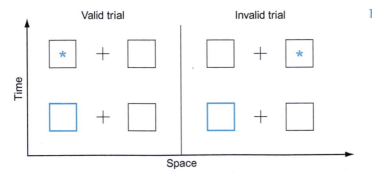

FIGURE 7.1   The Posner paradigm.

Try this code:

```
figure %Create a new figure
xlim([0 1]) %Set the range of values on the x-axis to (0 to 1)
ylim([0 1]) %Set the range of values on the y-axis to (0 to 1)
rectangle('Position', [0.2   0.6   0.5   0.2]) %Create a rectangle at the x-position 0.2,
%y-position 0.6 with an x-width of 0.5 and a y-height of 0.2
```

If you declare **rectangle** with a handle, you can change all properties of the rectangle. Try it. Rectangles have some interesting properties that can be changed.

Regarding data analysis, the most important function we can introduce at this point is the t-test. MATLAB uses **ttest2** to test the hypothesis that there is a difference in the mean of two independent samples.

Consider this:

```
A = rand(100,1); %Create a matrix A with 100 random elements in one column
B = rand(100,1); %Create a matrix B with 100 random elements in one column
[significant,p] = ttest2(A,B)
```

MATLAB should have returned:

```
significant = 0
p = 0.6689
```

This means that the null hypothesis was kept because you failed to reject it. You failed to reject it because the observed difference in means (given the null hypothesis is true) had a probability of about 0.67, which is far too high to reject the null hypothesis. This is what you should expect if the random number generator works. Now try this test:

```
B = B .* 2;
[significant,p] = ttest2(A,B)
significant = 1
p = 3.3467e-013
```

Now, the null hypothesis is rejected. As a matter of fact, the p-value is miniscule.

**Note on seeding the random number generator**: If you use the **rand()** function just as is, the SAME sequence of pseudorandom numbers will be generated each session. You can avoid this by seeding it first like this: **rand('state',number)**. It is important to note that the "random number generator" does no such thing. As a matter of fact, all numbers generated are perfectly deterministic, given the same seed number. We don't want to go on a tangent why this has to be the case or how to avoid this by relying on a genuinely random (at least as far as we can tell) natural process (such as radioactive decay). As long as you pick a different number as a seed each time, you should be fine, for all common intents and purposes. Hence, it is popular to make the number after the state argument dependent on the cpu-clock. In old versions of MATLAB (e.g., 7.04), this could be done as follows:

```
rand('state', sum(100*clock))
```

Newer versions of MATLAB (e.g., 8.1 onwards) rely on a very different system. Namely, the notion of a random number stream that underlies rand, randn, and randi. This random number stream is implemented as **randstream**.

To seed the generator in new versions of MATLAB, things are more complicated but also more versatile.

First type

**RandStream.list**

to get a list of available pseudorandom number generation methods. Mersenne twister with Mersenne prime $2^{19937}-1$ sounds appealing.

Now type

**s = Randstream('mt19937ar', 'seed', sum(100*clock))**

Note the value of "Seed".

Now type

**RandStream.setDefaultStream(s);**

These changes are due to the fact that MATLAB is becoming increasingly object oriented. We are now handling Randstream "objects." Expect to see more of this in the future. For the project in Section 7.4, it might be useful to know at least one other common data analysis function, namely **ANOVA** (analysis of variance). ANOVA generalizes the case of a two-sample t-test to many samples. For the purposes of this chapter, a one-way ANOVA will be sufficient:

**A = rand(100,5); %Generating 5 levels with 100 repetitions each.**
**anova1(A); %Do a one-way balanced ANOVA.**

In this case, there were no significant differences, as revealed by Table 7.1 and Figure 7.2.

Now try this:

**B = meshgrid(1:100); %Generate a large meshgrid**
**B = B(:,1:5); %We only need the first five columns**
**A = A .* B; %Multiply!**
**anova1(A); %Doing the one-way balanced ANOVA again**

This time, there can be no doubt that there is a positive trend, as you can see in Table 7.2 and Figure 7.3.

The **anova1** function assumes that different samples are stored in different columns and that different rows represent different observations in the same sample.

**TABLE 7.1**   ANOVA Table 1

| Source | SS | df | MS | F | Prob > F |
|---|---|---|---|---|---|
| Columns | 0.288 | 4 | 0.072 | 0.81 | 0.5221 |
| Error | 44.2525 | 495 | 0.0894 | | |
| Total | 44.5405 | 499 | | | |

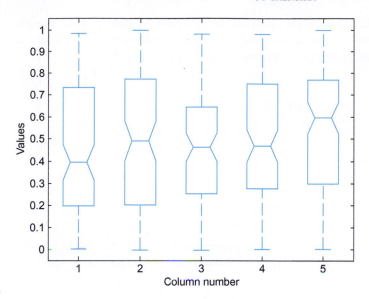

FIGURE 7.2   N.S.

TABLE 7.2   ANOVA Table 2

| Source | SS | df | MS | F | Prob > F |
|---|---|---|---|---|---|
| Columns | 207.519 | 4 | 51.8797 | 50.39 | 0 |
| Error | 509.66 | 495 | 1.0296 | | |
| Total | 717.179 | 499 | | | |

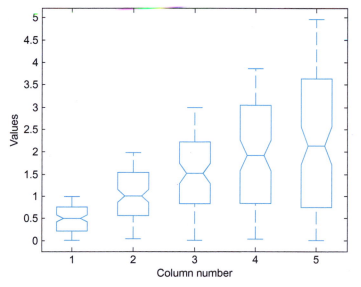

FIGURE 7.3   An effect.

Note that **anova1** assumes that there is an equal number of observations in each sample. For more generalized ANOVAs or unequal samples, see **anova2** or **anovan**. Their syntax is very similar. This, however, should not be necessary for the following project.

## 7.4 PROJECT

For this project, your task is to replicate a generalized version of the Posner paradigm. In essence, you will measure the speed of the "attentional spotlight" in the vertical versus horizontal directions. You need to create a program that allows you to gather data on reaction times in the Posner paradigm as described in the preceding sections. Most of the particular implementation is up to you (the nature of the cue, specific distances, etc.). However, be sure to implement the following:

- Cue and target must appear in one of 16 possible positions. See, for example, Figure 7.4.
- Make sure you have an equal number of valid and invalid trials. [If the trial is valid, the target should appear in the position of the cue. If the trial is invalid, the target position should be picked randomly (minus 1, the position of the cue).]
- Choose two temporal delays between cue and target: 100 ms and 300 ms. Make the delay an experimental condition.
- Collect data from 80 trials per spatial location of the cue (so that you have 20 for each combination of conditions: Valid/invalid, delay1/delay2). This makes for a total of 1280 trials. But they will go very, very quickly in this paradigm.
- Make sure that the picking of condition (valid/invalid, delay1/delay2, spatial location of cue) is random.
- After collecting the data, answer the following questions:
  1. Is there a difference in reaction times for valid versus invalid trials? (t-test)
  2. Is there a difference in reaction times for different delays? (t-test)
  3. Does the distance between target and cue matter? For this, use only invalid trials and plot reaction time as a function of
     a. Total distance of cue and target

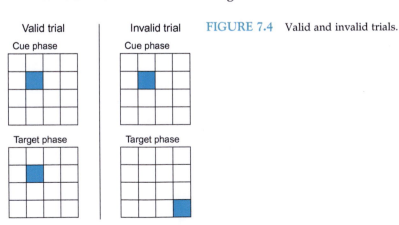

FIGURE 7.4    Valid and invalid trials.

    **b.** Horizontal distance of cue and target

    **c.** Vertical distance of cue and target

4. Related to this: Is there a qualitative difference in the slope of these lines? Is the scanner faster in one dimension than the other?

5. What is the speed of the attentional scanner? How many (unit of your choice, could be inches) does it shift per millisecond?

- Implement the project in MATLAB and answer the preceding questions. Illustrate with figures where appropriate.

- **\*If you are adventurous**: Use **anova2** or **anovan** to look for interaction effects between type of trial (valid/invalid, delay and spatial location of cue).

*Hints:*

- Start writing one trial and make sure it works properly.
- Be aware that you effectively have an experimental design with three factors: Cue position (16 levels), trial type (2 levels), and temporal delay (2 levels). However, you can break it up into four factors: Horizontal cue position (4 levels), Vertical cue position (4 levels), trial type (2 levels), and temporal delay (2 levels), which will make it easier to assess the x- versus y-speed of the scanner.
- If you can't produce a proper cue, try reviewing object handles (in figures).
- Write a big loop that goes through trials. Do this at the very end, if individual trials work.
- If you can't do everything, focus on subgoals. Implement one function after the other. Start with two conditions. If you are not able to implement all eight conditions, try to get as far as you can.

# MATLAB FUNCTIONS, COMMANDS, AND OPERATORS COVERED IN THIS CHAPTER

**randstream**
**rectangle**
**ttest2**
**state**
**clock**
**anova1**

# 8

# Psychophysics

## 8.1 GOALS OF THIS CHAPTER

In this chapter, you will learn how to use MATLAB® to do psychophysics. Once you master these fundamentals, you can use MATLAB to address any psychophysical question that might come to mind. While—in principle—all sensory modalities are open to psychophysical investigation, we will focus on visual psychophysics in this chapter.

## 8.2 BACKGROUND

Psychophysics deals with the nature of the quantitative relationship between physical and mental qualities. Today, the practice of psychophysics is ubiquitous in all fields of neuroscience that involve the study of behaving organisms, be they man or beast. Curiously enough, the origins of systematic psychophysics can be traced to a single individual: Gustav Theodor Fechner (1801−1887). Fechner's biography exhibits many telling idiosyncrasies. Born the son of a pastor, he studied medicine at the University of Leipzig, but never practiced it after receiving his degree. Mostly by virtue of translating chemistry and physics textbooks from French into German, he was appointed professor of physics at the University of Leipzig. In the course of studying afterimages by gazing into the sun for extended periods of time—himself being the primary and sole research participant—he almost lost his eyesight and went into deep depression in the early 1840s. This episode lasted for nearly a decade, a time which Fechner spent mostly within a darkened room. Emerging from this secluded state, he was overwhelmed by the sheer brilliant radiance of his surroundings, giving rise to his panpsychist worldview: he was now utterly certain that all things have souls, including inanimate objects such as plants and stones.

MATLAB® for Neuroscientists.
DOI: http://dx.doi.org/10.1016/B978-0-12-383836-0.00008-4

Determined to share his insights with the rest of humanity, he soon started publishing on the topic, formulating an "identity theory" stating that the physical world and the spiritual world are not separate entities, but actually the same—the apparent differences resulting from different perspectives (first versus third person) onto the same object. In his view, this reconciles the incompatible dominant philosophical worldviews of the 19th century: idealism and materialism. However, his philosophical treatises on subjects such as the soul-life of plants or the transcendence of materialism were poorly received by the scientific community of the day (Fechner, 1848). In order to convince his colleagues of the validity of his philosophical notions, he set out to devise methods that would allow him to empirically link physical and spiritual realms (Fechner, 1851). His rationale being that if it can be shown that mental and physical qualities are in a clear functional relationship, this would lend credence to the notion that they are actually metaphysically identical.

Publishing the results of empirical studies on the topic in his *Foundations of Psychophysics* in 1860, he showed that this is the case for several mental domains, such as the relationship between physical mass and the perception of heaviness or weight. Fechner formulated several methods to arrive at these results that are still in use today. Importantly, he expressed the results of his investigations in mathematical, functional terms. This allowed for the theoretical interpretation of his findings. Doing so, he introduced notions such as sensory thresholds quantitatively.

Ironically, inventing psychophysics did not help Fechner in convincing his philosophical adversaries of the merits of his identity theory. Few philosophers of the day renounced their idealistic or materialistic positions in favor of identity theory. Most of them simply chose to ignore Fechner, while the others mostly attacked him. Consequently, Fechner spent much of the remainder of his life fighting these real or imaginary adversaries, publishing two follow-up volumes in 1877 and 1882, chiefly focusing on the increasingly bitter

FIGURE 8.1    Gustav Theodor Fechner (1801–1887).

struggle against the philosophical establishment of Imperial Germany. Ultimately, these efforts had little tangible or lasting impact. Meanwhile, the first experimental psychologists, particularly the group around Wundt, pragmatically used these very same methods to create a psychology that was both experimental and empirical. It is not too bold to claim that they never stopped and that contemporary psychophysics derives in an unbroken line from these very roots.

The key to visual psychophysics (and psychophysics more generally) is to elicit relatively simple mental phenomena that lend themselves to quantification by presenting physical stimuli that are easily described by just a few parameters such as luminance, contrast, or spatial frequency.

It is imperative that the experimenter has complete control over these parameters. In other words, the visual stimuli that s/he is presenting have to be precise. One way to create these stimuli is to use commercially available graphics editors, most prominently Photoshop®. While this practice is very common, it comes at a cost. For example, the images created by Photoshop have to be imported by the experimental control software. It is more elegant to create the stimuli in the same environment in which they are used. More importantly, the experimenter surrenders some degree of control over the created stimuli, when using commercial graphics editors, because the proprietary algorithms to perform certain image functions are not always completely documented or disclosed. This problem is equally avoided by creating the stimuli in a controlled way within MATLAB.

## 8.3 EXERCISES

We need to introduce methods by which to create and present visual stimuli of any kind on the screen. Fortunately, MATLAB includes a large library of adequate functions. We will introduce the most important ones here.

By default, MATLAB visualizes images by triplets in a 256-element RGB space. Each element of the triplet has to be an integral value between 0 and 255. This corresponds to a range of 8 bits. Hence, these elements can be represented by variables of the type uint8. These values correspond to the intensity of red, green, and blue at a particular location in the image. Depending on the physical output device (typically cathode ray tube or LCD displays), these values effectively regulate the voltage of individual ray guns or pixels. Of course, MATLAB also supports much higher bit-depths. For the purposes of our discussion, 8 bit will suffice.

One important caveat is that the relationship between the assigned voltage values of the three individual ray guns in the cathode ray tube (0 to 255) and the perceived luminance of the screen is not necessarily linear. Visual scientists generally calibrate their monitors by "gamma correcting" them. This linearizes the relationship between assigned voltage or intensity values and the perceived luminance.

To be able to linearize the relationship, you need a photometer. Because we assume that those are—due to their generally high price—not readily available, we will forgo this step for the purposes of this book. However, we urge budding visual scientists to properly calibrate their monitors before doing an experiment in which the veracity of the data is crucial—as is the case if they are intended for publication. For more information on the issue of monitor calibration, see for example Carpenter and Robson (1999).

To begin, create a simple matrix with the following command:

```
>> test_disp = uint8(zeros(3,3,3))
```

This command creates the matrix **test_disp**, which is a three-dimensional matrix with three elements in each dimension. Importantly, it is of the data type **uint8**, which MATLAB assumes by default for its imaging routines. Of course, the function image can also image other matrices, but this would require to specify additional parameters. Imaging only matrices of the type **uint8** is the most straightforward thing to do for the purposes of this chapter.

Now type the following:

```
>> figure
>> subplot(2,2,1)
>> image(test_disp)
```

The function image compels MATLAB to interpret the values in the matrix as commands for the ray guns of the monitor and to display them on the screen. You should now be looking at a completely and uniformly dark (black) subplot 1.

Now, type the following:

```
>> test_disp(2,2,:) = 255
>> subplot(2,2,2)
>> image(test_disp)
```

The structure of the matrix and the function of the **image** command now become apparent.

The 0 values are interpreted as turning off all ray guns. The 255 values are interpreted as full power. As you maximally engage all three guns (represented by the third dimension of the matrix), the result is an additive mix of spectral information that is interpreted by the visual system as white. Now, try this:

```
>> subplot(2,2,3)
>> test_disp(2,2,1) = 0
>> image(test_disp)
```

The picture in subplot 2 is devoid of color from the red gun. It should appear cyan. Now try the following:

```
>> test_disp(2,2,:) = 0
>> test_disp(2,2,1) = 255
>> subplot(2,2,4)
>> image(test_disp)
```

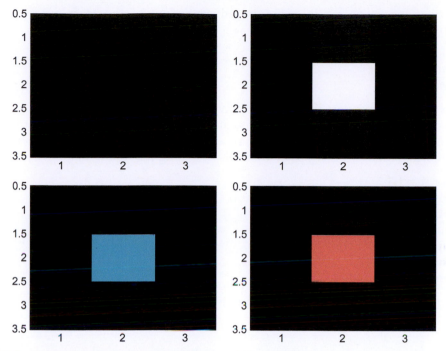

FIGURE 8.2   Testing the ray guns with matrices interpreted as images.

This code has the opposite effect, yielding a red inner pixel. The result should look something like Figure 8.2.

---

### SUGGESTION FOR EXPLORATION

Can you create arbitrary other colors? Can you create arbitrary shapes?

---

While it is hard to surpass this example of using just 9 pixels in clarity, it is also somewhat pedestrian. The true power of this approach becomes clear when considering natural stimuli, which are also increasingly used in visual psychophysics. To do this, you need to import an image into MATLAB. For this example, use the **imread** function:

```
>> temp = imread('UofC.jpg')
```

This command creates a large three-dimensional matrix of the **uint8** type (positive integral values from 0 to 255, as can be addressed by 8 bits).

Next, type the following to get a magnificent view of the Harper Library at the University of Chicago:

```
>> figure
>> subplot(2,2,1)
>> image(temp)
```

Now, you can manipulate this image in any way, shape, or form. Importantly, you will know exactly what you are doing, since you are the one doing it, which cannot be said for most of the opaque algorithms of image processing software. For example, you can separate the information in different color channels by typing the following:

```
>> for ii = 1:3
>> bigmatrix(:,:,:,ii) = zeros(size(temp,1),size(temp,2),3);
>> bigmatrix(:,:,ii,ii) = temp(:,:,ii);
>> subplot(2,2,ii + 1)
>> upazila = uint8(bigmatrix(:,:,:,ii));
>> image(upazila)
>> end
```

You get the picture shown in Figure 8.3 as a result.

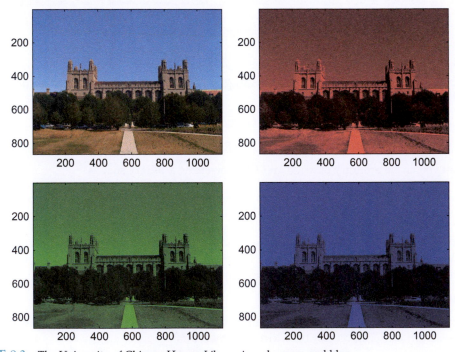

FIGURE 8.3    The University of Chicago Harper Library in red, green, and blue.

Note the use of the **upazila** helper variable. You have to use this because MATLAB interprets non-3D-**uint8** matrices differently when presenting images. Later you will learn how to make do without the **upazila** step. The different subplots illustrate the brightness values assigned to an individual ray gun (Upper left: all of them together. Upper right: red. Lower left: green. Lower right: blue). This allows you to assess the contribution of every single channel (red/green/blue) to the image. Another way of judging the impact of a particular channel is to leave it out. To do this, you add the other channels together, as follows:

```
>> bigmatrix2(:,:,:,1) = bigmatrix(:,:,:,1) + bigmatrix(:,:,:,2) + bigmatrix(:,:,:,3);
>> bigmatrix2(:,:,:,2) = bigmatrix(:,:,:,1) + bigmatrix(:,:,:,2);
>> bigmatrix2(:,:,:,3) = bigmatrix(:,:,:,1) + bigmatrix(:,:,:,3);
>> bigmatrix2(:,:,:,4) = bigmatrix(:,:,:,2) + bigmatrix(:,:,:,3);
>> figure
>> for ii = 1:4
>> subplot(2,2,ii)
>> image(uint8(bigmatrix2(:,:,:,ii)))
>> end
```

Doing so should yield the picture shown in Figure 8.4.

In this figure, the upper left is all channels. In the upper right, the blue channel is missing. In the lower left, the green channel is missing. In the lower right, the red channel is missing. Belaboring this point enhances an understanding of the relationship between the brightness values in the three-dimensional matrix and the appearance of the image.

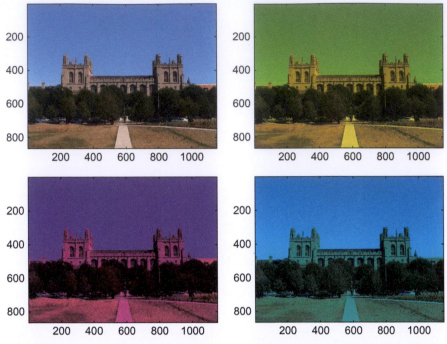

**FIGURE 8.4**   The University of Chicago Harper Library without red, green, and blue information.

You are now in a position to implement arbitrary changes to the image. For example, you can brighten it, increase the contrast, or selectively change the color balance. To explore this, start by changing the overall brightness by typing the following:

```
>> figure
>> subplot(2,2,1)
>> image(uint8(bigmatrix2(:,:,:,1)))
>> subplot(2,2,2)
>> image(uint8(bigmatrix2(:,:,:,1) + 50))
>> subplot(2,2,3)
>> image(uint8(bigmatrix2(:,:,:,1)-50))
>> subplot(2,2,4)
>> bigmatrix2(:,:,1,1) = bigmatrix2(:,:,1,1) + 100;
>> image(uint8(bigmatrix2(:,:,:,1)))
```

The result, shown in Figure 8.5, is a picture that has been somewhat brightened (upper right), darkened (lower left), and where the red channel has been turned (way) up in the lower right.

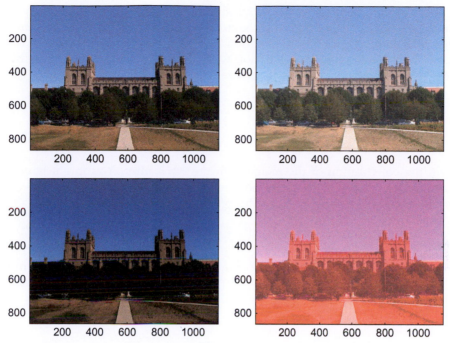

**FIGURE 8.5**    Brightening and darkening any or all ray guns has a profound effect.

---

## SUGGESTION FOR EXPLORATION

Increase the contrast of the image. Also try to image matrices that are not of the type **uint8**.

---

One of the most common image manipulations using image editors is the smoothing or sharpening of the image. The former is often performed to get rid of random noise or granularities in the image. Scientists might do this to simulate and understand the output of the visual system of another species. Importantly, these ends are typically achieved by low- or high-pass filtering of the original image. Unfortunately, most users don't really understand what is happening behind the scenes when using a commercially available image editor. Of course, this is unacceptable for doing psychophysics in particular or science in general.

Hence, we will now discuss how to perform these operations in MATLAB. First, import another image by typing the following. This image is more suited to making the effects of your manipulations more readily apparent.

```
>> pic = imread('filtering.jpg')
>> figure
>> subplot(2,2,1)
```

Look at the image. So far, so good. Now, slightly blur the image. To do so, you will convolve the image with a filter. Refer to Chapter 16, "Convolution," to understand precisely the underlying mathematics of the operation. For purposes of this example, it is enough to understand that the convolution operation will allow you to blur the image by blending brightness values of adjacent pixels. To create a small $3 \times 3$ filter, you type

```
>> filter = ones(3,3)
```

then

```
>> lp3 = convn(pic,filter)
```

to perform the convolution of the image with the $3 \times 3$ filter. You might have noted that the values are no longer in the range between 0 and 255. This is due to the multiplying and adding brought about by the convolution. Next, divide by the block size (9) to rectify this situation:

```
>> lp3 = lp3./9;
```

This operation creates floating-point values, so you have to be careful when imaging this:

```
>> subplot(2,2,2)
>> image(uint8(lp3))
```

This code creates a very slightly low-pass filtered version of the image. This result is most readily apparent when you look at the texture of the hat or hair in the image. Now, try a more radical low-pass filtering:

```
>> filter = ones(25,25)
>> lp25 = convn(pic,filter)
>> lp25 = lp25./625;
>> subplot(2,2,3)
>> image(uint8(lp25))
```

Looking at the image reveals significant blurring. This is the low-frequency component of the image. It is similar to what a typical nocturnal animal with relatively poor spatial acuity might see (sans the color). You arrive at the image by blurring a substantial number of pixels together.

---

### SUGGESTION FOR EXPLORATION

What happens if you use ever larger filters?

---

Note that the matrices you created with the convolution operation are slightly larger than the one that represents the initial picture (which had a format of $600 \times 800$). This is due to the nature of the convolution operation. It creates an artificial black rim not present in the original picture. You will understand why this happens and why this is a hard problem when reading Chapter 16. For now (we will encounter much more elegant ways in Chapter 16), and to (mostly) get rid of it and cut the image back to size, try the following:

>> lp25cor = lp25(13:612,13:812,:);

You might also have noted that the execution of the convolution operation took a significant amount of time. This might be important if you want to create your stimuli on the fly, as the observer does the experiment. To assess how your system stacks up against certain known benchmarks, type this command:

>> **bench**

Doing so makes MATLAB perform various typical operations and compare the speed of their execution to other benchmark systems. This is particularly crucial when you're running a time-sensitive program. Don't be surprised when receiving rather low benchmark values, particularly when running MATLAB over a network, despite basically fast hardware. To evaluate the reliability of the benchmark values, try running **bench** more than once, e.g. bench(5) runs it 5 times.

Let's get back to the filtering problem. The image in the lower left corresponds to what psychophysicists would call the "low spatial frequency" channel. It contains the low spatial frequency information in the image. Notably, it is mostly devoid of sharp edges. This information about edges in the image is contained in the "high spatial frequency" channel. How do you get there? By subtracting the low spatial frequency information from the original image. Try this:

>> subplot(2,2,4)
>> hp = pic-uint8(lp25cor);
>> image(hp)

The image in the lower right now contains the high spatial frequency information. It represents most of the textures and sharp edges in the original image.

Unfortunately, it is rather dark (due to the subtraction). To appreciate the full high spatial frequency information, add a neutral brightness level back in:

>> hp = hp + 127;
>> image(hp)

Much better. The final result should look something like that shown in Figure 8.6.

---

### EXERCISE 8.1

Use the information in the high spatial frequency channel to sharpen (enhance the edges) of the original image.

---

We have discussed a variety of image manipulations with MATLAB, namely the manipulation of form, color, and spatial frequency. One remaining major issue is the creation of

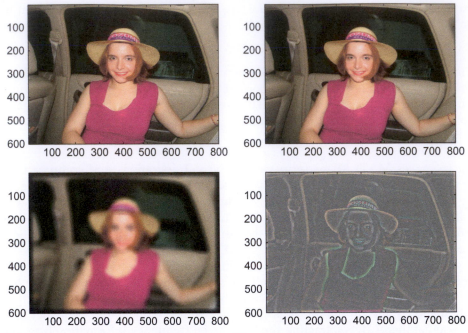

FIGURE 8.6    Information about texture is carried in different spatial frequency channels.

moving stimuli. There are many ways to do this in MATLAB. One of the most straightforward is to use the **circshift** function in combination with a frame capture function.

To use this function, type the following:

```
>> figure
>> pic3 = circshift(pic,[100 0 0]);
>> image(pic3)
```

The **circshift** function shifts all matrix values by the stated amount in the second argument—in this case, 100 in the direction of the first dimension, nothing in the others.

You can use this to create a movie:

```
>> figure
>> pic4 = pic;
>> for ii = 1:size(pic4,1)/10 + 1
>> image(pic4)
>> pic4 = circshift(pic4,[10 0 0]);
>> M(ii) = getframe;
>> end
```

There are other ways to create movie frames, for example, using the **im2frame** function. However, this version lets you preview the movie as you create it. You can play it by typing

```
>> movie(M,3,24)
```

This command plays the movie in matrix *M* three times, at 24 frames per second, in the existing figure. One caveat for movies is size in memory. These frames take up a considerable amount of space. You might get an error message indicating that the frames could not be created if you go beyond the available memory. The available memory depends on the computer and operating system. To assess the memory situation on the machine you are using, type

**memory**

There are several caveats when making movies with MATLAB. For example, the choppiness will depend on many factors, including machine speed, available memory, step size of the **circshift**, as well as frame rate. On most systems, it will be hard to avoid trade-offs to create movies that are reasonably smooth.

---

**EXERCISE 8.2**

Use this knowledge to create a movie of a single white dot (pixel) that moves from the far left of the screen to the far right. Then do it without circshift, simply updating the RGB values in the matrix for each frame with the right values. As most pixels don't change frame to frame, they don't need to be updated (circshift shifts all of them).

---

**SUGGESTION FOR EXPLORATION**

Import two pictures with the same size. Create a movie in which one morphs into the other. *Hint:* Over time, the numerical values in the matrix that represents the image should gradually shift from one to the next while you capture this process in frames.

---

As the color information is represented in the third dimension of the matrix, you can also use **circshift** to elegantly swap colors, as in this example:

```
>> figure
>> for ii = 1:3
>> subplot(1,3,ii)
>> image(pic)
>> axis equal
>> axis off
>> pic = circshift(pic,[0 0 1]);
>> end
```

The result should look something like Figure 8.7.

Often, when creating large numbers of stimuli, you might want to save them on the hard disk to free up some space in available memory. You can easily do this by using

FIGURE 8.7    You can use **circshift** to shift colors.

the **imwrite** function. For example, you can save the image in which the RGB values were swapped for BRG by typing **imwrite(pic,'BRG.jpg','jpg')**. This should have created the file BRG.jpg as a .jpg file in your working directory. You can now open it with other image editing software, put it online, etc. Similarly, you can save the movie by typing

**movie2avi(M,'upazila.avi','quality',100,'fps',24)**

which creates an .avi file named upazila at a frame rate of 24 and a quality of 100 in your working directory. From now on, this file will behave like any other movie file you might have on your hard disk.

At this point, we have explored several important image manipulation routines that should really give you a deep appreciation of the way MATLAB represents and displays images. Of course, many more image manipulations are possible in MATLAB. We will leave those for you to discover and return to the task of collecting psychophysical data, using this newfound knowledge. Because MATLAB represents images as brightness values in a three-dimensional matrix, you can manipulate them at will with any number of matrix operations. In principle, you could write your own Photoshop toolbox in MATLAB.

---

### EXERCISE 8.3

Can you rotate an image by 90°? Can you rotate by an arbitrary number of degrees?

---

### EXERCISE 8.4

Try adding different images together. For example, you can transmit secret information by embedding one image in another. Or create artificial stimuli. For example, in the attention community, it is popular to superimpose pictures of houses and faces.

> **SUGGESTION FOR EXPLORATION**
>
> Implement your favorite Photoshop routine in MATLAB.

Let's get back to psychophysics. Fechner formalized three fundamental methods to elicit the relationship between mental and physical qualities and introduced them to a wider audience. These methods are still in use today. You should recognize the *method of limits* from visits to the ophthalmologist investigating your vision or the otologist investigating your hearing. Basically, the observer is presented with a series of stimuli in increasing (or decreasing) intensity and asked to judge whether or not the stimulus is present. This method is extremely efficient because only a few stimuli are necessary to establish fairly reliable thresholds. Unfortunately, the method suffers from hysteresis; the threshold is path dependent, as participants exhibit a certain inertia (e.g., stating that the stimulus is still present even if they can't detect it, if coming from the direction of a stimulus being present). This problem can be overcome by counterbalancing (starting from different states). However, a better correction is the *method of constant stimuli*. In this method, the experimenter presents stimuli to be judged by the observer in random order, from a pre-determined set of values. The advantage of this method is that it yields very reliable and mostly unbiased threshold measurements. The drawback is that one needs to sample a relatively large range of stimuli (as one doesn't a priori know where the threshold will lie) and a large number of repetitions per conditions to reduce error. Hence, this method is usually not used where time is at a premium (such as in a doctor's office), but rather in research, where the time of undergrad or grad student observers is routinely sacrificed for increases in accuracy.

Finally, the *method of adjustment* lets the research participant manipulate a test stimulus that is supposed to match a given control. This method is particularly popular in color psychophysics. It is relatively efficient, but suffers from its own set of biases.

## 8.4 PROJECT

In this project, you will use the method of constant stimuli to determine the absolute threshold of vision, a classic experiment in visual psychophysics (Hecht, Shlaer, and Pirenne, 1942). Obviously, you will be able to do only a crude mock-up of this experiment in the scope of this chapter. The actual experiment was extremely well controlled and took a long time to carry out (not to mention specialized equipment).

Since you are unconcerned with publishing the results (these are extremely well established), you can pull off a "naïve" version in order to highlight certain features and principles of the psychophysical method. If you want to increase experimental control, perform the experiment in a dark room and wait 15 minutes (or better 30 minutes) before data collection. Also, try to keep a fixed distance from the monitor (e.g., 50 cm) throughout the data collection phase of the experiment.

However, before you can collect data, you need to write a stimulus control program utilizing the skills from the previous two chapters and the image manipulation skills introduced in this chapter. Here is a simple program that will do what is needed (make this an M-file). Note the somewhat obsolete use of the modulus function to order the stimuli. We could also do this with **randi** in the latest versions of MATLAB. On the other hand, the use of the modulus function allows you to have exactly the same number of trials per condition (as opposed to them having random frequencies).

```
clear all; %Emptying workspace
close all; %Closing all figures

temp = uint8(zeros(400,400,3)); %Create a dark stimulus matrix
temp1 = cell(10,1); %Create a cell that can hold 10 matrices

for ii = 1:10 %Filling temp1
    temp(200,200,:) = 255; %Inserting a fixation point
    temp(200,240,:) = (ii-1)*10; %Inserting a test point 40 pixels right
                                 %of it. Brightness range 0 to 90.
    temp1{ii} = temp; %Putting the respective modified matrix in cell
end %Done doing that

h = figure %Creating a figure with a handle h

stimulusorder = randperm(200); %Creating a random order from 1 to 200.
                               %For the 200 trials. Allows to have
                               %a precisely equal number per condition.

stimulusorder = mod(stimulusorder,10); %Using the modulus function to
                                       %create a range from 0 to 9. 20 each.

stimulusorder = stimulusorder + 1; %Now, the range is from 1 to 10, as
                                   %desired.

score = zeros(10,1); %Keeping score. How many stimuli were reported seen

for ii = 1:200 %200 trials, 20 per condition
image(temp1{stimulusorder(1,ii)}) %Image the respective matrix. As
                                  %designated by stimulusorder
ii %Give observer feedback about which trial we are in. No other feedback.
pause; %Get the keypress
temp2 = get(h,'CurrentCharacter'); %Get the keypress. "." for present,
                                   %"," for absent.
temp3 = strcmp('.', temp2); %Compare strings. If . (present), temp3 = 1,
                            %otherwise 0.
score(stimulusorder(1,ii)) = score(stimulusorder(1,ii)) + temp3; %Add up.
                                       % In the respective score sheet.
end %End the presentation of trials, after 200 have lapsed.
```

Note that these are relatively crude steps. In a real experiment, you might want to probe every luminance value and collect more samples per condition (50 or 100 instead of 20).

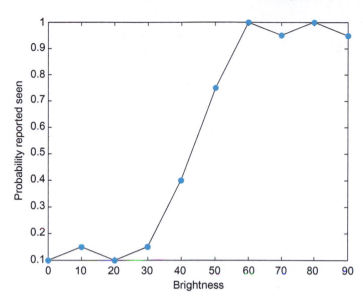

FIGURE 8.8 The psychometric curve reveals below-threshold regions, saturation regions, and a linear range.

Also, a time limit of exposure and decision time is usually used. But for now, this will do. When running this program yourself, make sure to focus on the central fixation dot. Don't get frustrated or bored. Psychophysical experiments are extremely intricate affairs, usually operating at the limits of the human sensory apparatus. Hence, they are rarely pleasant. So try to focus the firepower of your cortex on the task at hand. Also note that you will make plenty of errors. Don't get frustrated. That is the point of psychophysics. In a way, psychophysics amounts to a very sophisticated form of producing and analyzing errors. If you don't make any errors, there is no variance, and without variance, most of the psychophysical analysis methods fail—hence the large number of trials. Given enough trials, you can count on the statistical notion that truly random errors will average out, while retaining and strengthening the systematic trends in the data, revealing the properties of the system that produced it. As a matter of fact, you might want to throw in a couple of practice runs before deciding to analyze your data for real. Given that you are likely to be what is technically called an untrained observer there will be various dynamics going on during the experiment. At first, practice effects will enhance the quality of your judgments; then fatigue will diminish it. Also note that you are technically not a "naïve" observer, as you are aware of the purpose of the experiment. Don't let this discourage you for now. Doing so, we obtained the curve shown in Figure 8.8. This figure shows a fairly decent psychometric curve. It is obvious that we did not see the dot on the left tail of the curve (the observed variation represents errors in judgment). Similarly, it is obvious that we did always see the dot on the right of the curve, yet there is some variation in the reported instances seen. In other words, the points on the left are below threshold, whereas the points on the right are already saturated. In a real experiment, we would resample the range between the brightness values 20 and 70 much more densely, as it is clear that the date points outside this range add no information. However, this neatly illustrates one problem of the method of constant stimuli. We didn't know where the threshold would

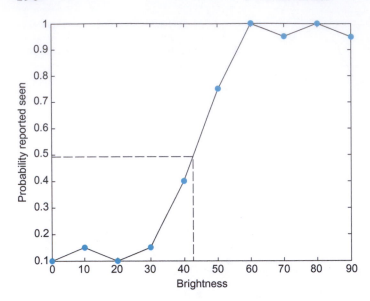

lie. Hence, we had to sample a broad range—undersampling the crucial range and oversampling regions of no interest. Even limiting the range from 0 to 90 was an educated guess. Strictly speaking, and without any previous knowledge, we would have had to sample the entire range of 0 to 255. Modern "staircase" procedures attempt to solve this problem, but are beyond the scope of this chapter. Psychophysicists like to boil down this entire dataset into one number: the *absolute threshold*. In this case, you can derive this value by interpolation. It is the x-value that corresponds to the intersection between the curve and the y-value of probability reported seen of 0.5, as shown in Figure 8.9. In other words, this analysis indicates an absolute threshold of a brightness value of 43. If you want to get a precise threshold, you would have to resample the range between 30 and 60 (or even between 40 and 50) very densely. Also note that this value of 43 is not inherently meaningful. Without having the monitor calibrated with a photometer, we don't know to how much physical light energy this corresponds to. Hence, we can't relate it to the minimum number of light quanta that can be detected. However, this threshold is meaningful in the context of a behavioral task: a shifted threshold under different conditions can give rise to conjectures about the structure and function of the physiological system producing these thresholds, as you will see when doing the exercises. Moreover, the absolute threshold is a stochastic concept. It is not true that lights below it are never seen. Of course, psychophysicists have very elaborate ways to analyze data like these. Most straightforwardly, they like to fit sigmoidal logistic curves to such data. We will go into the intricacies of psychophysical data analysis in the next chapters. Finally, we chose luminance values that worked on our monitor, yielding a decent psychometric curve, allowing us to determine the threshold. You might have to use a different range when working within your setup. For more background on psychophysical methods, read the classic *Elements of Psychophysics* by Fechner (1860) or, for a modern treatment of the use of these methods in visual neuroscience, *The Psychophysical Measurement of Visual Function* by Norton, Corliss, and Bailey (2002). In this project, you should specifically address the following issues: Compare thresholds in the periphery and center. You just did a parafoveal stimulus presentation (if you were honest and fixated the fixation point) or even a foveal presentation (if you looked at the stimulus directly). How does the threshold change in the periphery (putting the stimulus several hundred pixels away from the fixation point)?

Determine the thresholds for brightness values of the red, green, and blue guns individually. Which gun has the lowest threshold (is perceived as brightest)? Which gun has the highest threshold (is perceived as dimmest)? Can you account for the white threshold (as we did above) by adding the individual thresholds?

You just determined absolute thresholds. Another important concept in psychophysics is the relative threshold. To determine the relative threshold, put another test dot to the left of the fixation point. The task is now to indicate if the brightness of the right dot is higher (.) or lower (,). Does the relative threshold depend on the absolute brightness values of the dots? If so, can you characterize the relationship between relative threshold (difference in stimulus brightness values that gives a probability of 0.5) and absolute value of the stimuli?

When determining the relative threshold, can you reason why it makes sense to ask which of the two is brighter, instead of asking if they are the same or different (which might be more intuitive)?

## MATLAB FUNCTIONS, COMMANDS, AND OPERATORS COVERED IN THIS CHAPTER

**uint8**
**double**
**convn**
**circshift**
**image**
**bench**
**imread**
**imwrite**
**memory**
**movie**
**getframe**
**movie2avi**
**randperm**
**mod**
**spy**

# Psychophysics with GUIs

## 9.1 GOALS OF THIS CHAPTER

This chapter pursues dual goals. First, we want to build on the data collection with the psychophysical methods that were introduced in the last chapter. Second, and more importantly, this chapter will introduce the concept of a graphical user interface (GUI) within MATLAB® and demonstrate its gainful use.

## 9.2 INTRODUCTION AND BACKGROUND

A surprisingly large number of scientists pride themselves on their coding skills as a considerable source of self-esteem and identity. For them, it is bad enough that they are using a high-level interpreted language like MATLAB in the first place. They surely wouldn't be caught dead using a GUI on top of that, as giving up the command line would likely constitute a deadly blow to their street cred.

Nevertheless, there are legitimate uses for GUIs. These reasons are largely the same ones that made GUIs catch on in the community at large. Briefly put, they make things more accessible, and they require less user knowledge to operate. This is neatly illustrated by the success of Microsoft Windows, which made the use of computers conceivable for a mass audience that couldn't realistically be expected to learn how to profitably interact with a command line.

To illustrate this point within a MATLAB context, the following (true) example should suffice. I was once involved in a long-distance collaboration involving a question that was of theoretical interest to me. They had specialized equipment and would collect the data, but they were not trained in the arts of MATLAB, so we decided that I would do the data analysis. So far, so good. What I didn't realize is that there was a rather large amount of data files collected under a daunting number of experimental conditions that didn't seem to be organized or denoted in any way that made sense to me. Worse, they didn't seem to be able to communicate the structure to me, and on top of everything else, the organization of these data files seemed to keep changing. As you will learn in some of the later chapters, the proper organization of data is absolutely crucial when attempting to analyze complex datasets. I couldn't just send them the code so that they could adapt to their use, because they didn't have MATLAB, nor did they know how to use it. To make a long story short, after a few miserable and drawn-out attempts, the collaboration failed. It was my fault. What I should have done instead (short of physically going there, which I couldn't do, due to other commitments) was to write a self-contained GUI that encapsulated the data analysis itself. Such a GUI can be deployed on any machine, even if it doesn't have MATLAB, and—if properly set up—can be operated by anyone, even without any MATLAB knowledge. Put differently, I should have used a MATLAB GUI to create a data browser (and analyzer), then given it to those who understood the structure of the dataset (because they collected it). Instead of struggling with an impossible analysis, I should have invested my time in creating a purpose-built GUI. Luckily, you can learn from my mistake. This example serves—at least to me—as a very vivid cautionary tale of why one disregards GUIs at one's peril. They do have legitimate uses.

## 9.3  GUI BASICS

GUI stands for graphical user interface. It is therefore redundant to say "GUI interface." GUIs were pioneered by Xerox in the 1960s, along with the mouse. They were introduced to public use with the Apple Macintosh in 1984 and to widespread public use with Windows. In terms of MATLAB, early GUI functionality was introduced to MATLAB with version 4 in 1992.

For most purposes, the use of GUIs is not indicated. It takes effort to make them, and it is overkill if only the programmer will ever use the program, e.g., in most data analysis cases in research. This is particularly true if speed is of the essence, e.g., if you are rushing to meet a deadline. The two most common use cases that *do* warrant the construction of a GUI are:

- If the end user is not the programmer and not expected to know the intricacies of the program, e.g., for teaching, data collection, or analysis (as in the case of my ill-fated collaboration).
- If the program involves a lot of flags and parameters that need to be customized with every run. Instead of doing this on the command line, it is easier to press buttons (and not forget one). This is good if you, for instance, write a data browser or a simulation.

Having now discussed the conceptual history and defined use cases for GUIs, the obvious question is: How to make one?

This question is best answered not in the abstract, but with a practical example; which is why we cover that in the next section.

## 9.4  USING A GUI TO TRACK AN IP ADDRESS

In popular culture, GUIs are, perhaps unsurprisingly, steeped in mystery. For instance, in an episode of "CSI: New York," a character states, during an urgent crisis, that she will "create a GUI interface using Visual Basic to see if she can track an IP address" (http://goo.gl/0Dxv0). While this is plainly ridiculous on many levels (see use cases, previous section), this will do as a simple starter example that will allow you to grasp the basic functionality. Let's see if we can create a GUI (not a GUI interface!) to track an IP address. And we'll do it using MATLAB, not Visual Basic.

The first thing to note is that while you can, in principle, build a GUI by hand and from scratch, no one really does this anymore since MathWorks introduced the function **guide**.

GUIDE stands for "GUI design environment," and it provides exactly that, an editor that allows you to create GUIs in a point-and-click fashion. It allows you to quickly create functional GUIs, automatically taking care of most of the plumbing. In MATLAB, GUIs consist of a figure with associated code. The shell of all of this is created by **guide**, you are simply expected to add the functionality. You add elements like buttons, sliders, and lists in the figure, and you spell out what should happen once the users interact with these elements in the code. Let's try it.

After you type **guide**, you will be prompted with a quick start dialog window; see Figure 9.1. Note: This GUI was created on a Mac. If you use a PC, your directory structure will show backslashes, not slashes.

For now, please opt to create a new GUI, use the default "Blank GUI" template, and click "OK."

Once you do that, a figure opens that will allow you to build your GUI (see Figure 9.2). At this point, nothing is on it, but note in the bottom right the indication of a "current point" vector, which denotes the x and y position of your mouse cursor on the figure, as well as the "Position" vector. You can resize the figure by dragging the black rectangle on the bottom right corner of the gray figure canvas. Make sure to keep the screen dimensions of the machine that you want to deploy the GUI in mind. It would be annoying if the GUI you create so painstakingly wouldn't fit on the target screen.

On the left is a palette of tools you can insert in the figure, containing buttons and the like. Click on the button icon ("OK," below the arrow), and draw its outline on the canvas: Once you are done, it should look something like Figure 9.3.

The button reads "Push Button"; let's change that. Double-click on the button. This brings up the "Inspector." In this case, the inspector indicates that it is inspecting the uicontrol pushbutton1. You can also invoke this menu by typing "inspector" in the command line. You do need to give the handle of the uicontrol as an argument. For now, simply scroll

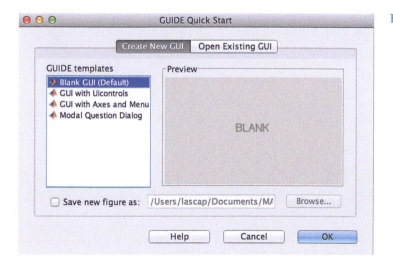

FIGURE 9.1    GUIDE quick start.

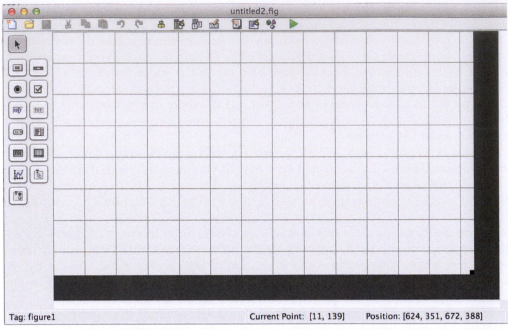

**FIGURE 9.2**   Blank GUI template.

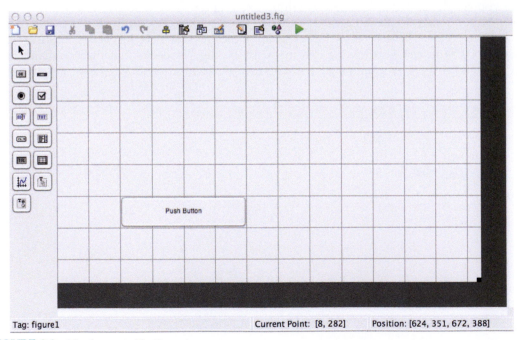

**FIGURE 9.3**   Newly created button.

down and take note of the different properties. At this point, the tag "pushbutton1" and the "String" are most relevant. The tag is how we will later address the button. We click on the text next to the string, change it to "TRACK!," and click OK; see Figure 9.4. Uicontrols are user interface objects such as buttons, sliders, and the like. They have a great many properties that can be set. They can be created both by dragging them onto the canvas within the guide, or programmatically. For now, we'll focus on the guide approach.

---

### EXERCISE 9.1

Add another button to the right of the existing button; label it "Own machine."

---

Now add an "Edit Text" control. After adding it, change its background color to white (via the inspector) and take note of its tag. Your figure should now look something like Figure 9.5.

Now add another "Edit Text" control, and label them with Static text controls as "Host" and "IP," as in Figure 9.6.

As you can see, the aesthetics of this GUI leave a lot to be desired, but we'll fix that later. For now, let's focus on functionality.

Click on the green arrow to run the GUI. If you do this for the very first time, you will be polled for a name. Call it iptracker. After you enter a filename, two things will open: The GUI as it will look to the end user and the code that was created by GUIDE; see Figure 9.7.

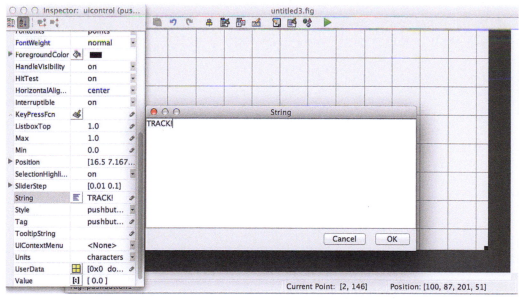

FIGURE 9.4    Renaming the button to "TRACK!".

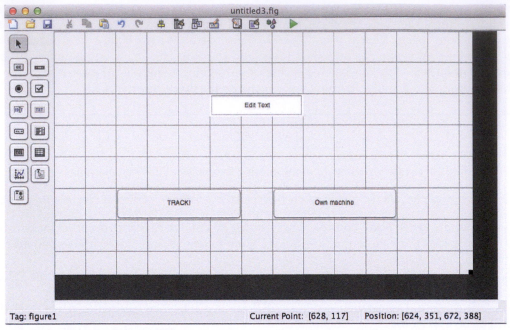

FIGURE 9.5    Edit Text control added.

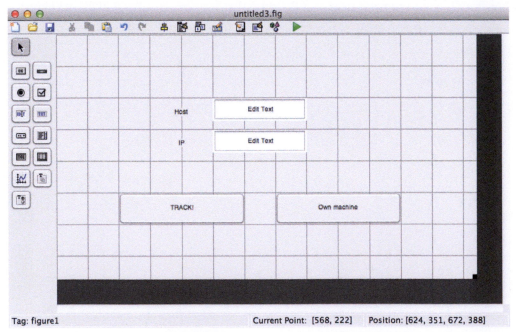

FIGURE 9.6    Host and IP Edit Text controls.

The code looks complicated. It seems like GUIDE already created over a hundred lines of code. But we'll look at that soon. First, click on the buttons. Nothing will happen. You can enter text into the text controls, but that's it. Clearly, we need to add functionality to our program.

That's what we'll do now. Close the running instance of the GUI, which should leave you with the GUI editor and the code. All existing code is stored in iptracker.m, and was created automatically by GUIDE. Analyzing the code, there are eight functional parts. First, a lengthy comment section explains the iptracker function (in principle). You can edit this if you want to. Then comes initialization code, which you should not edit until you know what you are doing; changes here can break the GUI. After that comes a function that executes at the opening of the GUI, before you can see it. Do not concern yourself with this either at this point. Then comes a function that handles potential output to the command line. Of particular interest for our purposes are the functions that come after that. Two of them, "pushbutton1_Callback" and "pushbutton2_Callback," govern what happens if the respective button is pressed. To make this explicit: If a given button (identified by the tag, which you can change via the inspector) is pressed, the code in the corresponding callback function is executed. The remaining functions govern the creation and updating of the two edit text boxes. These are the guts of the GUI. We need to define what the code should be doing by editing the "callback" functions. Speaking more generally, every time you invoke the uicontrol by moving a slider or pressing a button, the code within the associated "callback" function is executed. On a sidenote, you can add uicontrols and associated callback functions to any figure, without formally creating a GUI if you want the user to be able to interact with the figure directly, e.g., allowing to change line styles on the fly.

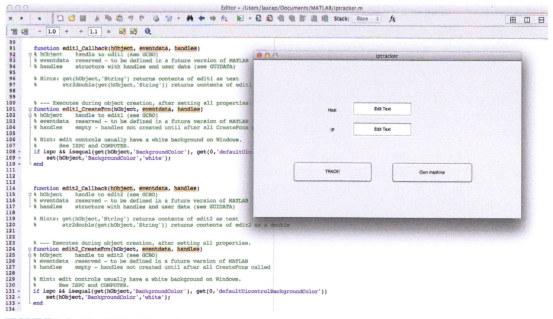

FIGURE 9.7   The GUI and its code.

---

### EXERCISE 9.2

Use the inspector to change the tag for the two buttons from pushbutton1 and pushbutton2 to "track" and "own," respectively. Similarly, change the tag of the two edit text elements to "ip" and "host." See how it changes automatically in the code after saving the figure by running it again.

This is recommended so that you don't lose track of what is what in GUIs with many elements.

The code still won't do anything, but at least the functions are now named properly.

---

Now, we are in a position to create the right functionality in the right places. There are actually only a few key structures and functions that govern the behavior of the GUI. These are:

- **handles**: this is the name of the structure that contains all data that is passed around within the GUI as its elements.
- **get**: this function is used to get the value of a GUI element.
- **set**: this function is used to set the value of a GUI element.
- **guidata**(hObject,handles): invoking this updates the handles data structure.

And that's basically it. The rest is details and commentary.

We now add the following code to the track callback function, which executes when the track button is pressed. Basically, we implement a three-step process. First, we read in the value of the "host" text field, then we use a special Java function provided by MATLAB (**getHostAddress**) to resolve the IP address. Third and finally, we put output this value into the ip text field, as follows:

handles.temp = get(handles.host,'String'); %Reading in the string in "host" text field, %putting it into temp.

ipaddress = char(getHostAddress(java.net.InetAddress.getByName(handles.temp))); %Resolving the IP address.

set(handles.ip,'String',ipaddress); %Updating the string field of the ip object with %the ip address.

At the end of this exercise, your function should look like Figure 9.8.

Now execute your code and try it. Input a URL into the host field, then click the track button (see Figure 9.9).

```
76      % --- Executes on button press in track.
77    □ function track_Callback(hObject, eventdata, handles)
78    □ % hObject     handle to track (see GCBO)
79      % eventdata  reserved - to be defined in a future version of MATLAB
80    └ % handles    structure with handles and user data (see GUIDATA).
81  -   handles.temp = get(handles.host,'String'); %Reading in the string in "host" text field, putting it into te
82  -   ipaddress = char(getHostAddress(java.net.InetAddress.getByName(handles.temp))); %Resolving the IP address.
83  - └ set(handles.ip,'String',ipaddress); %Updating the string field of the ip object with the ip address.
```

FIGURE 9.8   Track callback function.

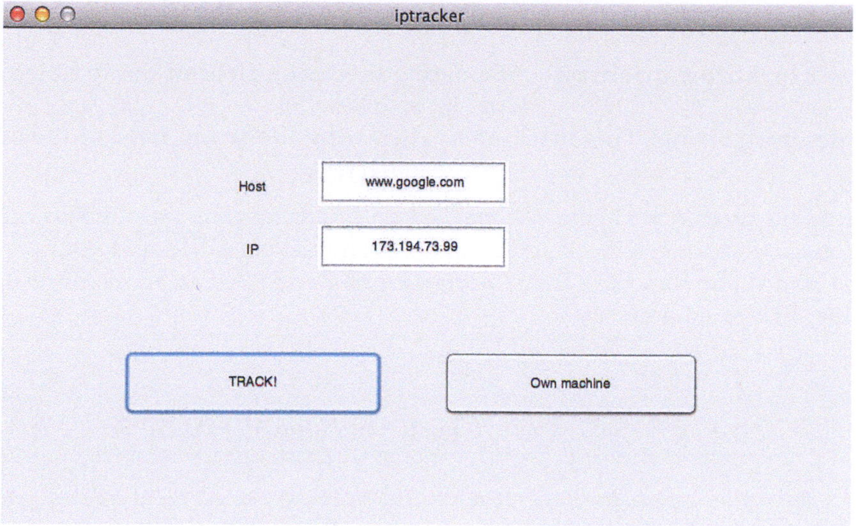

FIGURE 9.9 Success!

It works! Congratulations, you just executed your first MATLAB function from a GUI, and you tracked an IP address in the process.

Naysayers might complain that we didn't actually track an IP address, we just resolved one. OK, fine. Luckily, we anticipated this in the design of our program.

We'll use similar but slightly different logic as before, (as we now need to use functions to retrieve our own IP). To do that, add the following code to the callback function that corresponds to the "own machine" button (see Figure 9.10):

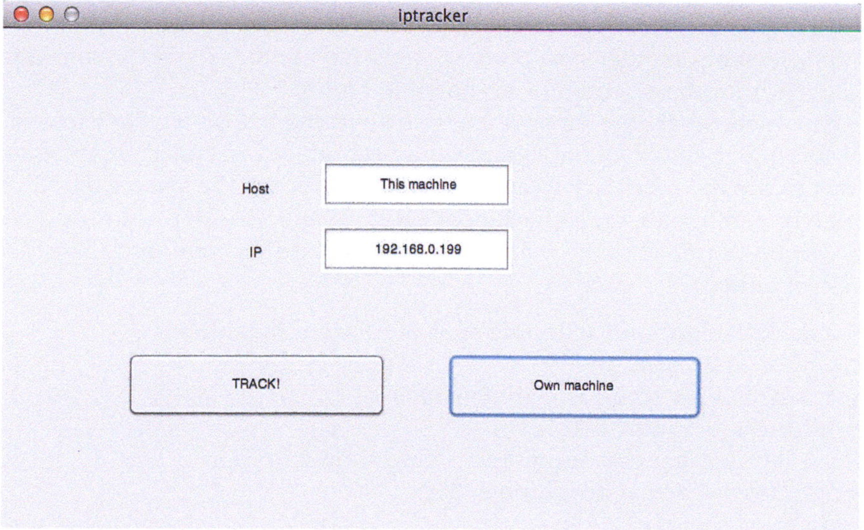

FIGURE 9.10 Success again!

ipaddress = char(getHostAddress(java.net.InetAddress.getLocalHost)); %Get own IP
%address.

set(handles.ip,'String',ipaddress); %Updating the string field of the ip object with %
%the ip address.

set(handles.host,'String','This machine'); %Updating the string field of the host
%object.

We taste sweet success yet again. We tracked an IP address (in real time!), even if it was our own. You can even toggle back and forth between tracking URLs and your own IP.

Now that you understand the basic mechanics of GUIs, we can move on to something more exciting, like psychophysics.

## 9.5  USING A GUI FOR PSYCHOPHYSICS

We won't reinvent the wheel here. If temporal precision is an issue and if you want to do advanced psychophysics, you should use the Psychophysics Toolbox or MGL, as explained in more detail in Appendix B. Nevertheless, you can use GUIs to nicely collect psychophysical data, and maybe even add a button to calculate thresholds, and so on.

As you will also learn in Appendix B, there is a MATLAB compiler that allows you to deploy a GUI without needing the machine of the end user to have MATLAB installed. This adds versatility if you need to collect data in the field.

People are very used to GUIs by now. Most participants in your experiments won't know much about MATLAB, but they will know how to use a GUI (if you designed it right).

For educational purposes, we won't start with a fresh GUI, but will continue in medias res. Let's add some things to the existing GUI that you will need for psychophysics.

The first thing to do is to resize it (make it bigger) to accommodate our changes.

Then, add an axes element (you can add as many axes as you want, but one will do for now), a slider element, another two buttons ("START" and "HAPPY"), and another text edit field (tagged brightness), roughly as shown in Figure 9.11.

As you know by now, these elements don't do anything yet, so we have to add the functionality by adding code. Once we execute the program by clicking on the green arrow, MATLAB will create the necessary wrappers/callbacks for us. We just have to fill them.

Appropriately, we'll start with the functionality of the "START" button. The point of this button will be to initialize the display in "axes." Put this code into the callback function for the start button:

```
%Create a dark background with one spot of random brightness
handles.X = zeros(500,500,3);
%Create a matrix with zeros, in 3 dimensions
actual_brightness = randi([0 255],1);
%Pick a random integer as a luminance value from 0 to 255.
actual_brightness = actual_brightness./255;
%Scale down
handles.X(250:259,200:209,:) = actual_brightness;
%Assign it to a 9 × 9 pixel square.
imagesc(handles.X,[0 255]); %Image it, scaled.
```

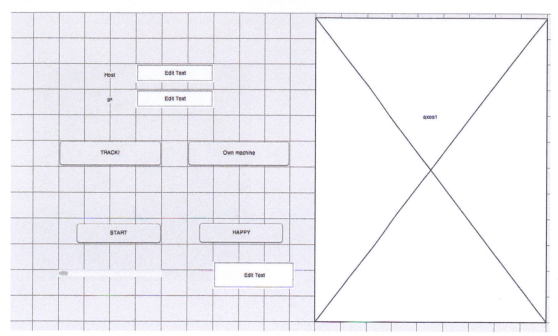

FIGURE 9.11   The expanded GUI.

**axis off; %Take off axes labels, etc.**
**axis square %Make it square**
**handles.actbright = actual_brightness; %Put the actual brightness in the handles**
**%structure.**

Now every time you execute the code, a bright square will be displayed on a dark background. Every time you press start again, a new, random brightness is picked (see Figure 9.12).

Now for the slider. The idea is that the participant can dial the brightness of a comparison square up and down. The goal is to match the brightness of the square set by MATLAB (once START is pressed).

Add this code to the slider function. It executes every time the slider is moved. The logic is that we first get the slider value, then add it to the matrix:

**handles.bright = get(handles.slider1,'Value'); %Getting the slider value**
**handles.X(250:259,300:309,:) = handles.bright; %Assign it to another 9 × 9 pixel**
**%square next to the other one.**
**imagesc(handles.X,[0 255]); %Image it, scaled.**
**axis off; %Take off axes labels, etc.**
**axis square %Make it square**

Run the code. Something weird will happen. In my case, the whole screen turns blue every time I move the slider (see Figure 9.13). Not quite the outcome I was hoping for. What is going on?

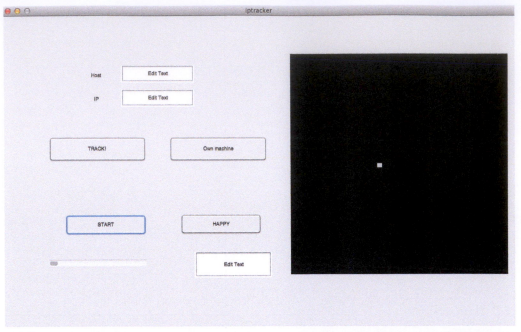

FIGURE 9.12   The START button at work.

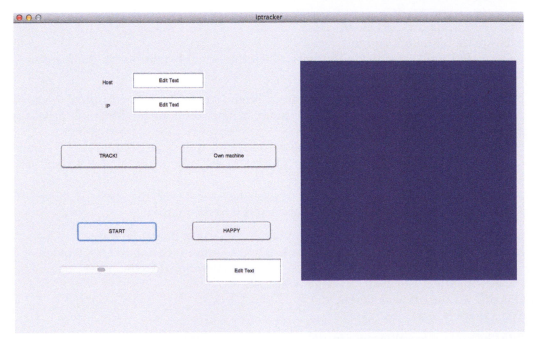

FIGURE 9.13   Not the desired outcome.

Bugs like these are hard to track down. In this case, it helps to remember what we discussed earlier in terms of the functions that implement virtually all basic GUI operations. The individual functions that make up the GUI don't by themselves have access to variables in other functions. They only do so via the handles structure. And that is only updated if the function **guidata** is invoked. This means there is an easy fix here. The slider function did not have access to the X matrix we created in the start button function, as handles hadn't yet been updated. To remedy this, add this code at the end of the start button code (and for good measure, put it at the end of the slider code as well):

**guidata(hObject, handles);**

That did it; see Figure 9.14. The user can adjust the brightness of the right patch at will, by moving the slider.

We now need to assign an end condition, a condition that allows the user to indicate that he is happy with the match and ready to move on to the next trial. That's where the "HAPPY" button comes in.

Add this code to the function that executes when the happy button is pressed:

**handles.diff = abs(handles.bright-handles.actbright).\*255; %Calculate the absolute**
**%difference between the values**
**set(handles.brightness,'String',num2str(handles.diff)); %Put it in the text field tagged**
**%"brightness"**
**guidata(hObject, handles); %Don't forget to update. Save your work.**

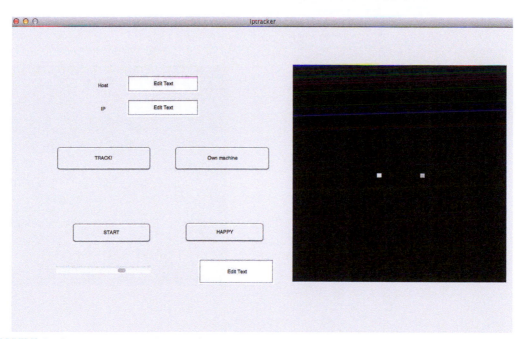

FIGURE 9.14  The brightness slider working properly.

It calculates the difference between the two brightness values, and outputs it in the text field we haven't used yet (see Figure 9.15).

Now, if you were adventurous, you could add code that starts a new trial in this very same function. You could also add code that calculates thresholds on the press of a button, you could display the data on the screen and prettify the design of this figure, etc.; you get the idea.

This is as far as we'll go for now. You can write GUIs of arbitrary complexity with hundreds of elements and multiple pages with the principles we covered here. If you are interested in more details and, in particular, how to build GUIs by hand (without using **guide**), we refer you to Smith (2006), although the book is somewhat dated by now.

Congratulations, you did "CSI: New York" one better. Not only did you create a GUI that tracks an IP address in real time, the same GUI also allows you to collect psychophysical data at the same time. Impressive.

---

### EXERCISE 9.3

The "method of production" sensu Fechner, in which the study participant controls a dial to match a given stimulus intensity, is particularly popular in color psychophysics. Create a GUI where the participant is presented with a given colored light and has to reproduce it by adjusting three sliders: one for the red gun, one for the green gun, and one for the blue gun of the screen.

---

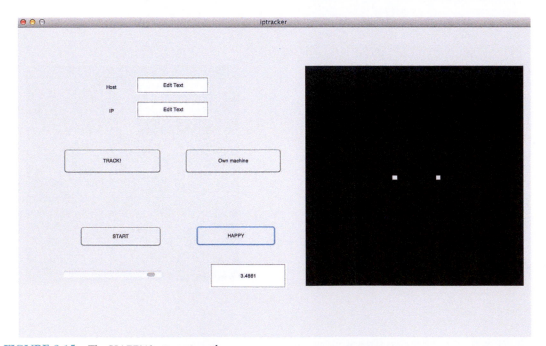

FIGURE 9.15   The HAPPY button at work.

## 9.6  PROJECT

This one is very straightforward. Put the psychophysics task that you created in the last chapter into a GUI! Make sure to add a button that allows you to calculate thresholds at the end. It doesn't have to calculate IP addresses, so start with a fresh GUI.

## MATLAB FUNCTIONS, COMMANDS, AND OPERATORS COVERED IN THIS CHAPTER

**guide**
**inspector**
**getHostAddress**
**guidata**
**randi**
**guidata**

# Signal Detection Theory

## 10.1 GOALS OF THIS CHAPTER

This chapter will mostly concern the use of signal detection theory to analyze data generated in psychophysical—and hypothetical neurophysiological—experiments. As usual, we will do this in MATLAB®.

## 10.2 BACKGROUND

At its core, signal detection theory (SDT) represents a way to optimally detect a signal in purely statistical terms without an explicit link to decision processes in particular or cognitive processes in general. However, in the context of our discussion, SDT provides a rich view of the problem of how to detect a given signal. It reframes the task as a decision process, adding a cognitive dimension to our understanding of this matter.

To illustrate the application of SDT in psychophysics, let us again consider the problem of reporting the presence or absence of a faint, barely visible dot of light, as in Chapter 8, "Psychophysics." In addition to the threshold, which is determined by the physical properties of the stimuli and the physiological properties of the biological substrate, there are cognitive considerations. In particular, observers have a *criterion* by which they judge (and report) whether or not a signal was present. Many factors can influence the criterion level and—hence—this report. You likely encountered some of those in the preceding chapters. For example, your criterion levels might have been influenced by doing a couple hundred trials, giving you an appreciation of what "present" and "absent" mean in the context of the given stimulus range (which is very dim overall). Moreover, motivational concerns might play a role when setting a criterion level. If research participants have an incentive to over- or under-report the presence of a signal—e.g., if they think that the experimenter expects this—they will in fact do so (Rosenthal, 1976). Interestingly, neuroeconomists utilize this effect by literally paying their research participants to prefer one alternative, in order to study the mechanisms of how the criterion level is set.

Of course, one could question the real-world relevance of these considerations, given that they arose in very particular and arguably often rather contrived experimental

*MATLAB® for Neuroscientists.*
DOI: http://dx.doi.org/10.1016/B978-0-12-383836-0.00010-2

settings. It is worth emphasizing that this first impression is extremely misleading. Today, signal detection theory constitutes a formal, stochastic way to estimate the probability by which some things are the case and others are not; by which some effects are real and others are not, and so on. As such, it has the broadest possible implications. Signal detection theory is used by pharmaceutical companies as well as oil prospectors, and it has even made its mark in public policy considerations. Of course, experimenters—psychophysicists in particular—also still use it.

The astonishing versatility and base utility of signal detection theory are likely owed to the fact that it goes to the very heart of what it means to be a cognitive organism or system, as we will now describe.

Consider the following situation. Let us assume you work for a company that builds and installs fire alarm systems. As these systems are ubiquitous in modern cities, business is good. However, you are confronted with a rather confounding problem: How sensitive should you make these alarms? The four possible cases are tabulated in a classic matrix, as shown in Table 10.1.

Let's peruse this matrix in detail as it is the foundation of the entire discussion to follow. The cell in the upper left represents the "desired" (as desired as it can be, given that there is a real fire in the building) case: there actually is a fire and the alarm does go off, urging the occupants of the building to leave and alerting the fire department to the situation. Ideally, you would like the probability of this event to be 1. In other words, you always want the alarm to go off when there is a fire present. This part should be fairly uncontroversial. The problem is that in order to reach a probability of 1 for this case, you need to set the criterion level for indicating "fire" by some parameters (usually smoke or heat or both) incredibly low. In fact, you need to set it so low that it will likely go off by levels of smoke or heat that can be reached without a fire being present. This puts you into the cell in the upper right. If this happens, you have a false alarm. From personal experience, you can probably confirm that the criterion levels of fire alarms are typically set in a hair-trigger fashion. Almost anything will set them off, and almost all alarms are therefore false alarms, given that the a priori probability of a real fire is very, very low. As is the case in modern cities. While this situation is better than having a real fire, false alarms are not trivial. Having them frequently is disruptive, can potentially have deleterious effects in case of a real fire (as the occupants of the building learn to stop taking action when the alarm goes off), and strains the resources of the firefighters (as a matter of fact, firefighters have been killed in traffic accidents on their way to false alarms). In other words, setting the sensitivity too high comes at a considerable cost. Hence, you want to lower the sensitivity enough to always be in a state that corresponds to one of the cells on the main diagonal of the matrix, either having a hit (if there is actually a fire present) or a "correct rejection," arriving in the lower right. The latter should be the most common case,

**TABLE 10.1**   The Signal Detection Theory Payoff Matrix

|                     | Real fire     | No fire (but possibly some smoke or heat) |
| ------------------- | ------------- | ----------------------------------------- |
| Alarm goes off      | Hit           | False alarm                               |
| Alarm does not go off | Miss        | Correct rejection                         |

indicating that there is no fire and the fire alarm does not go off. Unfortunately, if you drop the sensitivity too low, you arrive in the worst cell of all in terms of potential for damage and fatalities: having a real fire, but the fire alarm does not alert you to this situation. This is called a "miss," in the lower left.

In a way, this matrix illustrates what signal detection theory is all about: figuring out a way to set the criterion in a mathematically optimal fashion (in the applied version) and figuring out how and why people, organisms, and systems actually do set criteria when performing and solving cognitive tasks (in the pure research version).

If this description sounds familiar, it should. As contemporary science has largely adopted a stochastic view on epistemology, this fundamental situation of signal detection theory appears in many if not most experiments, disguised as the "p-value" problem.

You are probably well aware of this issue, so let us just briefly retrace it in terms of signal detection theory.

When performing an experiment, you observe a certain pattern of results. The basic question is always: How likely is this pattern, given there is no effect of experimental manipulation? In other words: How likely are the observed data to occur purely by chance? If they are too unlikely given chance alone, you reject the "null hypothesis" that the data came about by chance alone. That is—in a nutshell—the fundamental logic of testing for the statistical significance of most experimental data since Fisher introduced and popularized the concept in the 1920s (Fisher, 1925).

But how unlikely is too unlikely? Again, we face the fundamental signal detection dilemma, as illustrated in Table 10.2.

In science, the criterion level is conventionally set at 5%. This is called the *significance level*. If a certain pattern of data is less likely than 0.05 to have come about by chance, then you reject the null hypothesis and accept that the effect exists. Implicitly, you also accept that—at this level—5% of the published results will not hold up to replication (as they don't actually exist). It is debatable how conservative this standard is or should be. For extraordinary claims, a significance level of 1% or even less is typically required. What should be apparent is that the significance level is a social convention. It can be set according to the perceived consequences of thinking there is an effect when there is none (*alpha error*) or failing to discover a genuine effect (*beta error*), particularly in the medical community. The failure to find the (side-) effect of certain medications has cost certain companies (and patients) dearly. For a dissenting view on why the business of significance testing is a bad idea in the first place, see for example Ziliak and McCloskey (2008).

Regardless of this controversy, one can argue that any organism is—curiously—in a quite similar position. You will learn more about this in Chapters 21, "Neural Decoding: Discrete Variables," and 22, "Neural Decoding: Continuous Variables." For now, let us discuss the fundamental situation as it pertains to the nervous system (particularly the brain)

TABLE 10.2    Alpha and Beta Errors in Experimental Judgment

|  | Effect exists (H0 false) | Effect does not exist (H0 true) |
| --- | --- | --- |
| We conclude it exists | Discovery of effect | Alpha error (false rejection of H0) |
| We conclude it doesn't exist | Beta error (false retention of H0) | Failure to reject the null hypothesis |

of the organism. Interestingly, based on everything we currently assume to be true, the brain has no direct access to the status of the environment around it—as manifested in the values of physical parameters such as energy or matter. It learns about them solely by the pattern of activity within the sensory apparatus itself. In other words, the brain deduces the structure of the external world by observing the structural regularities of its own activity in response to the conditions in the outside world. For example, the firing of a certain group of neurons might be associated with the presence of a specific object in the environment. This has profound philosophical implications. Among them is the notion that the brain decodes its own activity in meaningful ways, as they were established by interactions with the environment and represent meaningful associations between firing patterns and states in the environment. In other words, the brain makes actionable inferences about the state of the external world by cues that are provided by activity levels of its own neurons. Of course, these cues are rarely perfectly reliable. In addition, there is also a certain level of "internal noise," as the brain computes with components that are not perfectly reliable either. This discussion should make it clear how the considerations about stochastic decision making introduced previously directly apply to the epistemological situation in which the brain finds itself. We will elaborate on this theme in several subsequent chapters. It should already be readily apparent that this is not trivial for the organism, as it has to identify predator and prey, along with other biologically relevant hazards and opportunities in the environment. In this sense, errors can be quite costly.

---

### SUGGESTION FOR EXPLORATION

The basic signal detection situation seems to reappear in different guises over and over again. What we call the two *fundamental errors* varies from situation to situation. Try framing the results of a diagnostic test for an arbitrary disease in these terms (which here appear as *false positive* and *false negative*). Also, try to model a case in the criminal justice system in this way.

---

## 10.3 EXERCISES

With this background in mind, it is now time to go back to MATLAB®. Let us discuss how you can use MATLAB to apply signal detection theory to the data generated in behavioral experiments.

Consider this situation. You run an experiment with 2000 trials. While running these trials, you record the firing rate from a single neuron in the visual cortex. In 1000 of the trials, you present a very faint dot. In the other 1000, you just present the dark background, without an added visual stimulus.

Let's plot the (hypothetical) firing rates in this experiment. To do so, you use the **normpdf** function. It creates a normal distribution. Normal distributions occur in nature when a large number of independent factors combine to yield a certain parameter. A normal distribution is completely characterized by just two parameters: its mean and variance.

Now, let us create a plausible distribution of firing rates. There is evidence that the baseline firing rate of many neurons in visual cortex in the absence of visual stimulation hovers around 5 impulses/second (Adrians). Moreover, firing rates cannot be negative. This makes our choice of a normal distribution somewhat artificial, as it does—of course— yield negative values.

With that in mind, the following code will produce a somewhat plausible distribution of firing rates for background firing in the absence of a visual stimulus:

```
x = 0:0.01:10;
y = normpdf(x,5,1.5)
plot(x,y)
```

The third parameter of **normpdf** specifies the variance. In this case, we just pick an arbitrary, yet reasonable value—for neurons in many visual areas, the variance of the neural firing rate scales with and is close to the mean. Other values would also have been possible. Note that strictly speaking, it would make more sense to consider only integral firing rates, but for didactic reasons, we will illustrate the continuous case. This will not make a difference for the sake of our argument, and it is the more general case.

Now consider the distribution where the stimulus is in fact present. However, it is very faint. It is sensible to assume that this will change the firing rate of an individual neuron only very modestly (as the neuron needs the rest of the firing range to represent the remaining luminance range). This assumes that the neuron changes its firing rate in response to luminance changes in the first place. Many—even in visual cortex—do not but are luminance invariant. Most do—however—modulate their firing in response to contrast. None of this is important for our didactic "toy" case. A plausible distribution will be created by:

```
z = (normpdf(x,6,1.5));
plot(x,z)
```

In other words, we assume that adding the stimulus to the background adds only—on average—one spike per stimulus in this hypothetical example. For the sake of simplicity, we keep the variance of the distribution the same, in reality it would likely scale with the increased mean.

---

## SUGGESTION FOR EXPLORATION

MATLAB has a large library of probability density functions. Try another one to model neural responses. A plausible starting point would be the Poisson distribution, as it yields only discrete and positive values, as is the case with integral neural firing rates. MATLAB offers the Poisson probability density function under the command **poisspdf**.

---

On a side note, this is a good point to introduce another class of MATLAB functions, namely cumulative distribution functions. They integrate the probability density of a given distribution function (e.g., the normal distribution).

These are used for many calculations, as they provide an easy way to determine the integrated probability density of a given distribution at a certain cutoff point.

For example, **normcdf** is often used to determine IQ-percentiles.

As IQ in the general population is distributed with a mean of 100 and a standard deviation of 15, we can type:

**normcdf(100,100,15)**

to get the unsurprising answer:

**ans =**
   **0.5000**

If we want to find out the percentile of someone with an IQ of 127, we simply type:

**>> normcdf(127,100,15)**
**ans =**
   **0.9641**

In other words, the person has an IQ higher than 96.41% of the population.

Back to SDT. If you did everything right, you can now cast the problem in terms of signal detection theory. It should look something like Figure 10.1.

The plot in Figure 10.1 contrasts the case of stimulus absence versus stimulus presence; firing rate in impulses per second is plotted on the x-axis, whereas probability or frequency is plotted on the y-axis. The thick vertical black line represents the criterion level we chose.

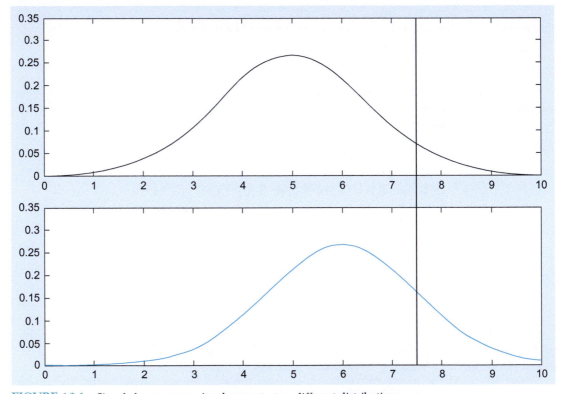

FIGURE 10.1    Signal absent versus signal present—two different distributions.

The upper panel represents the case of an absent stimulus. For the cases to the right of the black line, the neuron concluded "stimulus present," even in the absence of a stimulus. Hence, they are false alarms. Cases to the left of the black line are correct rejections. As you can see, at a criterion level of 7.5 impulses per second, the majority of the cases are correct rejections.

The lower panel represents the case of a present stimulus. For the cases to the left of the black line, the neuron concluded "stimulus absent," even in the presence of a stimulus. Hence, they are misses. Cases to the right of the black line represent hits. At a criterion level of 7.5 impulses per second, the majority of the cases are misses.

We are now in a position to discuss and calculate the receiver operating characteristic (ROC) curve for this situation, defined by the difference in mean firing rate, variance, and shape of the distribution. The exotic-sounding term *receiver operating characteristic* originated in engineering, in particular the study of radar signals and their interpretation.

Generally speaking, an ROC curve is a plot of the false alarms (undesirable) against hits (desirable), for a range of criterion levels. "Area under the ROC curve" is a metric of how sensitive an observer is, as will be discussed later. Given the conditions that you can generally assume, ROC curves are always monotonically non-decreasing curves. In the context of tests, you plot the hit rate (or true positive rate or sensitivity) versus the false positive rate (or 1-specificity) to construct the ROC curve. Keep this in mind for future reference. It will be important.

First, try to plot the ROC curve:

```
figure
for ii = 1:1:length(y) %Going through all elements of y
FA(ii) = sum(y(1,ii:length(y))); %Summing from ith element to rest → FA(ii)
HIT(ii) = sum(z(1,ii:length(y))); %Summing from ith element to rest → Hit(ii)
end
FA = FA./100; %Converting it to a rate
HIT = HIT./100; %Converting it to a rate
plot(FA,HIT) %Plot it
hold on
reference = 0:0.01:1; %reference needed to visualize
plot(reference,reference,'color','k') %Plot the reference
```

*Note:* This code could have been written in much more concise and elegant ways, but it is easier to figure out what is going on in this form.

To get the ROC curve, see Figure 10.2.

Note that the false alarms and hits are divided by 100 to get a false alarm and hit rate. The black line represents a situation in which hits and false alarms rise at the same rate — no sensitivity is gained at any point. As you can see, this neuron is slightly more sensitive than that, as evidenced by the deviation of its ROC curve from the black identity line. However, it rises rather gently. There is no obvious point where one should set the criterion to get substantially more hits than false alarms. This is largely due to the small difference in means between the distributions, which is smaller than the variance of the individual distribution. By experimenting with different mean differences, you can explore their effect on the ROC curves.

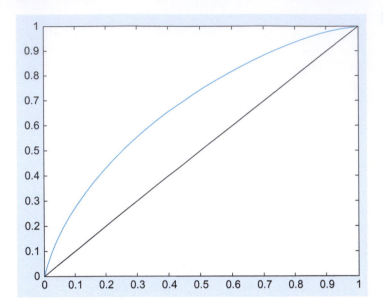

FIGURE 10.2    The ROC curve.

---

**EXERCISE 10.1**

Experiment with mean differences by yourself. The result should look something like Figure 10.3.

---

It becomes readily apparent that the ability to choose a criterion level that allows you to give a high hit rate without also getting a high false alarm rate is dependent on the difference between the means of the distributions. The larger the difference (relative to the variance) between the means, the easier it is to set a reasonable criterion level. For example, a mean difference of 5 allows you to get a hit rate of 0.9 virtually without any false alarms. This also gives a normative prescription to reduce false alarms: If you want to reduce false alarms, you should increase the difference in the means of the measured parameter between conditions of signal present (e.g., a fire) and signal not present (e.g., no fire). The clearer the parameters you choose to differentiate between these two cases, the better off you will be. A similar case can be made for the variance of the signals. The less variance (often noise) there is in the signals, the better off you will be, when you are trying to distinguish between them. Hence, in order to create highly sensitive tests that discriminate between two situations, one needs to measure parameters that can be measured reliably without much noise but which exhibit a large difference in the mean parameter value, given the different situations in question.

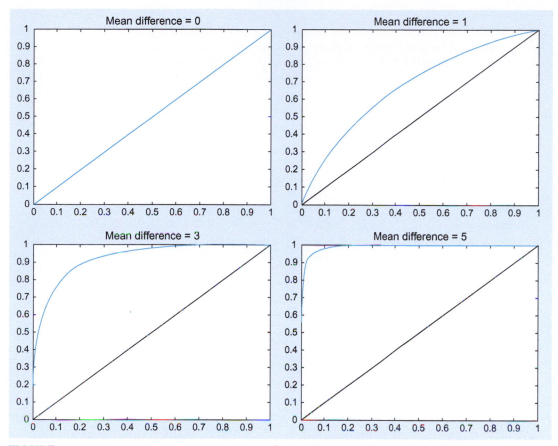

**FIGURE 10.3** The shape of the ROC curve is dependent on the mean difference of the distributions.

---

## SUGGESTION FOR EXPLORATION

How are the ROC curves affected by increasing the variance of the distribution, while keeping the absolute mean difference the same?

---

The concept of difference (or distance) between means relative to the variance is of central importance to signal detection theory. Hence, it received its own name: Discriminability index ($d'$ or *d prime*). $d'$ is defined as the distance between the means of the two distributions normalized (divided) by the joint standard deviation of the two distributions. Conceptually, it is an extension of the signal to noise ratio (SNR)—from means to mean differences. Here, it can be interpreted as a representation of signal strength relative to noise.

Importantly, $d'$ determines where an optimal criterion level should be set. For example, if $d'$ is very high, you can get 100% hits without any false alarms, by setting the criterion level properly. The situation is slightly more complicated when $d'$ is small, but there is a prescriptive solution for this case as well—it is discussed below.

This point makes intuitive sense. The errors derive from the fact that the "signal-present" and "signal-absent" distributions overlap. The more they overlap, the higher the potential for confusion. If the distributions don't overlap at all, you can easily draw a boundary without incurring errors or making mistakes.

---

### EXERCISE 10.2

Consider Figure 10.4. It represents two distributions: one for "stimulus absent" on the left and one for "stimulus present" on the right. At which x-value would you put the criterion level? Can you plot the corresponding ROC curve (mean difference = 5, variance = 0.5)?

---

### SUGGESTION FOR EXPLORATION

Create a movie that shows the evolution of the ROC curve as a function of increasing $d'$ (for added insight, try various degrees of variance in the distribution). You can also download this movie from the web site.

---

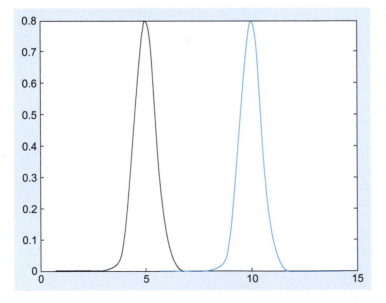

FIGURE 10.4    A case of high $d'$.

So far, so good. One central concept of signal detection theory that we are still missing is the notion of a likelihood ratio, or rather the use of likelihood ratios in signal detection theory.

While they sound rather intimidating, likelihood ratios are extremely useful because they are abstract enough to be powerful and flexible, yet specific enough to be of practical use. Hence, they are used in many fields, particularly diagnostics, but more generally in almost all of science. If you want to grasp the core of the concept, it is important to first strip off all these uses—some of which you might be already familiar with—and understand that it originally comes from statistics, or rather probability theory.

If you happen to appreciate analytical statistics, you might be appalled by the purely intuitive treatment of the likelihood ratio in this chapter. However, we deem this treatment appropriate for the purposes of our discussion.

Consider a situation in which you throw a fair and unbiased six-sided die. Each side has a probability of 1/6, which is about 0.1667. In other words, you expect the long-term frequency of a particular side to be 1 in 6. If you now want to know the probability that the die is showing one of the three lower numbers, you add the three individual probabilities and arrive at 0.5. Similarly, the probability that the die will show one of the three higher numbers is equally 0.5.

In other words, the ratio of the probabilities is 0.5/0.5 = 1.

If you ask what the probability ratio of the upper 4 versus the lower 2 numbers is, you arrive at (4*0.1667)/(2*0.1667) = 0.666/0.333 = 2/1 = 2. In other words, the ratio of the probabilities is 2 and—in principle—you could call this a likelihood ratio.

In practice, however, the term *likelihood ratio* has a specific meaning, which we will briefly develop here.

To do so, we have to do some card counting. Let's say a deck of cards contains eight cards valued 2 to 9. In each round, the dealer draws two cards from this deck (without showing them to you). There is an additional, special deck that contains only two cards: one that is valued 1 and one that is valued 10. In the same round, the dealer draws one card from this special deck—again without showing it to you. However, the dealer does inform you of the total point value of all three cards on the table. Your task is to guess whether the card from the special deck is a 1 or a 10.

While this may sound like a rather complicated affair, the odds are actually hugely in favor of the player once you do an analysis of the likelihood ratios. So don't expect to see this game offered in Vegas any time soon.

Instead, let us analyze this game—we happen to call it Chittagong—for educational purposes. The highest possible point value in the game is 27, and it can happen only if you get the 10 in the special deck and the 8 and 9 in the normal deck. So there is only one way to arrive at this value. Similarly, the lowest possible point value is 6—by getting 1 in the special deck as well as 2 and 3 in the normal deck. This case is also unique. Everything else falls somewhere in between. So let us construct a table where we explore these possibilities (see Table 10.3).

---

## SUGGESTION FOR EXPLORATION

Can you re-create Table 10.3 with MATLAB using the permutation functions?

TABLE 10.3   Exploring the Likelihood Ratios in the Chittagong Game

| Total points (TP) | Possible cases (=probability) in which special deck card is 10 | Possible cases (=probability) in which special deck card is 1 | Likelihood ratio (LR) |
|---|---|---|---|
| 6 | 0 | 1 | 0 |
| 7 | 0 | 1 | 0 |
| 8 | 0 | 2 | 0 |
| 9 | 0 | 2 | 0 |
| 10 | 0 | 3 | 0 |
| 11 | 0 | 3 | 0 |
| 12 | 0 | 4 | 0 |
| 13 | 0 | 3 | 0 |
| 14 | 0 | 3 | 0 |
| 15 | 1 | 2 | $1/2 = 0.5$ |
| 16 | 1 | 2 | $1/2 = 0.5$ |
| 17 | 2 | 1 | $2/1 = 2$ |
| 18 | 2 | 1 | $2/1 = 2$ |
| 19 | 3 | 0 | inf |
| 20 | 3 | 0 | inf |
| 21 | 4 | 0 | inf |
| 22 | 3 | 0 | inf |
| 23 | 3 | 0 | inf |
| 24 | 2 | 0 | inf |
| 25 | 2 | 0 | inf |
| 26 | 1 | 0 | inf |
| 27 | 1 | 0 | inf |

You can immediately see that vast regions of the table are not even in play. If the total value is below 15, you know that the special card had to be a 1. Moreover, if the total value is above 18, you know that the special card had to be a 10. Only four values are up to guessing, and even here, the odds are very good: As the player, you should guess 1 for 15 and 16, but 10 for 17 and 18. This state of affairs is due to the large difference between 1 and 10, relative to the possible range of normal values (5 to 17). In other words, $d'$ is very high in this game. This becomes immediately obvious when you plot the frequency distribution as histograms (10 = blue, 1 = black), as shown in Figure 10.5. This figure should look vaguely familiar (compare it to Figure 6.4).

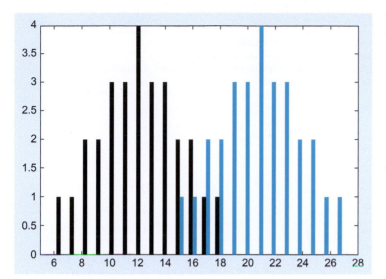

FIGURE 10.5  Histogram of frequency distributions.

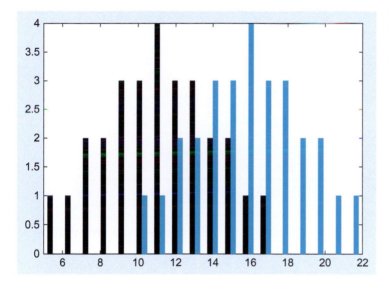

FIGURE 10.6  Histogram of frequency distribution with a smaller mean differences between the special cards (0 = black, 5 = blue).

Reducing the mean difference by 4 does change the distance between the distribution as well as the overall range. Suppose the cards in the special deck are replaced with two cards worth 0 and 5 points, respectively. What does the histogram of the frequency distributions look like now? (See Figure 10.6.)

The table of likelihood ratios, shown in Table 10.4, reflects this change.

As you can see, there is an intuitive and clear connection between likelihood ratio and $d'$. Of course, this relationship has been worked out formally. We will forgo the derivation here in the interest of getting back to neuroscience.

**TABLE 10.4**  Revisiting Likelihood Ratios

| Total points (TP) | Possible cases (=probability) in which special deck card is 5 | Possible cases (=probability) in which special deck card is 0 | Likelihood ratio (LR) |
|---|---|---|---|
| 5 | 0 | 1 | 0 |
| 6 | 0 | 1 | 0 |
| 7 | 0 | 2 | 0 |
| 8 | 0 | 2 | 0 |
| 9 | 0 | 3 | 0 |
| 10 | 1 | 3 | 1/3 = 0.33 |
| 11 | 1 | 4 | 1/4 = 0.25 |
| 12 | 2 | 3 | 2/3 = 0.66 |
| 13 | 2 | 3 | 2/3 = 0.66 |
| 14 | 3 | 2 | 3/2 = 1.33 |
| 15 | 3 | 2 | 3/2 = 1.33 |
| 16 | 4 | 1 | 4/1 = 4 |
| 17 | 3 | 1 | 3/1 = 3 |
| 18 | 3 | 0 | inf |
| 19 | 2 | 0 | inf |
| 20 | 2 | 0 | inf |
| 21 | 1 | 0 | inf |
| 22 | 1 | 0 | inf |

In this simple case, you can just set the criterion at the ratio between the probabilities. If this ratio is smaller than 1, guess 0. If it is larger, guess 5.

In the technical literature, the likelihood ratio takes more factors into account: the prior probability as well as payoff consequences. Let us illustrate this case. Suppose there are not 2, but 10 cards in the special deck: You know that 9 have a value of 5, 1 has a value of 0. Hence, there is an a priori chance of 9/10 that the card will have a value of 5, and this does influence the likelihood ratio, as it should. Taking payoff consequences into account makes good sense because not all outcomes are equally good or bad (see the discussion at the beginning of the chapter). A casino could still make money off this game by adjusting the payoff matrix. For example, it could make the wins very small (as they are expected to happen often in a game like this), but the rare losses could be adjusted such that they are rather costly. A player has to take these considerations into account when playing the game and setting an optimal criterion value.

To make this point more explicit, the likelihood ratio can be defined as follows:

$$l_{ij}(e) = \frac{p(e|s_i)}{p(e|s_j)} \tag{10.1}$$

So the likelihood ratio of an event $e$ is the ratio of two conditional probabilities. One is the probability of the event given state $s_i$; the other, the probability of the event given state $s_j$. $l_{ij}$ is always a single real number.

Moreover, we already discussed a more general situation where the likelihood ratio takes prior probabilities and payoffs into account:

$$l_{ij}(e) = \frac{stim\_frequency}{1 - stim\_frequency} * \frac{value\_of\_correct\_rejection - value\_of\_false\_alarm}{value\_of\_hit - value\_of\_miss} \tag{10.2}$$

This is particularly important for real life situations, where not all outcomes are equally valuable or costly.

As we alluded to before, the likelihood ratio is closely linked to ROC curves. Specifically, it is very important to characterize optimal behavior.

These considerations influence the likelihood ratio at which you should set your decision criterion. Importantly, there is a direct relationship between likelihood ratio and ROC curve: The slope of the ROC curve at a given point corresponds to the likelihood ratio criterion which generated the point (Green and Swets, 1966). In other words, an inspection of the slope can reveal where the criterion should optimally be set.

Let us illustrate these claims by revisiting the distributions introduced at the beginning of the chapter.

Use this code to plot the slope of the curves, analogous to Figure 10.2.

```
figure
x = 0:0.01:10;
%Note that x is ordered. If you start with empirical data, you will have to sort them
%first.
y = normpdf(x,5,1.5)
z = (normpdf(x,6.5,1.5));
subplot(2,1,1)
for ii = 1:1:length(x)
FA(ii) = sum(y(1,ii:length(y)));
HIT(ii) = sum(z(1,ii:length(y)));
end
FA = FA./100;
HIT = HIT./100;
plot(FA,HIT)
hold on
baseline = 0:0.01:1;
plot(baseline,baseline,'color','k')
subplot(2,1,2)
for ii = 1:length(x)-1
m1(ii) = FA(ii)-FA(ii + 1); %This recalls the
```

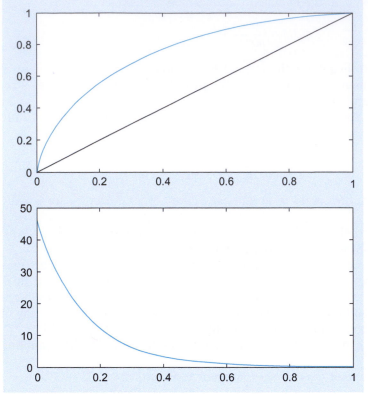

**FIGURE 10.7**    ROC curve with slope.

```
m2(ii) = HIT(ii)-HIT(ii + 1); %equation of a slope
end
m3 = m1./m2; %Dividing them
plot(m3)
```

The slope of the ROC curve is plotted in the lower panel; see Figure 10.7.

The philosophical implications of signal detection theory are deep. The message is that—due to the stochastic structure of the real world—infallibility is, in principle, impossible in most cases. In essence, in the presence of uncertainty (read: in all real life situations), errors are to be expected and cannot be avoided entirely. However, signal detection theory provides a precise analytical framework for optimal decision making in the face of uncertainty, while also being able to take into account subjective value judgments (such as preferring one kind of error over another).

As you might have noticed, we are really only scratching the surface here. Because situations in which a signal detection theory perspective is useful are truly ubiquitous—think of any kind of selection and quality control process, such as hiring decisions, admission decisions, marriage, dating, to say nothing of the myriad applications in materials science—signal detection theory has become a bottomless well. This should not be surprising, as it is arguably at the very heart of cognition itself. Yet, this led to a situation in which even specialists can be overwhelmed by the intricacies of the field. Hence, the point of this brief treatment was to cover the conceptual essentials and their application. We are confident that it is enough to get you started in applying signal detection theory with MATLAB to problems in neuroscience.

For further reading, we highly recommend the classical and elaborate *Signal Detection Theory and Psychophysics* by Green and Swets (1966); the latest edition is still available in print. It nicely highlights the role of signal detection theory in modern cognitive science with many colorful examples.

## 10.4 PROJECT

The project for this chapter is very straightforward. Many uses of signal detection in neuroscience involve the measurement of some "internal response" in addition to measuring a behavioral response (e.g., deciding whether a stimulus under the control of the experimenter is present or not). We assume that you do not currently have access to measure a "deep" internal response, such as firing rate of certain neurons that are presumably involved in the task. Instead, we ask you to redo the experiment in Chapter 8, but with a twist. Instead of just asking whether a faint stimulus is present or not, now elicit 2 judgments per trial: One whether the stimulus is present or not, the other how confident the observer is that it was present or not, on a scale from 1 (not certain at all) to 9 (very certain). Replot the data in terms of certainty. Get two distributions of certainty (one for situations where the stimulus was present, the other where it was not present). After doing so, please answer and explore the following questions:

Where does the internal criterion of the observer lie?
What is the $d'$ of the certainty distributions?
Construct the ROC curve for the data (including slope).
How sensitive is the observer? (Compare the area under the ROC curve with the area under the diagonal reference curve.)
Can you increase $d'$ by showing a different kind of stimulus?
Can you shift the position of the criterion by biasing the payoff matrix for your observer (e.g., rewarding the observer for hits)?

## MATLAB FUNCTIONS, COMMANDS, AND OPERATORS COVERED IN THIS CHAPTER

**normpdf**
**normcdf**
**sum**

# PART III

# DATA ANALYSIS WITH MATLAB

# Frequency Analysis Part I: Fourier Decomposition

## 11.1 GOALS OF THIS CHAPTER

This chapter introduces the most common method of decomposing a time series into frequency components, Fourier analysis. You will learn about the Fourier transform and the associated amplitude and phase spectra. The MATLAB® implementation of the fast Fourier transform (FFT), an efficient algorithm for calculating Fourier transformations, will be introduced and applied to the analysis of human speech sounds.

## 11.2 BACKGROUND

Figure 11.1 shows typical recordings of two human vowel sounds. How can you characterize these different sounds? Frequency analysis provides a way to examine the relative contributions of various frequencies to an overall signal. In the case of an auditory signal, a given frequency component would be termed *pitch*.

### 11.2.1 Real Fourier Series

Take some continuous function $f$. We can approximate such a function with a weighted series of sinusoids of various frequencies. Such a series is termed the real Fourier series:

$$f(t) = \frac{a_0}{2} + \sum_{n=1}^{\infty} a_n \cos(nt) + \sum_{n=1}^{\infty} b_n \sin(nt) \qquad (11.1)$$

Here, the coefficients $a_n$ and $b_n$ represent the relative strength of each frequency component $n/2\pi$. [$a_0$ represents the nonoscillatory component of $f(t)$.] So, given $f(t)$, determining the coefficients $a_n$ and $b_n$ allows for the representation of $f(t)$ as a series sum of sinusoids.

*MATLAB® for Neuroscientists.*
DOI: http://dx.doi.org/10.1016/B978-0-12-383836-0.00011-4

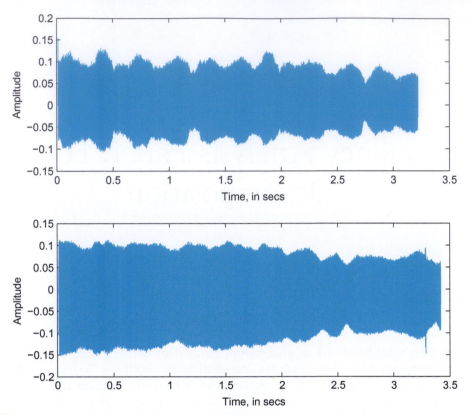

**FIGURE 11.1**    Acoustic time series representing two different human vowel sounds.

We will exploit two special properties of the sine and cosine functions to find the Fourier series coefficients $a_n$ and $b_n$. Over the interval $-\pi$ to $\pi$, cosine and sine functions with differing frequencies have the special property of *orthogonality*. The integral of the product of two mutually orthogonal functions evaluates to zero. So, the integral of the product of cosine or sine functions with differing frequencies results in zero over this interval. Another interesting property of sine and cosine is that the integral of the square of a cosine or sine function over this integral is $\pi$.

To find the strength, $a_m$, of a cosine component $m$, multiply by the corresponding cosine function and integrate:

$$\int_{-\pi}^{\pi} f(t)\cos(mt)dt = \int_{-\pi}^{\pi} \frac{a_0}{2}\cos(mt)dt + \sum_{n=1}^{\infty}\int_{-\pi}^{\pi}\cos(mt)a_n\cos(nt)dt + \sum_{n=1}^{\infty}\int_{-\pi}^{\pi}\cos(mt)b_n\sin(nt)dt$$

$$(11.2)$$

All terms on the right side except the cosine term where $m = n$ yield zero:

$$\int_{-\pi}^{\pi} f(t)\cos(mt)dt = a_m\int_{-\pi}^{\pi}\cos^2(mt)dt \qquad (11.3)$$

The right side integral evaluates to one over the integration range, yielding an expression for the Fourier series term coefficient:

$$\int_{-\pi}^{\pi} f(t) \cos{(mt)}dt = \pi a_m \tag{11.4}$$

$$a_m = \frac{1}{\pi} \int_{-\pi}^{\pi} f(t) \cos{(mt)}dt \tag{11.5}$$

In general, the interval of $f(t)$ will not be $-\pi$ to $\pi$. For an interval centered on $x$ with length $2L$, the expression becomes

$$a_m = \frac{1}{L} \int_{x-L}^{x+L} f(t) \cos{\left(\frac{\pi}{L}mt\right)}dt \tag{11.6}$$

A similar procedure using sine functions yields the coefficients for the sine terms of the Fourier series.

## 11.3 EXERCISES

---

**EXERCISE 11.1**

Write a MATLAB function to calculate coefficients for a real Fourier transform. *Hint:* The function will need to shift the interval so that the interval encompasses the entire time series. In other words, $x = 0$ and $L =$ half the range of $t$.

---

### 11.3.1 Complex Fourier Transform

Euler's identity,

$$e^{i\omega t} = \cos{\omega t} + i\sin{\omega t} \tag{11.7}$$

provides a straightforward way to formulate complex Fourier series representation for a given function, $f(t)$:

$$f(t) = \sum_{n=-\infty}^{\infty} c_n e^{int} \tag{11.8}$$

Similar to the real transform, coefficients for the complex Fourier transform can be found by

$$c_m = \frac{1}{2\pi} \int_{-\pi}^{\pi} f(t) e^{-imt} \, dt \tag{11.9}$$

for a given coefficient $m$ over the interval $-\pi$ to $\pi$. Over the interval $x - L$ to $x + L$, this becomes

$$C_m = \frac{1}{2L} \int_{x-L}^{x+L} f(t) e^{-(i\pi mt/L)} dt \tag{11.10}$$

---

### EXERCISE 11.2

Write a MATLAB function to calculate coefficients for a complex Fourier transform. This is essentially the discrete Fourier transform (DFT):

where $k$ ranges from 0 to $N - 1$, $N$ is the number of points, and $f_n$ is the value of the function at point $n$.

$$F_k = \frac{1}{N} \sum_{n=0}^{N-1} f_n e^{-i\frac{2\pi n}{N}k} \tag{11.11}$$

---

Let's look at how this method of the Fourier transform scales with $N$. Given a time series with $N$ values, this method requires a multiplication of the series and the corresponding Fourier component and subsequent sum for each coefficient. Assuming a number of coefficients equivalent to $N$, then you have a process that scales with $N^2$. In other words, as $N$ increases, the time required to compute the Fourier transform increases as $N^2$.

### 11.3.2 Fast Fourier Transform

With a few special tricks, a faster algorithm, the fast Fourier transform (FFT) that scales in $N \log N$ time can be formulated. One of these tricks involves taking advantage of datasets exactly $2^N$ elements long. The increase in processing speed has made the FFT ubiquitous in signal processing. While a complete derivation of the algorithm is beyond the scope of this book, invoking the MATLAB implementation of the FFT will be discussed.

MATLAB provides an FFT function **fft(X)**, where $X$ is a vector in time space. **fft** returns the frequency space representation of $X$.

To visualize the importance of the difference in scaling, execute the following code:

```
figure
hold on
N = 1:10 * 100;
plot(N, N.^2, 'b')
plot(N, N.*log(N), 'r')
```

---

### EXERCISE 7.3

If $N$ represents sample size, what can you observe about the benefits of scaling as $N$ grows? Where does the efficiency of the FFT algorithm benefit most, for large $N$ or small $N$?

---

## 11.3.3 The Inverse DFT

As you might imagine, there is an inverse to the DFT:

$$f_n = \sum_{k=0}^{N-1} F_k e^{i\frac{2\pi n}{N}k} \tag{11.12}$$

MATLAB provides **ifft()** to perform the inverse discrete Fourier transform.

---

### EXERCISE 7.4

Generate a single sine wave. Use **fft()** to generate the discrete Fourier transform. Use **ifft()** to retrieve the original sine wave from the DFT.

---

## 11.3.4 Amplitude Spectrum

Often when you are using Fourier analysis, the amplitude spectrum is one of the first analyses performed. The amplitude spectrum graphs amplitude against frequency. In terms of the Fourier series representation, the amplitude spectrum depicts the magnitude of the coefficients at various frequencies. As such, it depicts the relative strengths of the various frequency components.

The following code generates a time series composed of 10 sine waves whose frequencies and amplitudes vary systematically.

```
L = 1000;
X = zeros(1,L);
sampling_interval = 0.1;
t = (1:L) * sampling_interval;
for N = 1:10
X = X + N * sin (N*pi*t);
end
plot(t, X);
Y = fft(X)/L;
```

Now, the variable $Y$ contains the normalized FFT of $X$. Note the normalization factor $L$. Displaying the amplitude spectrum of $X$ requires plotting the amplitudes at various frequencies. Note that **fft** returns only a single value, the transform coefficients. Now, how do you determine the frequency scale?

The return value of the FFT assumes that frequency is evenly spaced, from 0 to a theoretical result called the *Nyquist limit*. Nyquist demonstrated that a discrete sampling of a continuous process can capture frequencies no higher than half the sampling frequency. Since the code above has the sampling interval, this **Nyquist limit** is half the inverse of the sampling interval.

The following code calculates the Nyquist limit for the time series:

**NyLimit = (1 / sampling_interval)/ 2;**

When viewing the FFT, it is important to remember that the result is the complex transform. Thus, simply using the result of the FFT as a set of real coefficients can cause a number of problems. To display the amplitude spectrum, the absolute value of the complex coefficients will be used. The values returned by **fft** are the coefficients for frequencies from the negative Nyquist limit to the positive Nyquist limit. If the time series data are purely real, then the resultant transform will have even symmetry. That is, the transform will be symmetrical across the abscissa. So, in this very frequent case, only the first half of the result of **fft** is used. The following code employs **linspace** to generate frequency values and plots the amplitude spectrum. **linspace** generates a linearly spaced sequence of values given initial and final values. Here, the initial and final values are 0 and 1, with a value count of L/2. The resultant vector is scaled by the Nyquist limit to generate the frequency vector.

**F = linspace(0,1,L/2)\*NyLimit;**
**plot(F, abs(Y(1:L/2)));**

### 11.3.5 Power

Power at a given frequency is defined as

$$\Phi(\omega) = |F(\omega)|^2 = F(\omega)F^*(\omega) \tag{11.13}$$

where $F^*$ is the complex conjugate of $F$. To do this in MATLAB, use the function **conj** to return the complex conjugate of a series of complex values.

Here is a plot of the power spectrum of the time series generated for the amplitude spectrum:

**plot(F, (Y(1:L/2).\*conj(Y(1:L/2))));**

### 11.3.6 Phase Analysis and Coherence

A power spectrum alone is not a complete representation of the information in the original signal. The various Fourier components can have various phases relative to one another, as illustrated in Figure 11.2.

You can plot relative phase by frequency by plotting the inverse tangent of the ratio between the imaginary component and the real component. Why is this the case? Imagine

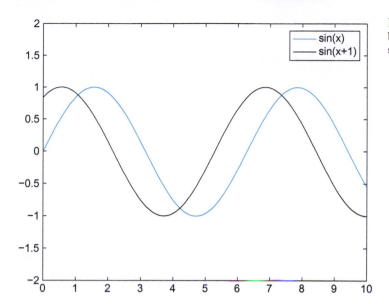

sin(x)
sin(x+1)

FIGURE 11.2 Phase difference between two sinusoids with the same frequency.

the complex plane, with pure real values along the abscissa (x-axis) and pure imaginary values along the ordinate (y-axis). Any complex value in your 1D Fourier transform can be represented with a coordinate pair. The magnitude of the value is simply the distance from the origin to the coordinates, or the complex modulus. The phase is the angle formed by the abscissa and the line passing through the origin and the complex point. Thus, using basic trigonometry, the phase angle is $\tan^{-1}\left(\frac{imag}{real}\right)$.

How can you represent this in MATLAB?

```
L = 1000;
X = zeros(1,L);
sampling_interval = 0.1;
t = (1:L) * sampling_interval;
for N = 1:10
X = X + N * sin (N*pi*t);
end
plot(t, X);
Y = fft(X)/L;
phi = atan(imag(Y)./real(Y));
F = linspace(0,1,L/2)*NyLimit;
plot(F, phi(1:L/2));
```

---

### EXERCISE 11.5

Compare the phase spectrum generated in the preceding exercises with the phase spectrum of the corresponding cosine function. Compare their power spectra.

---

**TABLE 11.1**   Average First and Second Formant Frequencies for Selected American English Vowels

| Vowel sound | First formant | Second formant |
|---|---|---|
| Bit | 342 | 2322 |
| But | 623 | 1200 |
| Bat | 588 | 1952 |
| Boot | 378 | 997 |

*(Data from Hillenbrand et al., 1995.)*

# 11.4 PROJECT

In this project, you will be asked to use Fourier decomposition to analyze vowel sounds produced by human speakers. On the companion website, you will find five examples of vowel sounds as produced by male American English speakers. Each sound corresponds to one of the vowel sounds in Table 11.1. The formant frequencies in Table 11.1 note the average formant frequencies as spoken by a male speaker of American English. You will use power spectra of these sounds to classify the recordings as one of these vowel sounds in the table.

To complete this project, you need to understand how formants relate to frequency analysis. The human vocal tract has multiple cavities in which speech sounds resonate. As such, most sounds have multiple strong frequency components. In classifying speech sounds, the lowest strong frequency band is termed the *first formant*. The next highest is termed the *second formant*, and so on.

Vowels lend themselves to a particularly simple characterization through their formants. Typically, vowel sounds have distinguishable first and second formants. Table 11.1 shows first and second formants for four vowel sounds in American English. Thus, the short "i" sound would have strong frequency representation at 342 Hz and at 2322 Hz.

# MATLAB FUNCTIONS, COMMANDS, AND OPERATORS COVERED IN THIS CHAPTER

**fft**
**ifft**
**conj**

# Frequency Analysis Part II: Nonstationary Signals and Spectrograms

## 12.1 GOAL OF THIS CHAPTER

The goal of this chapter is to extend Fourier analysis as covered in the previous chapter to nonstationary signals. The short-time Fourier transform will be introduced. Nonstationary examples will include applications to time-varying auditory signals and the EEG during sleep.

## 12.2 BACKGROUND

Figure 12.1 depicts the vocalizations of a zebra finch. How is this dissimilar from the sound signals you have examined thus far?

Note that different portions of the song have different envelopes with clearly defined breaks. If they are taken separately, you might imagine these subsections to have different Fourier spectra. In fact, they do. Figure 12.2 shows the Fourier spectrum for two subsections of song. The two subsections have very different distributions of power over frequency.

A Fourier transform of the full song returns the power distribution over the entire song. Any localization of frequency information to a time point or an interval is lost. Using the example of the bird song here, a Fourier transform of the entire song would eliminate any ability to associate frequency components with a given syllable. How, then, can you extend the techniques discussed earlier to such complex signals?

*MATLAB® for Neuroscientists.*
DOI: http://dx.doi.org/10.1016/B978-0-12-383836-0.00012-6

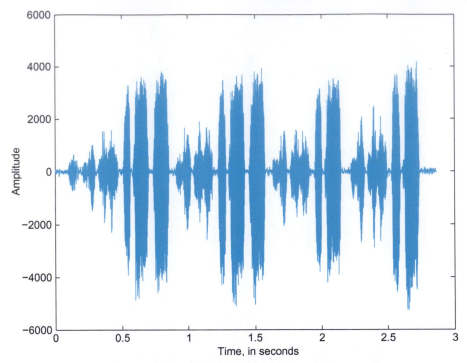

**FIGURE 12.1**    The sound amplitude of a zebra finch vocalization as a function of time.

## 12.2.1 The Fourier Transform: Stationary and Ergodic

When applied to a signal, the term *stationary* indicates that certain statistical properties of the signal are uniform throughout. In other words, a subset of the signal is sufficient for analysis of the entire signal. The distribution of power over frequency remains the same over the whole signal.

A similar idea is the concept of *ergodicity*. Imagine an ensemble of related signals. Going with the example of zebra finch vocalizations, an appropriate ensemble would be the set of vocalizations from a set of birds. An ergodic ensemble is one in which each sample and the ensemble approach the same mean. In other words, analyzing one sample or a subset of the signals from the group can approximate the analysis of the ensemble. Ergodicity and stationarity are independent qualities. Neither implies the other.

The Fourier transform assumes a stationary signal. Unfortunately, many biological signals, including the birdsong in Figure 12.1, are nonstationary.

## 12.2.2 Windows

How can you employ the Fourier transform to a nonstationary signal? If you assume that the Fourier spectrum will change relatively little over a small interval of the signal, you could divide the overall signal into windows and calculate the Fourier transform for

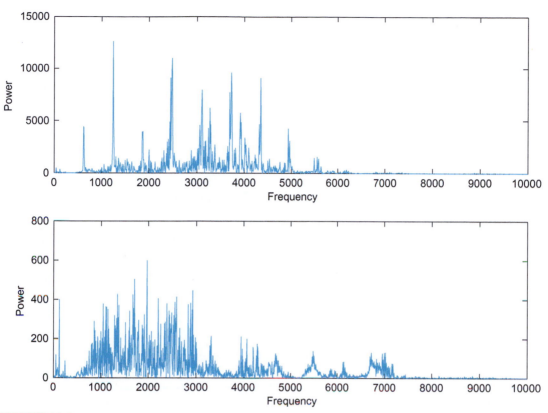

FIGURE 12.2   The power spectra of two portions of the zebra finch vocalizations depicted in Figure 12.1.

each window separately. If a signal is relatively stationary over short intervals, or quasistationary, this approach will often produce fruitful results. While many biological signals are not truly stationary, many are quasistationary and amenable to this approach.

However, this approach breaks down somewhat at the interval boundaries, due to the stationary assumptions of the Fourier transform. Choosing overlapping intervals mitigates this somewhat. This is the basis for the short-time Fourier transform (STFT).

While a simple flat subset of the original time series might be the most straightforward window, an appropriate choice of window shape can amplify or minimize characteristics of the time series. For example, windows with tapered ends are used to minimize artifacts from the edges of the window. This suggests a generalized window function, $w(t)$, which returns the value of the window at a given value of $t$. For values outside the window, $w(t)$ should return values equal to or close to 0.

Mathematically, the STFT is represented as:

$$X(\tau, \omega) = \int_{-\infty}^{\infty} x(t)w(t - \tau)e^{-j\omega t}dt \tag{12.1}$$

for a continuous signal. In this case, we are more interested in the discrete STFT,

$$X(m, \omega) = \sum_{n=-\infty}^{\infty} x(n)w(n - m)e^{-j\omega n} \tag{12.2}$$

As mentioned previously, there are many alternatives to the simple squared off window for calculating an STFT. We will briefly discuss three. The Hamming window function is commonly used:

$$w(n) = 0.53836 - 0.46164 \cos\left(\frac{2\pi n}{N - 1}\right) \tag{12.3}$$

where $N$ is the number of points and $n$ varies over the interval.

The Signal Processing Toolbox of the MATLAB® software provides the function **hamming**, which returns a Hamming window of the desired length:

```
L = 100;
w = hamming(L);
plot(1:L, w)
```

Note that the Hamming window has a high amplitude at the center and low amplitude at the ends. This attenuation reduces the artifacts from the edge of the window interval. Another window function is the Hann window, whose functional form is similar to the Hamming window:

$$w(n) = 0.5 - 0.5 \cos\left(\frac{2\pi n}{N - 1}\right) \tag{12.4}$$

Like the Hamming window, the shape of the Hann window is used to reduce artifacts introduced at the edges of the finite windows from the signal. Gaussian window functions are often used as well:

$$w(n) = e^{-n^2} \tag{12.5}$$

A short-time Fourier transform using a Gaussian window function is sometimes denoted as a *Gabor transform*. A Gabor function is the product of a sinusoid and a Gaussian function. The Gaussian function causes the amplitude of the sinusoid to diminish away from the origin, but near the origin, the properties of the sinusoid dominate. By applying a Gaussian window and a Fourier transform to the time series, you are, in effect, applying a Gabor function filter to the data.

## 12.3 EXERCISES

As a part of the Signal Processing Toolbox, MATLAB provides the function **spectrogram**, which calculates a short-time Fourier transform using a Hamming window. The data for Figure 12.1 is available on the companion web site. Download the file song1.wav, and load the file with **wavread** as follows:

**[amp, fs, nbits] = wavread('song1.wav');**

The function **wavread** loads a sound file in WAVE format and returns the data as amplitude information ranging from $-1$ to $+1$. Here, you store the amplitude information in the variable *amp*. The sampling rate is returned in *fs*, and the number of bits per sample (resolution) is stored in *nbits*.

Now type

**spectrogram(amp, 256, 'yaxis')**

You should see something like Figure 12.3. The default operation of spectrogram calculates power of the signal by dividing the whole signal into eight portions with overlap and windowing the portion with a Hamming window. Here, the specified window size was 256. The optional parameter *'yaxis'* specifies that frequency should be on the y-axis rather than x-axis. If no return values are specified, the default operation renders the power spectral density over time using "hotter" colors (red, yellow, etc.) to designate frequency bands of greater energy.

If a sampling frequency is not specified, the time scale will not be correct. To show the correct time space for the loaded song, type

>> **spectrogram(amp, 256, [ ], [ ], fs, 'yaxis')**

Here, the empty brackets signify that the default settings for the window overlap, and FFT size should remain.

Also, **spectrogram** can return the power spectral density:

>> **[S, F, T, P] = spectrogram(X);**
>> **mesh(P)**

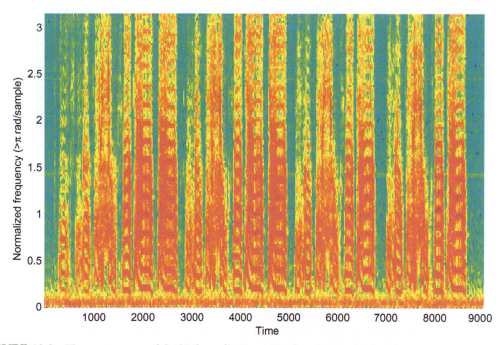

FIGURE 12.3    The spectrogram of the bird vocalization using the **spectrogram** function.

The preceding code generates a 3D plot of the spectrogram, where $z$ magnitude, rather than color, represents power.

---

### EXERCISE 12.1

In the bird song sample, try to determine where the sound changes using the time series data alone. Do the same with the STFT. Do your results agree?

---

### EXERCISE 12.2

Examine the result of the spectrogram with varying window sizes for the following time series:

```
>> t = 0:0.05:1;
>> X = [sin(5*t) sin(50*t) sin(100*t)];
```

Try values ranging from 16 to 1024 for the Hamming window width. How does the representation change with different Hamming window widths? Why might this occur?

---

## 12.3.1 Limitations of the STFT

The STFT is a fine resolution to the problem of determining the frequency spectrum of signals with time-varying frequency spectra. There are some limitations. Small frequency fluctuations are difficult to detect with the STFT because each subset of the signal is assumed to be stationary. Since the reported frequency distribution at a time point results from the analysis of the entire window, choosing a smaller window does allow for better localization in time. However, a smaller window allows for fewer samples in each Fourier transform, which ultimately reduces frequency resolution, especially for lower frequencies. In other words, a trade-off exists between frequency and time localization.

The STFT is best employed when the fluctuations in frequency occur over a fairly uniform time scale. This allows selecting a single window size without substantial loss of information.

## 12.4 PROJECT

Typical sleep in human adults includes the well-known REM sleep as well as four well-characterized stages of non-REM sleep, or NREM sleep. During wakefulness, alpha waves dominate the EEG, in the frequency range 8 to 13 Hz. As the person enters the first stage of non-REM sleep, the dominant wave type transitions from alpha waves to theta waves, in the range of 4 to 7 Hz. This is the first stage of non-REM sleep.

The second and third stages of non-REM sleep are characterized by sleep spindles, at 12 to 16 Hz, and the appearance of delta waves, ranging in frequency from 0.5 to 4 Hz. The fourth stage of sleep is characterized by a majority power distribution in the delta wave band. The third and fourth stages of NREM sleep are also termed slow wave sleep, to denote the prevalence of the low frequency delta waves in these two stages.

On the companion web site, you can find three EEGs from patients falling asleep. Using spectrogram and any other frequency analysis tools learned thus far, try to determine when the people enter each of the NREM stages of sleep.

# MATLAB FUNCTIONS, COMMANDS, AND OPERATORS COVERED IN THIS CHAPTER

**hamming**
**spectrogram**
**wavread**

## 13.1 GOALS OF THIS CHAPTER

In this chapter, you will be introduced to the use of wavelets and wavelet transforms as an alternative method of spectral analysis. We will discuss a number of common wavelets and introduce the Wavelet Toolbox of the MATLAB® software.

## 13.2 BACKGROUND

In Chapter 12, "Frequency Analysis Part II: Nonstationary Signals and Spectrograms," introduced the short-time Fourier transform (STFT) to decompose the frequency composition of nonstationary signals. Under certain situations, though, the STFT results in a less-than-optimal breakdown of frequency as a function of time. With increased precision in frequency distribution, localization in time becomes less precise. In other words, there is a time–frequency precision tradeoff. The reverse is also true: better temporal localization reduces the precision of the frequency distribution. This may bring to mind the well-known relationship of position and momentum of the Heisenberg uncertainty principle.

One of the benefits of the STFT is that the transform window can be chosen to optimize the resolution of frequency or localization of that frequency in time. A larger window allows for better frequency resolution, and a smaller window allows for better temporal resolution. However, for the STFT, the window size is constant throughout the algorithm. So, while the STFT can optimize for frequency or time in a given signal, the choice in the time-frequency tradeoff holds for the entire signal. This can pose a problem for some nonstationary signals. The wavelet transform provides an alternative to the STFT that often provides a better frequency/time representation of the signal.

### 13.2.1 What is a Wavelet?

A *wavelet* is a function that satisfies at least the following two criteria:

**1.** The integral of the function $\psi(x)$ over all $x$ is 0.

$$\int_{-\infty}^{\infty} \psi(x)\mathrm{d}x = 0 \tag{13.1}$$

**2.** The square of $\psi(x)$ has integral 1. A function adhering to this property is called *square-integrable*.

$$\int_{-\infty}^{\infty} \psi(x)\mathrm{d}x = 1 \tag{13.2}$$

Fulfilling the first criterion mandates that the wavelet function has an equal area above and below zero. Fulfilling the second criterion mandates that the function approach zero at positive and negative infinity. Because of this second criterion, the function decays away from the origin, unlike sinusoidal or other infinite waves (thus, wave*let*).

### 13.2.2 The Continuous Wavelet Transform

The continuous wavelet transform (CWT) is analogous to the continuous Fourier transform:

$$W(s,t) \equiv \int_{-\infty}^{\infty} x(u)\psi_{s,t}(u)\mathrm{d}u \tag{13.3}$$

Here, the parameter $t$ is the typical $t$ in the time series $x(t)$. The parameter $s$ is called *scale* and is analogous to frequency for Fourier transforms. The wavelet function itself varies with both $s$ and $t$:

$$\psi_{s,t}(x) \equiv \frac{1}{\sqrt{s}}\psi^*\left(\frac{x-t}{s}\right) \tag{13.4}$$

The inclusion of $t$ and $s$ allows the function to be scaled and translated (shifted) for different values of $s$ and $t$. The original wavelet function (untranslated and unscaled) is often termed the *mother wavelet*, since the set of wavelet functions is generated from that initial function.

The scaling provides a significant benefit over the short-time Fourier transform. The multiple scales of the wavelet transform permit the equivalent of large- or small-scale transform windows in the same time series. The preceding transform can be approximated for a discrete time series.

### 13.2.3 Choosing a Wavelet

A number of wavelet functions are commonly used in data analysis. Here are two used primarily for spectral analysis.

*Morlet wavelet* (for large $\omega_0$):

$$\psi(t) = \pi^{-\frac{1}{4}}e^{-\frac{1}{2}t^2}e^{-i\omega_0 t} \tag{13.5}$$

The Morlet wavelet was originally developed to analyze signals with short, high-frequency transients and long, low-frequency transients (see Figure 13.1).

*Mexican hat wavelet*:

$$\psi(t) = \frac{1}{\sqrt{2\pi\sigma^3}} \left(1 - \frac{t^2}{\sigma^2}\right) e^{\frac{-t^2}{2\sigma^2}}$$

(13.6)

The Mexican hat wavelet has poorer frequency resolution than the Morlet wavelet, but often better temporal resolution.

## 13.2.4 Scalograms

The *scalogram* depicts the strength of a particular wavelet transform coefficient at a point in time. As such, it is the wavelet analog of the spectrogram.

The scalogram in Figure 13.2 shows the continuous wavelet transform of the following signal with a Morlet wavelet (sigma = 10). This code generates a time series with three long blocks of time at 100, 500, and 1000 Hz. At every half second, a 0.05 transient at 1000 Hz is inserted.

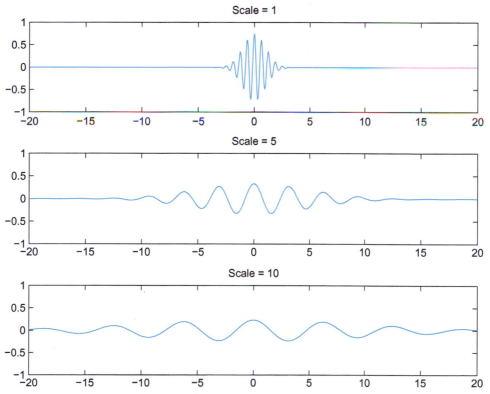

FIGURE 13.1   Morlet wavelet at various scales

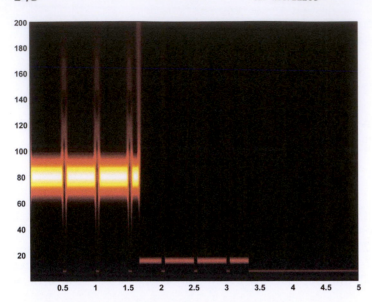

FIGURE 13.2   Scalogram for sinusoid + transient signal in text.

```
Fs = 5000;
total_time = 5;
t = (1/Fs):(1/Fs):(total_time/3);
f = [100   500   1000];
x = [cos(f(1)*2*pi*t) cos(f(2)*2*pi*t) cos(f(3)*2*pi*t)];
t = (1/Fs):(1/Fs):total_time;
%add short transients
trans_time = 0:(1/Fs):0.05;
trans_f = 1000;
for secs = 0.5:0.5:4
trans = cos(trans_f*2*pi*trans_time);
x((secs*Fs):(secs*Fs + length(trans) − 1)) = trans;
end
```

Be aware that the relationship between scale and frequency is an inverse one and that frequency increases with *decreasing* scale. Also, note how the frequency resolution improves for the higher frequency band in the later third of the series. This corresponds to the 1000 Hz section of the time series.

The code to generate and plot the CWT follows.

In **my_cwt.m**:

```
function coefs = simple_cwt(t, x, mother_wavelet, max_wavelet, scales, params)
% Generates coefs for a continuous wavelet transform
% t, x are time and data points for time series data
% mother_wavelet is a function, taking parameters (t, params),
% where the value of params depends on the specific function used
```

```
% max_wavelet is the maximum range of the wavelet function (beyond which
% the wavelet is essentially zero)
% scales is a vector of desired scales
% params is the parameter for the mother wavelet function
max_t = max(t);
dt = t(2)-t(1);
full_t = − (max_t/2):dt:(max_t/2);
coefs = zeros(length(scales), length(x));
points = length(x);
t_scale = linspace( − max_wavelet, max_wavelet, points);
dt = (max_wavelet*2)/(points − 1);
mom_wavelet = feval(mother_wavelet, t_scale, params);
row = 1;
for scale = scales
time_scale = [1 + floor([0:scale*max_wavelet*2]/(scale*dt))];
wavelet = mom_wavelet(time_scale);
w = conv(x,wavelet)/sqrt(scale);
mid_w = floor(length(w)/2);
mid_x = floor(length(x)/2);
w = w((( − mid_x:mid_x) + mid_w));
scale % print scale to show progress
coefs(row,:) = abs(w);
row = row + 1;
end
```

In **my_morlet.m**:

```
function m = morlet(t, params)
sigma = params(1);
m = pi^ − 0.25*exp( − i*sigma.*t − 0.5*t.^2);
```

In **plot_cwt.m**:

```
function plot_cwt(t, coefs, scales)
imagesc(t, scales, coefs);
colormap(hot);
axis xy;
end
```

Here, **imagesc** generates an imagemap from two vectors of data. Given parameters $x$, $y$, and $c$, **imagesc** generates a colored area of *color(n,m)* centered at $x(n)$ and $y(m)$. So, here in **plot_cwt**, at values of $t$ and *coefs*, the corresponding scales value is used to assign a color.

To generate the scalogram, type:

```
scales = 1:200;
coefs = my_cwt(t, x, @my_morlet, 10, scales, [10]);
plot_cwt(t, coefs, scales);
```

## 13.2.5 The Discrete Wavelet Transform

In addition to the continuous wavelet transform, there is a transformation termed the *discrete wavelet transform* (DWT). However, the DWT is not merely a discretized continuous wavelet transform. Instead, the discrete wavelet transform calculates only a subset of the possible scales, usually dyadic values (successive values in $2^n$, i.e., 1, 2, 4, 8, 16, 32, etc.). Moreover, the DWT is usually calculated using an algorithm called the *pyramid algorithm*, in which the data series is recursively split in two and reprocessed.

An exploration of the pyramid algorithm is beyond the scope of this chapter. For a thorough discussion, see Percival and Walden (2000). The DWT has been used to denoise signals and to cluster neural spikes for sorting (Quiroga, Nadasdy, and Ben-Shaul, 2004).

## 13.2.6 Wavelet Toolbox

Wavelet Toolbox provides an implementation of the DWT and a number of appropriate wavelets. Analyses using the discrete wavelet transform use different wavelets than analyses with the continuous transform. The Haar wavelet and the Daubechies wavelet are among the most widely used.

In the following commands, *'wname'* corresponds to the name of a specific wavelet included in Wavelet Toolbox. Possible choices are *'dbN'* for Daubechies N, *'haar'* for Haar, *'morl'* for Morlet, and *'mexh'* for the Mexican hat. To view all supported wavelets, use **help waveform.**

```
coefs = cwt(S, SCALES, 'wname')
```

The function **cwt** performs a continuous wavelet transform on the dataset *S*. The scales given as *SCALES* are used, and the wavelet is given by *'wname'*. The function **cwt** will also automatically plot the scalogram if given the parameter *'plot'* at the end:

```
coefs = cwt(x, 1:200, 'morl', 'plot')
[cA. cD] = dwt(X, 'wname')
X = idwt(cA, cD, 'wname')
```

The functions **dwt** and **idwt** perform a single level decomposition and synthesis given the wavelet name.

```
[C, L] = wavedec(X, N, 'wname')
X = waverec(C, L, 'wname')
```

The functions **wavedec** and **waverec** perform multilevel decomposition and synthesis given wavelet name and level *N*. Note that *N* cannot be greater than the exponent of the largest power of 2 less than the size of *X*. The *C* vector contains the transform, and the *L* vector contains bookkeeping information used by **wavedec** and **waverec** to find the position of the parts of the transform in *C*.

Here is an example plotting scales 2 through 7 for a Debauches 4 wavelet:

```
% here size(s) = 128
[C, L] = wavedec(s, 7, 'db4');
for scale = 2:7
subplot(7,1,scale)
c_sub = (2^(scale − 1)):(2^scale);
t_sub = linspace(1, time, time/size(c_sub));
plot(t_sub, C(c_sub))
end
wavedemo
```

The **wavedemo** function opens an automated tour of Wavelet Toolbox, showing various transforms and functions provided by the toolbox.

## 13.3 EXERCISES

### EXERCISE 13.1

Which of the following MATLAB functions can be wavelet functions? Why or why not? **function x = f_one(t)**

```
x = cos(t);
end
function x = f_two(t)
if (x < 0 or x > pi/2)
x = 0;
else
x = cos(t);
end
```

```
end
function x = f_three(t)
x = sqrt(2) * t * exp( − t^2/2) / pi^4;
end
function x = f_four(t)
x = sqrt(2) * t^2 * exp( − t^2/2) / pi^4;
end
function x = f_five(t)
x = (x > −1 && x < 0) * −1 + (x > 0
&& x < 1);
end
```

### EXERCISE 13.2

Generate the scalogram in Figure 13.2. Generate a spectrogram and compare. How clearly does each render the transients? The primary frequencies?

---

### EXERCISE 13.3

Write a Mexican hat mother wavelet function compatible with the previous continuous wavelet transform code. Generate a scalogram of the sinusoid + transient signal used in Figure 13.2. Compare Mexican hat transform to the Morlet transform.

---

### EXERCISE 13.4

Download the EEG signal wavelet, **eeg**, from the companion website. Generate scalograms using the Mexican hat and Morlet wavelet transforms. Compare to a spectrogram generated with **spectrogram()**.

## 13.4 PROJECT

In Chapter 12, "Frequency Analysis Part II: Nonstationary Signals and Spectrograms," you used the short-time Fourier transform to look for sleep state transitions. Here, you will be asked to examine the same data files using the continuous wavelet transform and Morlet and Mexican hat wavelets. Compare and contrast your findings with what you found using only the STFT.

## MATLAB FUNCTIONS, COMMANDS, AND OPERATORS COVERED IN THIS CHAPTER

**cwt**
**dwt**
**idwt**
**wavedec**
**waverec**

# Introduction to Phase Plane Analysis

## 14.1 GOAL OF THIS CHAPTER

The goal of this chapter is to examine the cone and horizontal cell system using a qualitative visualization technique called *phase plane analysis*; this system will be discussed further in Chapter 28, "Models of the Retina." The techniques presented here will be used again in Chapter 15, "Exploring the Fitzhugh-Nagumo Model."

## 14.2 BACKGROUND

In this chapter you will be studying a retinal feedback model; this model is described further in Chapter 28, "Models of the Retina." The system is represented as follows:

$$\frac{d\tilde{C}}{dt} = \frac{1}{\tau_C}(-\tilde{C} - k\tilde{H}) \tag{14.1}$$

$$\frac{d\tilde{H}}{dt} = \frac{1}{\tau_H}(-\tilde{H} + \tilde{C}) \tag{14.2}$$

Typical values for these parameters are $\tau_C = 0.025$ sec, $\tau_H = 0.08$ sec, and $k = 4$. Now assume that the light intensity is $L = 10$ (i.e., daylight). For your initial conditions, choose that $C(0) = H(0) = 0$. Finally, be aware that:

$$\tilde{C} = C - \frac{L}{k+1} \quad \text{and} \quad \tilde{H} = H - \frac{L}{k+1} \tag{14.3}$$

For further details of the basic biology of this system, see Chapter 28, "Models of the Retina." In that chapter, we will examine in more detail the model of retinal feedback between cone cells and horizontal cells of the retina, shown in Equations 14.1 and 14.2. Although the explicit solutions determined in that chapter are more informative, many more complicated systems (such as the Fitzhugh-Nagumo system presented in Chapter 15, "Exploring the Fitzhugh-Nagumo Model") can only be qualitatively described. When we describe a system

*MATLAB® for Neuroscientists.*
DOI: http://dx.doi.org/10.1016/B978-0-12-383836-0.00014-X

qualitatively, we look for steady-state values of the solutions (often called *fixed points*) and try to classify the dynamics of the solution that led to these steady-state values. In Chapter 28, "Models of the Retina," we will consider the following system in more detail:

$$\frac{dx}{dt} = x + y \tag{14.4}$$

$$\frac{dy}{dt} = 4x + y \tag{14.5}$$

which has eigenvalues $-1$ and $3$ and has the solution:

$$x(t) = C_1 e^{3t} + C_2 e^{-t} \tag{14.6}$$

$$y(t) = 2C_1 e^{3t} - 2C_2 e^{-t} \tag{14.7}$$

This solution can be described qualitatively. If you wait long enough, then this system will approach one of two states. If $C_1 = 0$, then:

$$\lim_{t \to \infty} x(t) = \lim_{t \to \infty} y(t) = 0 \tag{14.8}$$

Therefore, one says that $(x, y) = (0, 0)$ is a steady-state or fixed point of the system. For $C_1 \neq 0$, then:

$$\lim_{t \to \infty} x(t) = \lim_{t \to \infty} y(t) = \infty \tag{14.9}$$

Therefore, the only finite steady-state solution to this system is $(x, y) = (0, 0)$. Regardless of how you choose $C_1$ and $C_2$, there are no other steady-state values for this system. Since the initial conditions determine $C_1$ and $C_2$, then those initial conditions that lead to $C_1 = 0$ will have solutions that steadily tend toward the fixed point $(0, 0)$, while all others will steadily tend toward infinity (i.e., away from the fixed point at the origin). A fixed point with this property—that is, with some initial conditions leading to the fixed point and others leading away from it—is called a *saddle point*. This simple qualitative description of identifying the steady state(s) of the solution, the dynamics of what initial conditions lead to the steady state(s), and how it is reached steadily or in an oscillatory fashion can all be determined from a phase plane analysis of the system.

The first step in phase plane analysis is to set up a phase plane. The axes for the plane represent the state variables characterizing the system. In the preceding example, the phase plane is constructed with $y$ as the ordinate and $x$ as the abscissa. Next, the *-x-* and $y$-nullclines are plotted. The $x$-nullcline is the curve in the $x$-$y$ plane, where:

$$\frac{dx}{dt} = 0$$

A similar definition applies for the $y$-nullcline. Intersections of these nullclines represent points where:

$$\frac{dx}{dt} = \frac{dy}{dt} = 0$$

so $x$ and $y$ are no longer changing with time. In other words, these intersections represent steady-state values or fixed points of the system. Next, a vector field is constructed by assigning the following vector to every point on the $x$-$y$ plane:

$$\left[ \frac{dx}{dt} \quad \frac{dy}{dt} \right]^T$$

Notice that this vector field can be determined without knowing the solution to the system. Since the slope of these vectors is:

$$m = \frac{dy}{dt} \bigg/ \frac{dx}{dt} = \frac{dy}{dx} \tag{14.10}$$

by the chain rule, the vector field must be tangent to any solution $(x, y)$ of the system. This allows you to use the vector field to calculate the solution of the system for any initial condition $(x_o, y_o)$. Such a solution when plotted on the phase plane is called a *trajectory*. The phase plane, nullclines, vector field, and several trajectories are shown in Figure 14.1 for the system in Equations 14.4 and 14.5.

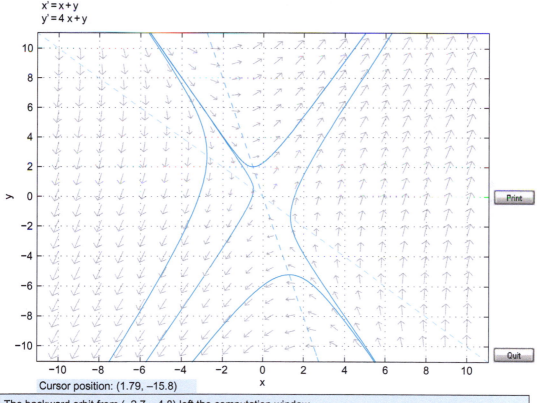

FIGURE 14.1 Phase plane of a linear system showing saddle node stability.

In the figure, the nullclines are plotted as dashed lines. Notice that these nullclines intersect at the point $(x, y) = (0, 0)$ indicating that this is the steady-state of the system in agreement with what was predicted by considering the explicit solutions (Equations 14.6 and 14.7). Any linear system of ordinary differential equations described by a matrix with real eigenvalues of opposite sign (recall that the eigenvalues for this system are $-1$ and $3$) will have a saddle point at the intersection of its nullclines.

If the matrix describing the linear system has real eigenvalues that are both negative, then the fixed point is called a *nodal sink*. The classic phase portrait of a nodal sink is shown in Figure 14.2.

If the matrix describing the linear system has real eigenvalues that are both positive, then the fixed point is called a *nodal source*. The classic phase portrait of a nodal source is shown in Figure 14.3. Notice the difference in the direction of the arrows in the vector field in this figure.

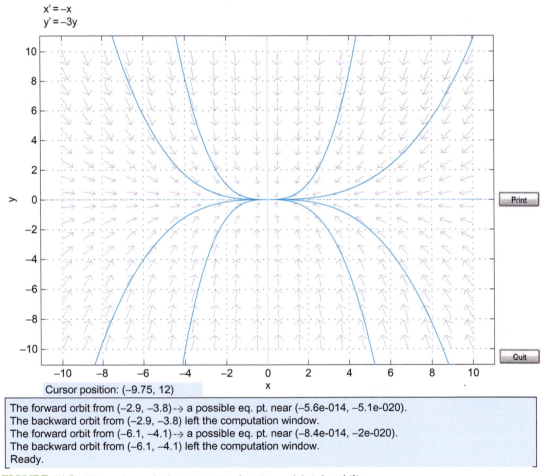

FIGURE 14.2    Phase plane of a linear system showing nodal sink stability.

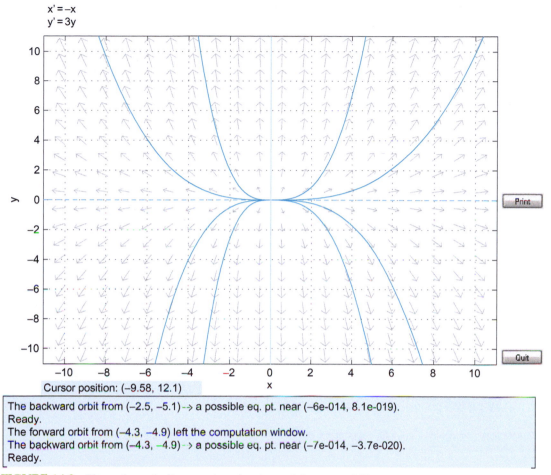

FIGURE 14.3    Phase plane of a linear system showing nodal source stability.

If the matrix describing the linear system has imaginary eigenvalues that have negative real parts, then the fixed point is called a *spiral sink*. The classic phase portrait of a spiral sink is shown in Figure 14.4.

If the matrix describing the linear system has imaginary eigenvalues that have positive real parts, then the fixed point is called a *spiral source*. The classic phase portrait of a spiral source is shown in Figure 14.5.

These five types of equilibria are collectively known as the *generic equilibria*. There are also five nongeneric equilibria. The most important nongeneric equilibrium is called a *center*. It occurs when the eigenvalues of the matrix are purely imaginary. The classic phase portrait of a center is shown in Figure 14.6.

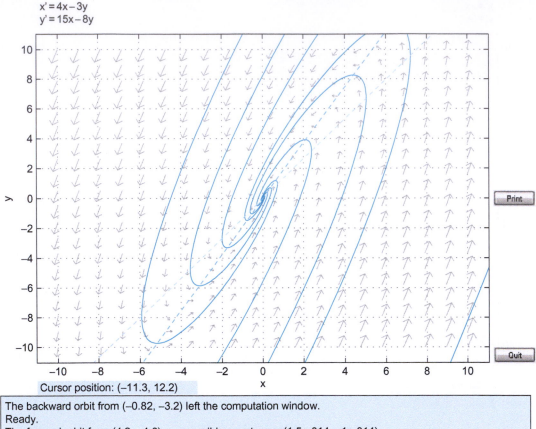

**FIGURE 14.4**    Phase plane of a linear system showing spiral sink stability.

## 14.3 EXERCISES

The phase portraits in the preceding section were drawn using a downloadable M-file called pplane7.m. The phase plane consists of three basic features: the nullclines intersecting at the fixed point of the system, the vector field showing how the solutions change over time, and trajectories showing how the solution approaches its steady-state from a given initial condition. The first exercise of this chapter will involve writing a simple version of pplane7. Several functions built into MATLAB will aid in coding each of the basic features of the phase plane mentioned previously.

Plotting the nullclines of the system requires no more than the basic plotting commands used throughout previous chapters. Plotting the vector fields can be greatly aided by the functions **meshgrid()** and **quiver()**. The function **meshgrid** takes two vector arguments $x$ and $y$, and returns two square matrices $X$ and $Y$ such that each row of $X$ is a copy of the

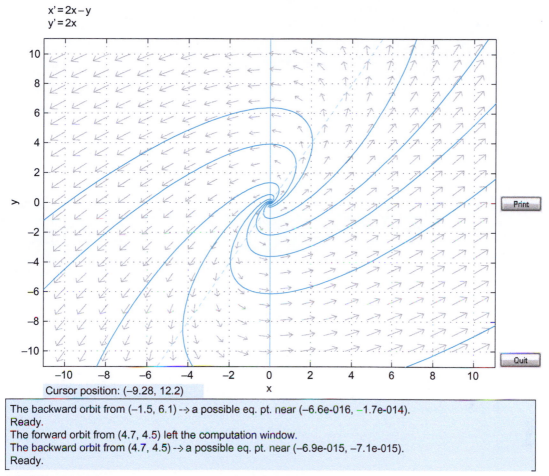

$x' = 2x - y$
$y' = 2x$

Cursor position: (−9.28, 12.2)

The backward orbit from (−1.5, 6.1) --> a possible eq. pt. near (−6.6e-016, −1.7e-014).
Ready.
The forward orbit from (4.7, 4.5) left the computation window.
The backward orbit from (4.7, 4.5) --> a possible eq. pt. near (−6.9e-015, −7.1e-015).
Ready.

**FIGURE 14.5**   Phase plane of a linear system showing spiral source stability.

vector $x$, and each column of $Y$ is a copy of the vector $y$. This function is useful for evaluating functions of two variables. For example, suppose you wanted to evaluate the function $f(x,y) = x + y$. You could do this using **for** loops; for example, you could type

```
>> x = [0:0.1:10];
>> y = x;
>> for ii = 1:length(x)
>> for jj = 1:length(y)
>> f(ii,jj) = x(ii) + y(jj);
>> end;
>> end;
```

which produces the same results as the commands

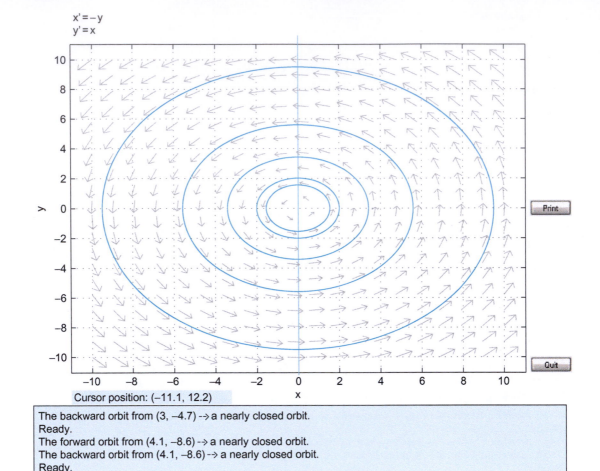

**FIGURE 14.6**     Phase plane of a linear system showing center stability

```
>> x = −10:10;
>> y = x;
>> [X, Y] = meshgrid(x,y);
>> f = X + Y;
```

Evaluating functions of two variables is important in this chapter because the model you wish to study (Equations 14.1 and 14.2) expresses the derivatives as functions of the two variables: $\tilde{C}$ and $\tilde{H}$. By comparing the matrix $f$ as defined in the preceding code to Equation 14.1, you see that $f$ holds the values of the derivative $\dfrac{dx}{dt}$ for several values of $x$ and $y$. You could define a matrix $g$ that holds the $y$-derivative by using the following command:

```
>>g = 4X + Y;
```

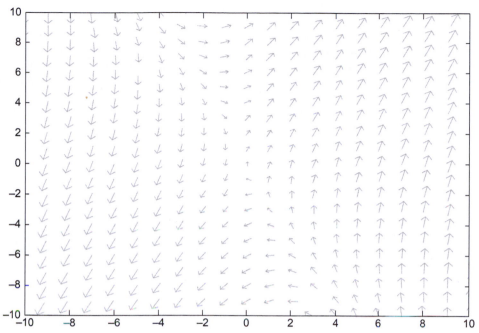

FIGURE 14.7    Vector field created by the **quiver()** command.

Once you have evaluated these derivatives using **meshgrid,** we can plot a vector field using the **quiver** command. If you have the matrices $X$, $Y$, $f$, and $g$ defined as shown here, then type this command:

>> quiver(X,Y,f,g);

You should get the result shown in Figure 14.7.

The function **quiver** works by plotting a vector on the plane at points $(x, y)$ with components $(f, g)$.

You can plot the trajectory of a system given an initial condition in several ways. One method is to use a numerical solver such as the **ode_euler()** or **RK4()** functions you will write in Chapter 25, "Voltage-Gated Ion Channels," to solve for $x$ and $y$ given some initial condition and then plot $x$ versus $y$. Another method would be to calculate the derivatives of $x$ and $y$ at the initial condition, move the system a short distance in the direction indicated by the derivative, and then repeat over many time steps. Either method will work, and both can be done with no more than the basic functions introduced in Chapter 2, "MATLAB Tutorial."

---

### EXERCISE 14.1

Write a function **phase_plane(A, init)** that takes a matrix and performs a phase
plane analysis for the linear system, $\mathbf{u}' = \mathbf{Au}$. The function should plot a phase

plane with axes $x$ and $y$, plot the nullclines, create the vector field, and plot the phase plane trajectory that passes through the initial condition. Finally, the program should output the type of equilibrium point, saddle point, spiral sink, etc. If the equilibrium point is nongeneric, then the program can just output nongeneric as the class type. This is quite a complicated program, so you may want to write several smaller functions that can be called within **phase_plane**—for example, a separate function that will simply classify the fixed point and then another that will create a vector field, etc. *Hint:* The Boolean function **isreal()** will return 0 if the argument is not a real number and 1 if it is. This function might be useful for deciding whether or not the eigenvalues of $A$ are real, so that the fixed point of the system can be classified.

## 14.4 PROJECT

Use your **phase_plane** program to analyze the retinal model described at the beginning of this chapter. The matrix describing this linear system is:

$$\begin{bmatrix} \dfrac{-1}{\tau_C} & \dfrac{-k}{\tau_C} \\ \dfrac{1}{\tau_H} & \dfrac{-1}{\tau_H} \end{bmatrix}$$

Identify what kind of behavior the fixed point exhibits. Repeat using the parameters for dim light:

$$\tau_C = 0.1 \text{ sec, } \tau_H = 0.5 \text{ sec, and } k = 0.5$$

What is the behavior of the fixed point now?

## MATLAB FUNCTIONS, COMMANDS, AND OPERATORS COVERED IN THIS CHAPTER

**isreal**
**quiver**
**eig**

# Exploring the Fitzhugh-Nagumo Model

## 15.1 GOAL OF THIS CHAPTER

In this chapter we will use the techniques of phase plane analysis to analyze a simplified model of action potential generation in neurons known as the Fitzhugh-Nagumo (FN) model. Unlike the Hodgkin-Huxley model, which has four dynamical variables (see Chapter 27, "Modeling of a Single Neuron"), the FN model has only two, so the full dynamics of the FN model can be explored using phase plane methods.

## 15.2 BACKGROUND

The FN model can be created from the Hodgkin-Huxley model by combining the variables $V$ and $m$ into a single variable $v$ and combining the variables $n$ and $h$ into a single variable $r$. The four equations of the Hodgkin-Huxley model then become the two-equation system (Fitzhugh, 1961)

$$\frac{dv}{dt} = c(v - \frac{1}{3}v^3 + r + I) \tag{15.1}$$

$$\frac{dr}{dt} = -\frac{1}{c}(v - a + br) \tag{15.2}$$

where $a$, $b$, $c$, and $I$ are parameters of the model.

In Chapter 14, "Introduction to Phase Plane Analysis," we analyzed a system of linear differential equations that had the following general form:

$$\frac{dx}{dt} = ax + by \tag{15.3}$$

*MATLAB® for Neuroscientists.*
DOI: http://dx.doi.org/10.1016/B978-0-12-383836-0.00015-1

$$\frac{dy}{dt} = cx + dy \tag{15.4}$$

In the current chapter we would like you to consider more complicated differential equations (such as those of the FN model). Suppose that you have a system of differential equations of the form:

$$\frac{dx}{dt} = f(x, y) \tag{15.5}$$

$$\frac{dy}{dt} = g(x, y), \tag{15.6}$$

where $f$ and $g$ are more complicated functions of $x$ and $y$. You begin by plotting the $x$- and $y$-nullclines, which are given by $f(x, y) = 0$ and $g(x, y) = 0$, respectively. These nullclines may intersect never, once, or more than once. If the nullclines never intersect, then the system has no finite steady-state solutions. If there is one point of intersection, then there is only one steady-state solution. Linear systems have at most one steady-state solution (unless they are degenerate). Nonlinear systems, however, can have any number of steady-state values. This will be important in your understanding the trajectories, which may be seen in nonlinear systems. A vector field and trajectories given initial conditions can be calculated for nonlinear systems in the exact same manner as calculated for linear systems. Lastly, you can classify the fixed points (steady-state values) as you did in Chapter 14, "Introduction to Phase Plane Analysis." You perform this by linearizing the functions $f$ and $g$ about each fixed point. You assume that the functions $f$ and $g$ have Taylor expansions, so:

$$f(x, y) = f(x_{ss}, y_{ss}) + \frac{\partial f(x_{ss}, y_{ss})}{\partial x}(x - x_{ss}) + \frac{\partial f(x_{ss}, y_{ss})}{\partial y}(y - y_{ss}) + higher\ order\ terms, \tag{15.7}$$

and

$$g(x, y) = g(x_{ss}, y_{ss}) + \frac{\partial g(x_{ss}, y_{ss})}{\partial x}(x - x_{ss}) + \frac{\partial g(x_{ss}, y_{ss})}{\partial y}(y - y_{ss}) + higher\ order\ terms. \tag{15.8}$$

As you approach the fixed points, the higher order terms tend to zero since $x - x_{ss}$, $y - y_{ss} \ll 1$. Additionally, $f(x_{ss}, y_{ss}) = g(x_{ss}, y_{ss}) = 0$, so:

$$f(x, y) \approx \frac{\partial f(x_{ss}, y_{ss})}{\partial x}(x - x_{ss}) + \frac{\partial f(x_{ss}, y_{ss})}{\partial y}(y - y_{ss}) \quad and \tag{15.9}$$

$$g(x, y) \approx \frac{\partial g(x_{ss}, y_{ss})}{\partial x}(x - x_{ss}) + \frac{\partial g(x_{ss}, y_{ss})}{\partial y}(y - y_{ss}). \tag{15.10}$$

Substituting these equations into Equations 15.1 and 15.2 yields:

$$\frac{dx}{dt} = \frac{d(x - x_{ss})}{dt} = \frac{\partial f(x_{ss}, y_{ss})}{\partial x}(x - x_{ss}) + \frac{\partial f(x_{ss}, y_{ss})}{\partial y}(y - y_{ss}) \tag{15.11}$$

$$\frac{dy}{dt} = \frac{d(y - y_{ss})}{dt} = \frac{\partial g(x_{ss}, y_{ss})}{\partial x}(x - x_{ss}) + \frac{\partial g(x_{ss}, y_{ss})}{\partial y}(y - y_{ss}). \tag{15.12}$$

Expressing this system as a matrix equation gives:

$$\begin{bmatrix} (x - x_{ss})' \\ (y - y_{ss})' \end{bmatrix} = \begin{bmatrix} \dfrac{\partial f(x_{ss}, y_{ss})}{\partial x} & \dfrac{\partial f(x_{ss}, y_{ss})}{\partial y} \\ \dfrac{\partial g(x_{ss}, y_{ss})}{\partial x} & \dfrac{\partial g(x_{ss}, y_{ss})}{\partial x} \end{bmatrix} * \begin{bmatrix} (x - x_{ss}) \\ (y - y_{ss}) \end{bmatrix}. \tag{15.13}$$

If you let:

$$u = \begin{bmatrix} (x - x_{ss}) \\ (y - y_{ss}) \end{bmatrix} \quad \text{and} \quad J = \begin{bmatrix} \dfrac{\partial f}{\partial x} & \dfrac{\partial f}{\partial y} \\ \dfrac{\partial g}{\partial x} & \dfrac{\partial g}{\partial y} \end{bmatrix} \tag{15.14}$$

then you can write Equation 15.13 as:

$$u' = J|_{(x_{ss}, y_{ss})} * u. \tag{15.15}$$

The matrix $J$ is called the *Jacobian matrix*. It is a very important matrix in the mathematics of multivariable calculus. Equation 15.15 tells you that to a first-order approximation the nonlinear system in Equations 15.5 and 15.6 can be approximated by the linear system of Equation 15.15. The eigenvalues of the Jacobian matrix (evaluated at the fixed point) allow you to classify the fixed point as a saddle point, spiral sink, etc. Equation 15.15 is an approximation to the nonlinear system. You might wonder at what point the approximation breaks down. There is a theorem that we will state without proof which says that when the dynamics of the fixed point of the linear system in Equation 15.12 is a generic fixed point, then the fixed point of the nonlinear system in Equations 15.1 and 15.2 has the same dynamics. If the linear system has a nongeneric fixed point such as a center, then no conclusion can be drawn about the dynamics of the fixed point of the nonlinear system. See Chapter 14, "Introduction to Phase Plane Analysis," for a review of generic and nongeneric equilibria.

Note that information about the dynamics of the fixed point applies only to a limited neighborhood centered about the fixed point. A spiral source, for example, can spiral out to infinity or spiral out and approach a circular orbit. The latter case is called a *limit cycle*. Nonlinear systems in higher dimensions (three or more) can have even more complicated dynamics, not all of which have currently been discovered. The best studied dynamics of higher order nonlinear systems include Lorenz attractors and chaos.

## 15.3 EXERCISES

In this chapter we will explore the **pplane** program written by Dr. John C. Polking of Rice University. This program was used to make the figures in the Background section of Chapter 14, "Introduction to Phase Plane Analysis," and the latest version can be

FIGURE 15.1    **pplane8** Setup window.

downloaded free at http://math.rice.edu/~dfield/. After downloading the script, you can run it by typing the following:

```
>> pplane8;
```

Entering this command will open the pplane8 Setup window shown in Figure 15.1.

Set up the FN model by changing the variables $x$ and $y$ to $v$ and $r$ according to Equations 15.1 and 15.2. The parameter values you can use for now are $a$ 0.7, $b$ 0.8, $c$ 3, and $I$ 0. Set the display window such that $v$ ranges from $-3$ to 3 and $r$ ranges from $-2$ to 4. Leave the other settings the same. If you have done this correctly, the Setup window will look like the one in Figure 15.2.

Now click the Proceed button, and a pplane8 Display window will come up, as shown in Figure 15.3.

Next, open the Solutions menu and click Show nullclines. This will display the $v$-null-cline in magenta and the $r$-nullcline in red. The phase plane will now look like the one shown in Figure 15.4.

Next, open the Option menu, select Solution Direction, and then select Forward. This will ensure that when an initial condition is provided to the system, the trajectory will be plotted only as time moves forward. Finally, open the Solutions menu and select Find an Equilibrium Point. This will turn the mouse pointer into a crosshair. Place the crosshair near the intersection of the nullclines and click. An Equilibrium point data window will open, revealing that the equilibrium is located at $(v, r) = (1.1994, -0.62426)$. If you would like to enter in an initial condition to see a trajectory in the phase plane, you have two options. First, you can open the Solutions menu and then click Keyboard Input. This will

FIGURE 15.2 **pplane8** Setup window with FN model.

allow you to enter the initial conditions. After you click Compute, a trajectory in blue is depicted on the phase plane in the pplane8 Display window. Alternatively, you can click Solutions and then select Plot several solutions. Again, the mouse pointer is converted to a crosshair. You can now click on the phase plane at the point representing the initial condition and press Enter. Several trajectories are shown in the phase plane in Figure 15.5.

Finally, you can obtain the voltage trace from the phase plane by opening the Graph menu and selecting *v vs t*. This will again convert the mouse pointer into a crosshair. Use the crosshair to select any trajectory on the phase plane. A pplane8 *t*-plot such as the one in Figure 15.6 will appear.

The plot in Figure 15.6 shows that if you change the membrane potential of the neuron to 2.4, it decays back down to the equilibrium value 1.1994 as previously determined. This is analogous to giving a neuron a subthreshold depolarizing stimulus. After the brief depolarizing stimulus, the neuron's membrane potential will exponentially relax back down to its equilibrium resting potential.

---

**EXERCISE 15.1**

Is the equilibrium point in the preceding model system stable (i.e., are trajectories attracted to this point or repelled from it)?

---

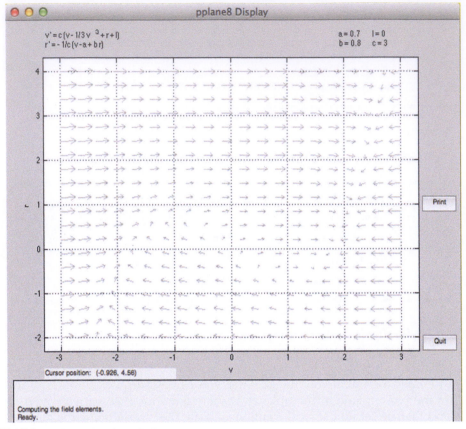

FIGURE 15.3    **pplane8** Display window.

# 15.4 PROJECT

In this project, you will explore the Fitzhugh-Nagumo model that you setup with **pplane8** by injecting different levels of current and examining how the behavior of the model neuron mimics that of a real neuron. Specifically, you should do the following:

Change the injected current value to $I = -0.2$ in the Setup window and click Proceed. Follow the previous instructions to display the nullclines. Calculate a trajectory in the Forward direction with the initial condition $(v, r) = (1.1994, -0.62426)$. Is this point still stable?

Determine what $v$ versus $t$ looks like for a trajectory on this phase plane. Would you classify the injected input of $-0.2$ as a superthreshold or subthreshold stimulus? Does this neuron exhibit subthreshold oscillations for this value of injected current?

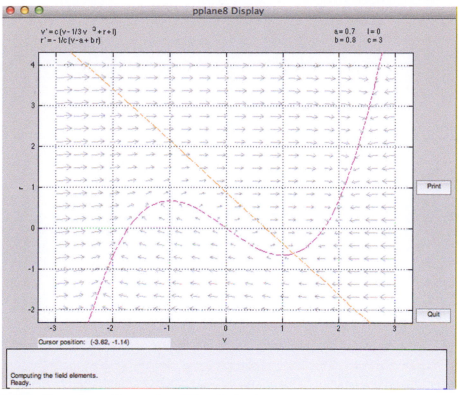

**FIGURE 15.4** Phase plane with $v$- and $r$-nullclines depicted.

Change the injected current value to $I = -0.4$ in the Setup window and click Proceed. Follow the previous instructions to display the nullclines. Calculate a trajectory in the Forward direction with the initial condition $(v, r) = (1.1994, -0.62426)$. Is this point still stable? Plot several trajectories on this phase plane. Since the nullclines intersect at only a single point, there are no other equilibrium points for this system, but trajectories may be attracted to some other closed orbit—for example, a circular orbit. Are these trajectories attracted to a closed orbit?

Determine what $v$ versus $t$ looks like for a trajectory that is attracted to a closed orbit, also called a *limit cycle*. Would you classify this injected stimulus as a superthreshold or subthreshold stimulus?

Finally, repeat the analysis for $I = -1.6$ and examine $v$ versus $t$. Does this neuron spike continuously as it did before? Neurons are known to exhibit a phenomenon called *excitation block*, whereby increasing the current injection can often repress repetitive firing behavior.

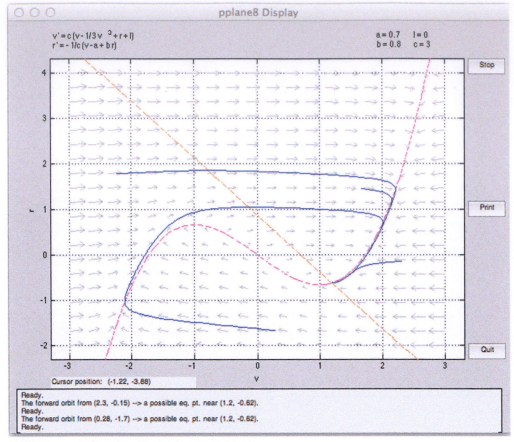

FIGURE 15.5    Phase plane with sample trajectories.

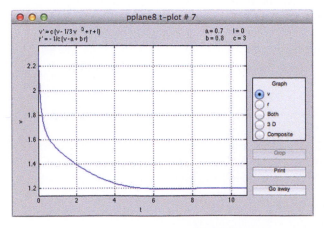

FIGURE 15.6    **pplane8** *t*-plot showing voltage over time.

# MATLAB FUNCTIONS, COMMANDS, AND OPERATORS COVERED IN THIS CHAPTER

**pplane8** (free script by John C. Polking available at http://math.rice.edu/~dfield/)

# Convolution

## 16.1 GOALS OF THIS CHAPTER

The purpose of this chapter is to familiarize you with the convolution operation. You will use this operation in the context of receptive fields in the early visual system as input response filters whose convolution with an input image approximates certain aspects of your perception. Specifically, you will reproduce the Mach band illusion and explore the Gabor filter as a model for the receptive field of a simple cell in the primary visual cortex.

## 16.2 BACKGROUND

A *convolution* is the mathematical operation used to find the output $y(t)$ of a linear time-invariant system from some input $x(t)$ using the impulse response function of the system $h(t)$, where $h(t)$ is defined as the output of a system to a unit impulse input. It is defined as the following integral:

$$y(t) = h(t) * x(t) = \int_{-\infty}^{\infty} h(\tau)x(t - \tau)d\tau \qquad (16.1)$$

This can be graphically interpreted as follows. The function $h(\tau)$ is plotted on the $\tau$-axis, as is the flipped and shifted function $x(t - \tau)$, where the shift $t$ is fixed. These two signals are multiplied, and the signed area under the curve of the resulting function is found to obtain $y(t)$. This operation is then repeated for every value of $t$ in the domain of $y$. It turns out that it doesn't matter which function is flipped and shifted since $h * x = x * h$.

You can also define a convolution for data in two dimensions:

$$y(k, t) = h(k, t) * x(k, t) = \int_{-\infty}^{\infty} \int_{-\infty}^{\infty} h(\tau, K)x(k - K, t - \tau)dKd\tau \qquad (16.2)$$

*MATLAB® for Neuroscientists.*
DOI: http://dx.doi.org/10.1016/B978-0-12-383836-0.00016-3

Basically, you take a convolution in one dimension to establish the $k$ dependence of the result $y$ and then use that output (which is a function of $k$, $t$ and $\tau$) to perform another convolution in the second dimension. This second convolution provides the $t$ dependence of the result, $y$. It is important that you understand how to apply this to a two-dimensional data function because in this chapter you will be working with two-dimensional images. In the MATLAB® software, since you are working with discrete datasets, the integral becomes a summation, so the definition for convolution in 2D at every point becomes

$$y(n_1, n_2) = \sum_{k_1 = -\infty}^{\infty} \sum_{k_2 = -\infty}^{\infty} h(k_1, k_2) x(n_1 - k_1, n_2 - k_2) \tag{16.3}$$

Again, this is easier to understand pictorially. What you are doing in this algorithm is taking the dataset $x$, which is a matrix; rotating it by 180 degrees; overlaying it at each point in the matrix $h$ that describes the response filter; multiplying each point with the underlying point; and summing these points to produce a new point at that position. You do this for every position to get a new matrix that will represent the convolution of $h$ and $x$.

## 16.2.1 The Visual System and Receptive Fields

In this section we discuss in general the anatomy of the visual system and the input response functions that explain how different areas of the brain involved in this system might "perceive" a visual stimulus.

Light information from the outside world is carried by photons that enter the eyes and cause a series of biochemical cascades to occur in rods and cones of the retina. This biochemical cascade causes channels to close which leads to a decrease in the release of neurotransmitter onto bipolar cells. In general, there are two fundamental varieties of bipolar cells. On-bipolar cells become depolarized in response to light and off-bipolar cells become hyperpolarized in response to light. The bipolar cells then project to the ganglion cells which are the output cells of the retina. The response to light in this main pathway is also influenced by both the horizontal and amacrine cells in the retina. There are many types of retinal ganglion cells that respond to different visual stimuli.

A stimulus in the visual field will elicit a cell's response (above the background firing rate) only if it lies within a localized region of visual space, denoted by the cell's *classical receptive field*. In general, the ganglion cells have a center-surround receptive field due to the types of cells that interact to send information to these neurons. That is, the receptive field is essentially two concentric circles, with the center having an excitatory increase (+) in neuronal activity in response to light stimulus and the surround having an inhibitory decrease (−) in neuronal activity in response to light stimulus, or vice versa. The response function of the ganglion cells can then be modeled using a *Mexican hat* function, also sometimes called a *difference of Gaussians* function.

In the main visual pathway, the ganglion cells send their axons to the lateral geniculate nucleus (LGN) in the thalamus, which is in charge of regulating information flow to the cortex. These cells also are thought to have receptive fields with a center-surround architecture. LGN cells project to the primary visual cortex (V1). In V1, simple cells are thought to receive information from LGN neurons in such a way that they respond to bars of light at certain orientations and spatial frequencies. This can similarly be described as a Gabor function—a two-dimensional Gaussian filter whose amplitude is modulated by a sinusoidal function along an axis at a given orientation. Thus, different simple cells in V1 respond to bars of light at specific orientations with specific widths (this represents spatial frequency; see Dayan and Abbott, 2001). These and other cells from V1 project to many other areas in the cortex thought to represent motion, depth, face recognition, and other fascinating visual features and perceptions.

## 16.2.2 The Mach Band Illusion

Using your knowledge of the receptive fields or the response functions of the visual areas can help you understand why certain optical illusions work. The *Mach band illusion* is a perceptual illusion seen when viewing an image that ramps from black to white. Dark and light bands appear on the image where the brightness ramp meets the black and white plateau, respectively. These bands are named after Ernst Mach, a German physicist who first studied them in the 1860s. They can be explained with the center-surround receptive fields of the ganglion or LGN cells (Ratliff, 1965; Sekuler and Blake, 2002); we will use this model in this chapter although alternative explanations exist (for example, see Lotto, Williams, and Purves, 1999).

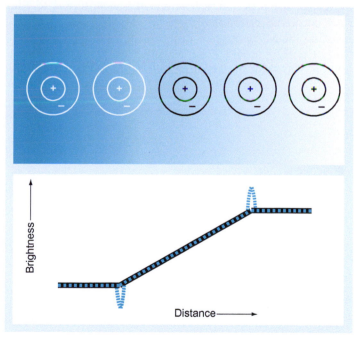

**FIGURE 16.1**    The Mach band illusion. Top of figure: the visual stimulus with various center-surround receptive fields superimposed. Bottom of figure: the actual brightness of the visual stimulus (black solid line) and the perceived brightness of the optical illusion (blue dotted line).

III. DATA ANALYSIS WITH MATLAB

The illusion is demonstrated in Figure 16.1. At the initiation of the stimulus brightness ramp, a dark band, darker than the dark plateau to the left, is usually perceived. At the termination of the brightness ramp, a light band is perceived brighter than the light plateau to its right. Figure 16.1 shows the center-surround receptive fields of sample neurons, represented by concentric circles, superimposed on the stimulus image. The center disk is *excitatory*, and the surrounding annulus is *inhibitory*, as indicated by the plus and minus signs. When the receptive field of a neuron is positioned completely within the areas of uniform brightness, the center receives nearly the same stimulation as the surround; thus, the excitation and inhibition are in balance. A receptive field aligned with the dark Mach band has more of its surround in a brighter area than the center, and the increased inhibition to the neuron results in the perception of that area as darker. Conversely, the excitation to a neuron whose receptive field is aligned with the bright Mach band is increased, since more of its center is in a brighter area than the surround. The decreased inhibition to such a neuron results in a stronger response than that of the neuron whose receptive field lies in the uniformly bright regime and thus the perception of the area as brighter.

## 16.3 EXERCISES

The goal for this chapter is to reproduce the Mach band optical illusion. First, you will create the visual stimulus. Then you will create a center-surround Mexican hat receptive field. Finally, you will convolve the stimulus with the receptive field filter to produce an approximation of the perceived brightness.

You begin by creating the M-file named ramp.m that will generate the visual input (see Figure 16.2). The input will be a $64 \times 128$ matrix whose values represent the intensity or brightness of the image. You want the brightness to begin dark, at a value of 10, for the first

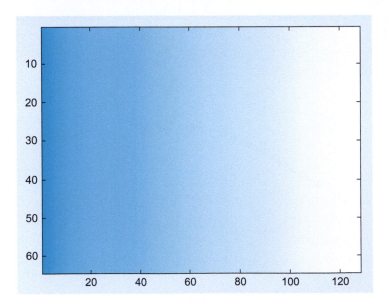

FIGURE 16.2 The brightness ramp stimulus used as visual input.

32 columns. In the next 65 columns, the value will increase at a rate of one per column, and the brightness will stay at the constant value of 75 for the rest of the matrix. Open a new blank file and save it under the name ramp.m. In that file enter the following commands:

```
%ramp.m
% This script generates the image that creates the Mach band visual illusion.
In = 10*ones(64,128); %initiates the visual stimulus with a constant value of 10
for ii = 1:65
    In(:,32 + ii) = 10 + ii;
    %ramps up the value for the middle matrix elements (column 33 to column 97)
end
In(:,98:end) = 75; %sets the last columns of the matrix to the final brightness value of 75
figure
imagesc(In); colormap(bone); set(gca, 'fontsize',20) %view the visual stimulus
```

Notice how the function **imagesc** creates an image whose pixel colors correspond to the values of the input matrix *In*. You can play with the color representation of the input data by changing the colormap. Here, you use the colormap *bone*, since it is the most appropriate one for creating the optical illusion, but there are many more interesting options available that you can explore by reading the help file for the function **colormap**.

You've just created an M-file titled ramp that will generate the visual stimulus. Note, however, that you use a **for** loop in ramping up the brightness values. Although it doesn't make much of a difference in this script, it is good practice to avoid using **for** loops when programming in MATLAB if possible, and to take advantage of its efficient matrix manipulation capabilities for faster run times (see Chapter 4.4.5.1, "Vectorizing Matrix Operations"). How might you eliminate the **for** loop in this case? One solution is to use the function **cumsum**. Let's see what it can do:

```
>> z = ones(3,4)
z =
   1   1   1   1
   1   1   1   1
   1   1   1   1
>> cumsum(z)
ans =
   1   1   1   1
   2   2   2   2
   3   3   3   3
```

The function will cumulatively add the elements of the matrix by row, unless you specify that dimension along which to sum should be the second dimension, or by column:

```
>> cumsum(z,2)
ans =
   1   2   3   4
   1   2   3   4
   1   2   3   4
```

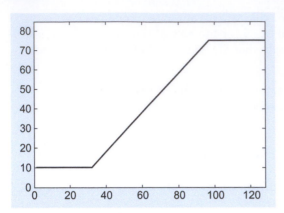

FIGURE 16.3 The brightness values in a slice through the ramp stimulus shown in Figure 16.2.

You will want this cumulative sum by columns for this ramp function. Now rewrite the code in proper style for MATLAB without the **for** loop:

**%ramp.m**
**% This script generates the image that creates the Mach band visual illusion.**
**In = 10*ones(64,128); %initiates the visual stimulus with a constant value of 10**
**% now ramp up the value for the middle matrix elements using cumsum**
**In(:,33:97) = 10 + cumsum(ones(64,65),2);**
**In(:,98:end) = 75; %sets the last columns of the matrix to the end value of 75**
**figure; imagesc(In); colormap(bone); set(gca, 'fontsize',20) %view the visual stimulus**

You can look at how the values of the brightness increase from left to right by taking a slice of the matrix and plotting it, as shown in Figure 16.3. Look at the $32^{nd}$ row in particular.

**>>  plot(In(32,:),'k','LineWidth',3); axis([0 128 0 85]); set(gca,'fontsize',20)**

Next, you will create a script titled mexican_hat.m that will generate a matrix whose values are a difference of Gaussians. For this exercise, you will make this a $5 \times 5$ filter, as shown in Figure 16.4.

**% mexican_hat.m**
**% this script produces an N by N matrix whose values are**
**% a 2 dimensional Mexican hat or difference of Gaussians**
**%**
**N = 5; %matrix size is NXN**
**IE = 6; %ratio of inhibition to excitation**
**Se = 2; %variance of the excitation Gaussian**
**Si = 6; %variance of the inhibition Gaussian**
**S = 500;%overall strength of Mexican hat connectivity**
**%**
**[X,Y] = meshgrid((1:N)-round(N/2));**
**% − floor(N/2) to floor(N/2) in the row or column positions (for N odd)**
**% − N/2 + 1 to N/2 in the row or column positions (for N even)**

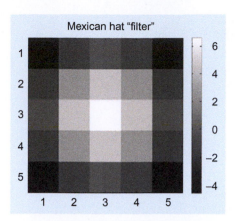

FIGURE 16.4 A 5 × 5 Mexican hat spatial filter.

```
%
[THETA,R] = cart2pol(X,Y);
% Switch from Cartesian to polar coordinates
% R is an N*N grid of lattice distances from the center pixel
% i.e. R = sqrt((X).^2 + (Y).^2) + eps;
EGauss = 1/(2*pi*Se^2)*exp(-R.^2/(2*Se^2)); % create the excitatory Gaussian
IGauss = 1/(2*pi*Si^2)*exp(-R.^2/(2*Si^2)); % create the inhibitory Gaussian
%
MH = S*(EGauss-IE*IGauss); %create the Mexican hat filter

figure; imagesc(MH) %visualize the filter
title('mexican hat "filter"','fontsize',22)
colormap(bone); colorbar
axis square; set(gca,'fontsize',20)
```

Now take a second look at some of the components of this script. The function **meshgrid** is used to generate the $X$ and $Y$ matrices whose values contained the $x$ and $y$ Cartesian coordinate values for the Gaussians:

```
>> X
X =
    -2  -1  0  1  2
    -2  -1  0  1  2
    -2  -1  0  1  2
    -2  -1  0  1  2
    -2  -1  0  1  2
>> Y
Y =
    -2  -2  -2  -2  -2
    -1  -1  -1  -1  -1
     0   0   0   0   0
     1   1   1   1   1
     2   2   2   2   2
```

The function **cart2pol** converts the Cartesian coordinates $X$ and $Y$ into the polar coordinates $R$ and $THETA$. You use this function to create the $5 \times 5$ matrix $R$ whose values are the radial distance from the center pixel:

```
>> R
R =
    2.8284   2.2361   2.0000   2.2361   2.8284
    2.2361   1.4142   1.0000   1.4142   2.2361
    2.0000   1.0000        0   1.0000   2.0000
    2.2361   1.4142   1.0000   1.4142   2.2361
    2.8284   2.2361   2.0000   2.2361   2.8284
```

The $THETA$ variable is never used; however, it gives the polar angle in radians:

```
>> THETA
THETA =
   -2.3562   -2.0344   -1.5708   -1.1071   -0.7854
   -2.6779   -2.3562   -1.5708   -0.7854   -0.4636
    3.1416    3.1416        0        0        0
    2.6779    2.3562    1.5708    0.7854    0.4636
    2.3562    2.0344    1.5708    1.1071    0.7854
```

Finally, you're ready to generate the main script called mach_illusion.m to visualize how the Mexican hat function/center-surround receptive field of the neurons in the early visual system could affect your perception. In this simple model, the two-dimensional convolution of the input image matrix (generated by the ramp.m M-file) with the receptive field filter (generated by the mexican_hat.m M-file) gives an approximation to how the brightness of the image is perceived when filtered through the early visual system. This operation should result in a dip in the brightness perceived at the point where the brightness of the input just begins to increase and a peak in the brightness perceived at the point

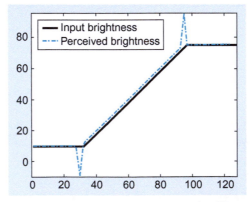

**FIGURE 16.5**    The Mach band illusion generated using the Mexican hat filter on the ramp input.

where the brightness of the input just stops increasing and returns to a steady value, consistent with the perception of Mach bands (see Figure 16.5). For a first pass, use the two-dimensional convolution function, **conv2**, that is built into MATLAB. As described in detail in the help section, this function will output a matrix whose size in each dimension is equal to the sum of the corresponding dimensions of the input matrices minus one. The edges of the output matrix are usually not considered valid because the value of those points have some terms contributing to the convolution sum which involved zeros padded to the edges of the input matrix. One way to deal with the problem of such edge effects is to reduce the size of the output image by trimming the invalid pixels off the border. You accomplish this by including the option **'valid'** when calling the **conv2** function:

```
%mach_illusion.m
clear all; close all
mexican_hat %creates the Mexican hat matrix, MH, & plots
ramp %creates image with ramp from dark to light, In, & plots
A = conv2(In,MH,'valid'); %convolve image and Mexican hat
figure; imagesc(A); colormap(bone) %visualize the "perceived" brightness
%create plot showing the profile of both the input and the perceived brightness
figure; plot(In(32,:),'k','LineWidth',5); axis([0 128 −10 95])
hold on; plot(A(32,:),'b-.','LineWidth',2); set(gca,'fontsize',20)
lh = legend('input brightness','perceived brightness',2); set(lh,'fontsize',20)
```

Make sure that the mexican_hat.m and ramp.m M-files are in the same directory as the mach_illusion.m M-file. Note that the size of the output is indeed smaller than the input:

```
>> size(A)
ans =
    60   124
```

For fun, you can learn more about how the convolution works by changing the **'valid'** option in the **conv2** function call to either **'full'** or **'same'** and see how the output matrix $A$ changes. One way to minimize the edge effects of convolution is to pad the input matrix with values that mirror the edges of the input matrix before performing the two-dimensional convolution and returning only the valid part of the output, which will now be the size of the original input matrix. The function **conv2mirrored.m** will do just this trick. It has been written in a generic form to accept matrices of any size:

```
%conv2_mirrored.m
function sp = conv2_mirrored(s,c)
% 2D convolution with mirrored edges to reduce edge effects
% output of convolution is same size as leading input matrix
[N,M] = size(s);
[n,m] = size(c); %% both n & m should be odd
%
% enlarge matrix s in preparation for convolution with matrix c
```

```
%via mirroring edges to reduce edge effects.
padn = round(n/2) - 1;
padm = round(m/2) - 1;
sp = [zeros(padn,M + (2*padm)); zeros(N,padm) s zeros(N,padm); zeros(padn,M +
(2*padm))];
sp(1:padn,:) = flipud(sp(padn + 1:2*padn,:));
sp(padn + N + 1:N + 2*padn,:) = flipud(sp(N + 1:N + padn,:));
sp(:,1:padm) = fliplr(sp(:,padm + 1:2*padm));
sp(:,padm + M + 1:M + 2*padm) = fliplr(sp(:,M + 1:M + padm));
%
% perform 2D convolution
sp = conv2(sp,c,'valid');
```

---

### EXERCISE 16.1

Put the figures generated by the mach_illusion.m script into a document, explain each figure, and give a short summary of the Mach band illusion as you understand it.

---

### EXERCISE 16.2

Rather than **cumsum**, you could have also used the function **meshgrid** to efficiently ramp up the brightness values from dark to light when creating the matrix *In*. Read the help file for **meshgrid** and rewrite the ramp.m script using **meshgrid** rather than **cumsum**.

---

### EXERCISE 16.3

Create the function **conv2_mirrored.m** using the code provided previously and place it in the same directory as your other files. Learn how the mirroring of the edges of the input matrix is accomplished by reviewing the help files on the functions **flipud** and **fliplr**. What determines the size of the mirrored-edge padding necessary and why? Rewrite your main script mach_illusion.m to use this convolution function rather than the **conv2** function. Check that your output matrix *A* is now the same size as the input matrix *In*.

---

### EXERCISE 16.4

Change the slope of the ramp without changing the beginning or ending values of the input image. [*Hint:* The command **linspace** can be useful to find values of the ramp that will go from 10 to 75 in, say, 30 steps rather than 65: **linspace(10,75,30)**.] How does increasing or decreasing the slope affect the strength of the illusion?

---

### EXERCISE 16.5

Convert the M-file named mexican_hat.m into a function where the inputs are the size of the matrix, the ratio of excitation to inhibition, the variance of excitatory and inhibitory Gaussians, and the overall strength of the filter. Also make the appropriate changes to the main script that calls this function, mach_illusion.m.

---

## 16.4 PROJECT

The receptive fields of simple cells in V1 reflect the orientation and spatial frequency preference of the neurons. One way to model this is to use the Gabor function, which is basically a two-dimensional Gaussian modulated by a sinusoid, as shown in Figure 16.6.

1. Observe how the receptive fields of simple cells in V1 modeled as Gabor functions with various spatial frequency and orientation preferences filter an image of a rose, which

FIGURE 16.6 A Gabor function modeling the oriented receptive field of a V1 neuron.

can be downloaded from the companion web site. Create two files, the gabor_filter.m function and the gabor_conv.m script (using the following code), in the same directory as the conv2_mirrored.m file. Also, place the rose.jpg image file in the same directory. Now, run the gabor_conv.m script. It will take a convolution between the rose image with a Gabor function of a given orientation (OR) and spatial frequency (SF). The input parameters OR and SF will determine the orientation and spatial frequency of the filter. Thus, you will essentially "see" how simple cells in V1 with a given orientation and spatial frequency preference perceive an image. Try values of SF = 0.01, 0.05, and 0.1, and OR = 0, pi/4, and pi/2. Put the resulting figures into a document and explain the results. Try changing the Gabor filter from an odd filter to an even filter by using **cos** instead of **sin**. How does this affect the output?

```
% gabor_filter.m
function f = gabor_filter(OR, SF)
% Creates a Gabor filter for orientation and spatial frequency
% selectivity of orientation OR (in radians) and spatial frequency SF.
%
% set parameters
sigma_x = 7;% standard deviation of 2D Gaussian along x-dir
sigma_y = 17;% standard deviation of 2D Gaussian along y-dir
%
% create filter
[x,y] = meshgrid(-20:20);
X = x*cos(OR) + y*sin(OR); %rotate axes
Y = -x*sin(OR) + y*cos(OR);
f = (1/(2*pi*sigma_x*sigma_y)).*exp(-(1/2)*(((X/sigma_x).^2) + ...
((Y/sigma_y).^2))).*sin(2*pi*SF*X);
```

```
%gabor_conv.m
clear all; close all
I = imread('rose.jpg');
OR = 0; SF = .01;
G = gabor_filter(OR,SF);
figure
subplot(1,3,1); imagesc(G); axis square; colorbar; title ('Gabor function')
subplot(1,3,2); imagesc(I); title('original image')
subplot(1,3,3); imagesc(conv2_mirrored(double(I),G));
colormap(bone); title(['Convolved image OR = ',num2str(OR),' SF = ', num2str(SF)])
```

2. Now you can have some fun times with image processing and convolutions. Choose any image and convolve it with a function or filter of your choosing. To avoid edge problems, you can use the conv2_mirrored function provided, the **conv2** function with the **'valid'** option (as in the exercises), or you can use the function **imfilter** from the Image Processing Toolbox built into MATLAB, which does an operation similar to convolution. You can create your own filter or choose a predesigned filter in MATLAB using the **fspecial** function, also from the Image Processing Toolbox. You can learn more about these functions through the online help. Hand in your code and picture of before and after the filtering, along with an image of the filter used in the convolution.

# MATLAB FUNCTIONS, COMMANDS, AND OPERATORS COVERED IN THIS CHAPTER

**imagesc**
**colormap**
**cumsum**
**meshgrid**
**cart2pol**
**conv2**
**flipud**
**fliplr**

# Neural Data Analysis I: Encoding

## 17.1 GOALS OF THIS CHAPTER

The primary goal of this chapter is to introduce you to the fundamental methods of analyzing spike trains of single neurons used to characterize their encoding properties: raster plots, peri-event time histograms, and tuning curves. While there are prepackaged tools available for these methods, in this chapter you will program these tools yourself and use them to analyze behavioral data recorded from a motor area of a macaque monkey.

## 17.2 BACKGROUND

In general, neuroscientists are interested in knowing what neurons are doing. More specifically, neuroscientists are often interested in neural *encoding*—how neurons represent stimuli from the outside world with changes in their firing properties. Let's say you are studying a neuron from a visual area. You would first present a research participant with controlled visual stimuli with a number of known properties—orientation, luminance, contrast, etc. Using standard electrophysiological techniques, you then record the response of the neuron to each stimulus. You can repeat the presentation of a given stimulus and then see how similar (or different) the neuronal responses are. A *raster plot* is a simple method to visually examine the trial-by-trial variability of these responses. You can examine what features these responses have in common by averaging over all responses to create a *peri-event time histogram*. Finally, to capture how the average response of the neuron varies with some sensory feature, you can generate a *tuning curve* that maps the feature value onto the average response of the neuron.

*MATLAB® for Neuroscientists.*
DOI: http://dx.doi.org/10.1016/B978-0-12-383836-0.00019-9

# 17.3 EXERCISES

## 17.3.1 Raster Plot

Because action potentials are stereotyped events, the most important information they carry is in their timing, as opposed to their size or shape. A raster plot replaces each action potential with a tick mark that corresponds to the time where the raw voltage trace crosses some threshold.

Load the dataset for this chapter from the companion web site. Contained within that dataset is a variable *spike*, which contains the firing times (in seconds) of a single neuron for 47 trials of the same behavioral task. Here, you are examining a recording from a cell in the motor cortex, and the task involves moving the hand from the same starting position to the same ending position. For each trial, the spike times are centered so that the start of movement coincides with a timestamp of 0 seconds. Because the neuron did not fire the same number of times for each trial, the data are stored in a *struct*, which is a data structure that can bundle vectors (or matrices) of different lengths. To access the spike times for the first and second trials, type

```
t1 = spike(1).times;
t2 = spike(2).times;
```

If you look at the workspace, you can verify that the vectors *t1* and *t2* are not the same length. Now plot the first trial as a raster (remember, you don't have to type the comments marked with "%"):

```
figure                          %Create a new figure
hold on                         %Allow multiple plots on the same graph
for ii = 1:length(t1)           %Loop through each spike time
    line([t1(ii) t1(ii)], [0 1])   %Create a tick mark at x = t1(ii) with height of 1
end
ylim([0 5])                     %Reformat y-axis for legibility
xlabel('Time (sec)'); ylabel('Trial #')
```

Even when you're looking at one trial, it appears that the neuron fires sparsely at first but then ramps up its firing rate a few hundred milliseconds before the start of movement. Now plot the next trial:

```
for ii = 1:length(t2)
    line([t2(ii) t2(ii)], [1 2])
end
```

Your results should look like those in Figure 17.1.

The relationship between the firing rate and start of movement is not nearly as clear in the second trial as in the first trial. However, in this chapter's final project, you will want to visualize data from all trials at once. One way is to simply write a loop to plot the raster for each trial as above. Another way is to take advantage of a built-in MATLAB® function called **histc**. This simply computes a histogram, meaning it counts how many values in a vector fall in within a discrete set of intervals, or bins. If we select a small enough bin width (say 5 ms), it will be very unlikely that we will have more than one spike in a given

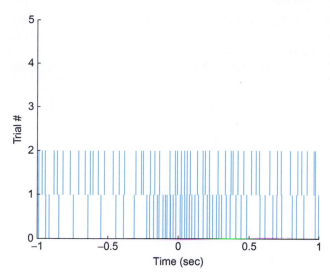

FIGURE 17.1 A raster plot of spike times
of the sample neuron for the first two trials.

bin, so we can convert our collection of spikes times into a matrix of zeros and ones, indicating whether or not a spike is present for a given trial in a given time range. We can then use the image plotting function **imagesc** (introduced in Chapter 16, "Convolution") to plot that matrix. This is less precise than plotting each spike time as a line as before, but it serves for most purposes. Try the code below:

```
raster = zeros(47,401);              %Initialize raster matrix
edges = [-1:.005:1];                 %Define bin edges
for jj = 1:47                        %Loop over all trials
                                     %Count # of spikes in each bin

    raster(jj,:) = histc(spike(jj).times,edges);
end
figure                               %Create figure for plotting
imagesc( ~raster )                   %'~' inverts 0s and 1s
colormap('gray')                     %Zero plotted as black, one as white
```

## 17.3.2 Peri-Event Time Histogram

The raster shows us the trial to trial variability, but it would also be nice to see what the response of an "average trial" looks like. This average neural response is captured by the *peri-event time histogram*, which is abbreviated PETH. *Peri-event* means that all the trials are centered relative to some relevant event—in this case, the start of movement. If our data were from a sensory array, the relevant event would be whatever stimulus we presented. This is why a PETH is sometimes also referred to as a *peri-stimulus time histogram*, or PSTH. However, in a motor system, where neural firing precedes the event we measure, this term is a little awkward, so we will stick with the more general term — peri-event time histogram.

*Time histogram* means you divide the time period into a series of bins (0 to 100 ms, 100 to 200 ms, etc.) and count how many spikes fall in each bin for all trials. Luckily, we just saw that MATLAB has a function that makes this easy: **histc**. To look at all trials, you will initialize the PETH with zeros and then sequentially add each trial's results. Try the following:

```
edges = [-1:0.1:1];            %Define the edges of the histogram
peth = zeros(21,1);            %Initialize the PETH with zeros
for jj = 1:47                  %Loop over all trials
                               %Add current trial's spike times

    peth = peth + histc(spike(jj).times,edges);
end
bar(edges,peth);              %Plot PETH as a bar graph
xlim([-1.1 1])                %Set limits of X-axis
xlabel('Time (sec)')          %Label x-axis
ylabel('# of spikes')         %Label y-axis
```

Your results should look like those in Figure 17.2.

Now the pattern in neuronal activity is clear: the firing rate begins to increase about half a second before movement start and returns to baseline by half a second after movement start. Of course, for the y-axis to indicate firing rate in spikes per second, you would need to divide each bin's spike count by both the bin width and the number of trials.

### 17.3.3 Tuning Curves

Many neurons respond preferentially to particular values of a stimulus. Typically, this activity gradually falls off from a maximum (corresponding to the preferred stimulus) along some stimulus dimension (e.g., orientation, direction). By plotting the stimulus

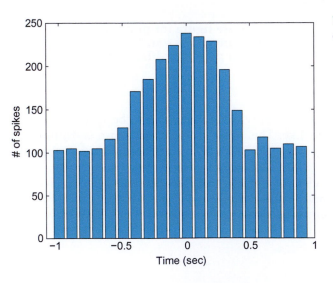

**FIGURE 17.2**   A peri-event time histogram centered on the start of movement.

dimension on the x-axis and the neural activity (typically a firing rate) on the y-axis, you can determine the preferred stimulus of a neuron. Figure 17.3 shows a tuning curve of a neuron from area MT, which is a part of the visual cortex that aids in the perception of motion. As you can tell, the neuron prefers upward motion (motion toward 90°).

## 17.3.4 Curve Fitting

Typically, tuning curves like this are fit to a function such as a Gaussian curve or a cosine function. Because all measurements made in the real world come with errors, it is usually impossible to describe empirical data with a perfect functional relationship. Instead, you fit data with a curve that represents a model of the proposed relationship. If this curve fits the data well, then you conclude that your model is a good one.

The simplest relationship you will typically look for is a linear one. Many neurons are thought to encode stimuli linearly. For example, ganglion cells in the limulus (horseshoe crab) increase their firing rate linearly with luminance of a visual stimulus (Hartline, 1940). You can simulate this relationship as follows:

```
x = 1:20;            %Create a vector with 20 elements
y = x;               %Make y the same as x
z = randn(1,20);     %Create a vector of random numbers
y = y + z;           %Add z to y, introducing random variation
plot(x,y, '.' )      %Plot the data as a scatter plot
xlabel('Luminance')
ylabel('Firing rate')
```

MATLAB contains prepackaged tools for fitting linear relationships. Just click on the figure, select Tools, and then select Basic Fitting. Check the boxes for Linear and Show

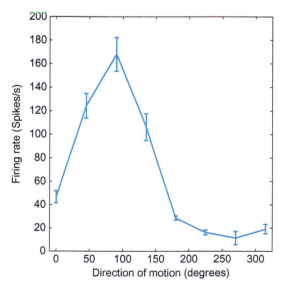

FIGURE 17.3 A tuning curve for a neuron from area MT.

equations, and you will see the line and equation that best fit your data. However, you might also like to be able to do this yourself. The command in MATLAB to fit data to a polynomial is **polyfit**. For example:

**p = polyfit(x,y,1)  %Fits data to a linear 1ˢᵗ degree polynomial**

The first value in *p* is the slope and the second value is the y-intercept. If you plot this fitted line, your result should be similar to Figure 17.4:

```
hold on                  %Allows 2 plots of the same graph
yFit = x*p(1) + p(2);    %Calculates fitted regression line
plot(x,yFit)             %Plots regression
```

Because MATLAB has a number of curve-fitting functions, there are a number of ways to perform this regression. One function worth mentioning is the function **regress**, because this can perform multiple linear regression, where a dependent variable is a function of a matrix of multiple independent variables. For this example, we assume the firing rate is a function of the luminance plus some baseline firing rate. We simply need to bundle the luminance with a vector of ones (representing the baseline) before performing the regression. In the code below, note the use of the **transpose** function (using the apostrophe as a shortcut) to convert row vectors into column vectors.

```
predictor = [x' ones(20,1)];  %Bundle predictor variables together into a matrix
p = regress(y',predictor)     %Perform regression
yFit = predictor*p;           %Calculate fit values
```

Now you will fit data to a more complicated function—a cosine. First, generate some new simulated data:

```
x = 0 : 0.1 : 30;    %Create a vector from 0 to 10 in steps of 0.1
y = cos (x);         %Take the cosine of x, put it into y
z = randn(1,301);    %Create random numbers, put it into 301 columns
y = y + z;           %Add the noise in z to y
figure               %Create a new figure
plot (x,y)           %Plot it
```

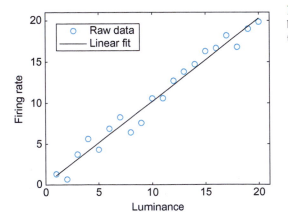

FIGURE 17.4 A linear fit of the relationship between the firing rate of a simulated ganglion cell and the luminance of the stimulus.

MATLAB does not have a built-in function for fitting this to a cosine-tuning function, but it does have a nonlinear curve-fitting function: **nlinfit**. You will need to specify the details of the fit. Here, you will use a cosine function with the y-offset, amplitude, and phase as free parameters. You can define this function "inline," which means it can be used by other functions in MATLAB in the same session or M-file.

Type this command to define a generic cosine function:

**mystring = 'p(1) + p(2) * cos ( theta - p(3) )'; %Cosine function in string form**

Here, *p(1)* represents the y-offset; *p(2)*, the amplitude; and *p(3)*, the phase. You can assume the frequency is 1. Now enter the following:

**myfun = inline ( mystring, 'p', 'theta' ); %Converts string to a function**

This function accepts angles *theta* and parameter vector *p* and transforms them using the relationship stored in *mystring*.

**p = nlinfit(x, y, myfun, [1 1 0] ); %Least squares curve fit to inline function "myfun"**

The first parameter of **nlinfit** is a vector of the x-values (the angle *theta* in radians). The second parameter is the observed y-values. The third parameter is the name of the function to fit, and the last parameter is a vector with initial guesses for the three free parameters of the cosine function. If the function doesn't converge, use a different initial guess. The **nlinfit** function returns the optimal values of the free parameters (sorted in *p*) that fit the data with the cosine function, as determined by a least squares algorithm.

Instead of defining a function inline, you can also save a function in an M-file. In that case, you will need to include an @ (at) symbol before the function name, which will allow MATLAB to access the function as if it were defined inline:

**p = nlinfit(x, y, @myfun, [1 1 0] ); %Least squares curve fit to function "myfun.m"**

You can use the inline function to convert the optimized parameters into the fitted curve. After plotting this, your result should look similar to Figure 17.5.

```
hold on                 %Allows 2 plots of the same graph
yFit = myfun(p,x);      %Calculates fitted regression line
plot(x,yFit,'k')        %Plots regression
```

We introduced the **nlinfit** function because it can be used to fit any arbitrary relationship you are interested in, whether it is linear or not. However, there are a couple of drawbacks to using to fitting a cosine-tuning function. The preferred direction isn't necessarily restricted to a reasonable range (say, from $-\pi$ to $\pi$). Worse, the value may be off by $\pi$ if the amplitude is found to be negative. A solution is possible because the cosine-tuning function can be reformulated as linear regression of the sine and cosine of the movement direction (review your trigonometric identities to see why). So you can again use the multiple regression function **regress** to find the preferred direction:

```
predictor = [ones(301,1) sin(x)' cos(x)'];  %Bundle predictor variables
p = regress(y',predictor)                   %Linear regression
yFit = predictor*p;                         %Calculate fit values
theta = atan2(p(2),p(3));                   %Find preferred direction from fit weights
```

## 17.4 PROJECT

The data that you will use for your project were recorded from the primary motor cortex (abbreviated MI) of a macaque monkey (data courtesy of the Hatsopoulos laboratory). MI is so named because movements can be elicited by stimulating this area with a small amount of electricity. It has also been shown that MI has direct connections to the spinal cord. MI plays an important role in the control of voluntary movement (as opposed to reflexive movements). This doesn't mean that MI directly controls movement, because other areas in the basal ganglia and the brainstem are important as well. Animals with a lesioned MI can still make voluntary movements, albeit less dexterously than before. However, it is clear that MI contains information about voluntary movement, usually a few hundred milliseconds before it actually happens. There is also a somatotopic map in MI, meaning that there are separate areas corresponding to face, arm, leg, or hand movements. These data are recorded from the arm area.

The behavioral data were collected using a manipulandum, which is an exoskeleton that fits over the arm and constrains movement to a 2D plane. Think of the manipulandum as a joystick controlled with the whole arm. The behavioral task was the center-out paradigm pioneered by Georgopoulos and colleagues (1982). The animal first holds the cursor over the center target for 500 ms. Then a peripheral target appears at one of eight locations arranged in circle around the center target. In this task there is an instructed delay, which means that after the peripheral target appears, the animal must wait 1000–1500 ms for a go cue. After the go cue, the animal moves to and holds on the peripheral target for 500 ms, and the trial is completed.

There are two interesting time windows here. Obviously, MI neurons should respond during a time window centered around the go cue, since this is when voluntary movement begins. However, MI neurons also respond during the instructed delay. This result is somewhat surprising because the animal is holding still during this time. The usual interpretation is that the animal is imagining or preparing for movement to the upcoming target. This means that MI is involved in planning as well as executing movement.

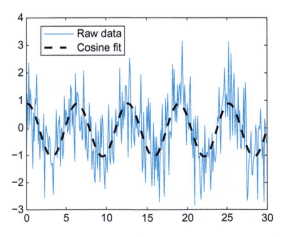

FIGURE 17.5    A nonlinear fit of a simulated, noisy cosine relationship.

If you treat the direction to the peripheral target as the "stimulus," you can arrange the neuronal responses in a tuning curve. These can be described with the same cosine curve used before, where the phase of the fitted cosine corresponds to the preferred direction of the neuron.

In this dataset, the neuronal spiking is stored in a struct called *unit*. Information for unit #1 is accessed with *unit(1)*. Spike times are stored in *unit(1).times*. There are three more important variables: the instruction cue times are stored in *instruction*, the go cue times are stored in *go*, and the direction of peripheral target is stored in *direction* (1 corresponds to 0 degrees, 2 corresponds to 45 degrees, etc.).

In this project, you are asked to do the following:

1. Make raster plots and PETHs for all the neurons for both time periods: instruction cue to 1 second afterward, and 500 ms before the movement onset to 500 ms afterward. Which neurons are the most responsive? Print out a few examples. Do you think the PETHs are a good summary of the raster plots? How does the time course of the responses differ between the two time periods?
2. Create tuning curves and fit a cosine tuning curve to the firing rates of all neurons for each time period. Report the parameters of the fit for each neuron and save this information for later chapters. How good of a description do you think the cosine curve is? Do the tuning curves differ between the two time periods? If so, why do you think this is?

Figures 17.6 and 17.7 show examples of what your results might look like. The locations of the smaller plots correspond to the locations of their associated peripheral targets. Here, a timestamp of 0 corresponds to the start of movement. You can use the command **subplot** to subdivide the plotting area. For example, the command **subplot(3,3,i)** makes the *i*th square in a 3 × 3 grid the active plotting area.

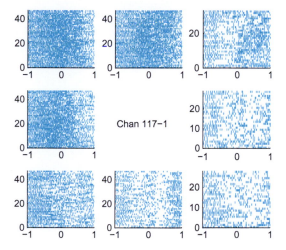

**FIGURE 17.6** An example of a full raster plot for the first neuronal unit recorded from electrode #117.

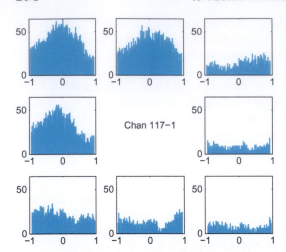

**FIGURE 17.7** An example of a full peri-event time histogram for the first neuronal unit recorded from electrode #117.

# MATLAB FUNCTIONS, COMMANDS, AND OPERATORS COVERED IN THIS CHAPTER

**histc**
**randn**
**bar**
**polyfit**
**regress**
**nlinfit**
**subplot**

# Neural Data Analysis II: Binned Spike Data

## 18.1 GOALS OF THIS CHAPTER

Previously, you used a simple linear encoding model to predict a neuron's firing rate, obtained by averaging over many trials for a range of stimuli. However, linear models have limitations for modeling observed neural data. In this chapter, we will examine a simple nonlinear encoding model which can model discrete, non-negative data obtained by counting the number of spikes occurring in a given time bin. Luckily, the nonlinear encoding model introduced here can be fit using a built-in function available in MATLAB®.

## 18.2 BACKGROUND

In the last chapter, you saw that a cosine tuning model does a good job of describing the firing rate (averaged across many trials) of a neuron as a function of direction. In particular, you saw that neurons have a baseline firing rate, a preferred direction where the firing rate is at its maximum, and an anti-preferred direction where the firing is at its minimum. When the cosine tuning model is expressed as a sum of a sine and cosine (see the end of Section 17.3.4), then the model can be fit using linear regression and the MATLAB function **regress**. Linear regression may not always be the best choice, as it makes certain assumptions about the raw data. For example, linear regression assumes data are continuous. However, our data consists of timestamps, and our binned data are counts of the number of spike timestamps that fall within a given bin. What does this mean? To start with, we know that a count is always non-negative.

Load the data for this chapter. This data was also collected during the eight-target center-out task, and the data is formatted the same as the data from Chapter 17. Start by

*MATLAB® for Neuroscientists.*
DOI: http://dx.doi.org/10.1016/B978-0-12-383836-0.00020-5

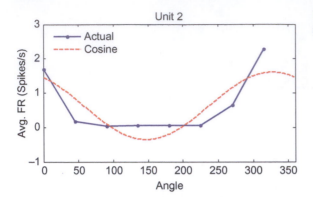

FIGURE 18.1 The cosine tuning model poorly fits data from a sparsely firing neuron.

constructing the empirical tuning curve for neuron #2, as we did in Chapter 17, and fit it to the standard cosine tuning model. If you do this, you will end up with a plot like Figure 18.1. The interesting thing about this neuron is that it has a lower baseline firing rate than previous neurons we have examined. The neuron does increase its firing rate for its preferred direction (around 315 degrees), but because it cannot fire less than zero spikes per second, it fires minimally for a wide range of directions (90, 135, 180, and 225 degrees). When the standard cosine tuning model is fit to this data, it does not fit the peak firing rate well, and it predicts a negative firing for the anti-preferred direction of 135 degrees. Obviously, this is not an adequate model for this neuron.

### 18.2.1 Exponential Function

What we need is a function which can only predict nonnegative firing rates. One such function is the exponential function, $e^x$. If $x$ is one, the value of this function is simply $e$, or approximately 2.718. If $x$ is zero, the function is 1, and as $x$ becomes more negative, the function will approach but never reach zero. In MATLAB, the exponential function is **exp**. To deal with low-firing neurons, we simply need to apply the exponential function to the cosine tuning model we used in the last chapter:

**mystring = 'exp( p(1) + p(2)\*cos(theta/180\*pi-p(3)) )';**

The mean firing rate can be fit as before, and now the prediction is quite good (see Figure 18.2).

However, we are doing something a little odd here by averaging the firing rates before fitting the function. It means that we are weighting data from each direction equally, even though the number of trials completed in each direction is not exactly equal (there are 71 trials for direction 4, but 48 for direction 5). A better approach would be to fit the raw data from each single trial, which consists of counts of how many times the neuron spiked in certain time windows. Thus the raw data is nonnegative and discrete. Using linear regression on discrete data is usually not appropriate, because linear regression assumes that the data is continuous, and that the relationship is disturbed by Gaussian noise.

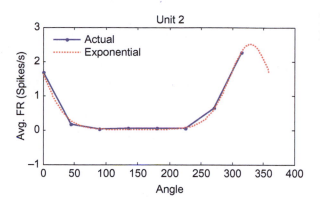

FIGURE 18.2 An exponential cosine model provides a better fit to this neuron's data.

FIGURE 18.2 An exponential cosine model provides a better fit to this neuron's data.

### 18.2.2 Poisson Distribution

A better choice is to assume that a discrete count follows a Poisson distribution. This is a simple but useful discrete distribution, which is good for modeling the number of events which occur in small amount of time or space. For example, the Poisson distribution might be used to model how many phone calls a call center might receive in an hour, or the number of grass seeds which sprout in a small patch of earth. Here, we use it to model the number of spikes detected in a given time bin. The use of the Poisson distribution for modeling spike trains is discussed in more detail in Chapter 33. We will compare the distribution of spikes we actually observe to what is expected with the Poisson distribution. We first need to collect the raw spike count in a 2-second window centered on the go cue for each trial:

```
neuronNum = 2;                                        %Select which neuron we want
numT = length(direction);                             %Count number of trials
spikeCount = zeros(numT,1);                           %Initialize count vector
for ii = 1:numT
  centerTime = go(ii);                                %Find go cue for given trial
  allTimes = unit(neuronNum).times-centerTime;        %Center spike times on go
  spikeCount(ii) = sum(allTimes > -1 & allTimes < 1); %2 seconds window
end
```

We now have a vector of the raw spike counts for each trial. Take a look at the distribution of spike counts for direction two:

```
dirNum = 2;                              %Select the direction we want
indTemp = find(direction == dirNum);     %Find appropriate trials
spikeTemp = spikeCount(indTemp);         %Pick out counts
edges = [0:4];                           %Bin edges
b = histc(spikeTemp,edges);              %Make histogram
bar(edges,b,'g')                         %Plot histogram
```

To compare this to the Poisson distribution, we can use the built-in MATLAB function **poisspdf,** which gives the probability mass function for the Poisson distribution. This gives the probability that a Poisson random variable is equal to a given count. The Poisson only takes the mean as a parameter, unlike the Gaussian distribution, which takes the mean and the variance. This is because the Poisson distribution has the property that its mean is equal to its variance. To see what the histogram should be if our neural data followed a Poisson distribution, we evaluate **poisspdf** with a mean that matches our neural data, and multiply this by the number of observations to get the expected number of counts.

```
y = poisspdf(edges,mean(spikeTemp));   %Match mean of Poisson dist. to data
yCount = y*length(indTemp);            %Multiply by number of trials
hold on
plot(edges,yCount,'r.-')
```

As shown in Figure 18.3, the Poisson distribution matches the actual data fairly well.

### 18.2.3 Log-Linear Models

In linear regression, the assumption is that the dependent variable $\mu$ (the firing rate) is equal to the linear function of the predictor variables (cosine and sine of direction). In matrix format, the firing rate is the product of the matrix of predictor variables $X$ and a vector of coefficients $b$ (see equation below). We previously found this vector of coefficients using the MATLAB function **regress**.

$$\mu = Xb \tag{18.1}$$

While Poisson-distributed data should not be fit using a linear model, they can be fit using a generalized linear model (GLM). Luckily, the procedure for fitting a GLM is already built into MATLAB with the function **glmfit**. GLM still predicts the firing rate using a linear function of the predictor variables, but since the mean of the Poisson distribution must be nonnegative, this prediction must be transformed. We have already seen that this transformation can be accomplished with the exponential function, and in fact in GLM the mean of Poisson-distributed data are expressed as the exponential of a predictor matrix $X$ times the coefficients $b$. Equivalently, the natural logarithm of the firing rate is expressed as a linear function of predictor variables (see Equation 18.2). Because of this, the model used in Poisson regression is referred to as a log-linear model. In GLM, the

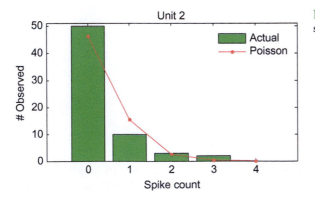

FIGURE 18.3　The Poisson distribution models sparse spike count data well.

natural logarithm is known as the "link function," tying the data to the linear function of the observed variables. If the data follows a different distribution (such as the binomial distribution), GLM requires a different link function.

$$\mu = \exp(Xb) \quad \log(\mu) = Xb \tag{18.2}$$

How do we apply this to our data? Instead of fitting the exponential cosine model to the mean firing rates (as in Section 18.2.1), we will fit a model to the raw spike counts and matrix of the predictor variables.

```
rad = [0:pi/4:2*pi-pi/4]';              %Match direction # with radians
                                        %Matrix of predictors variables:
predictor = [ones(numT,1) sin(rad(direction)) cos(rad(direction))];
coeff = glmfit(predictor(:,2:3),spikeCount,'poisson');   %Fit log-linear model
spikeFit = exp(predictor*coeff);        %Predict firing rate.
```

One difference from the **regress** function is that **glmfit** will automatically add a coefficient for a constant term. This means that vector of 1 does not need to be included in **glmfit** (which is why the predictors are supplied as "predictor(:,2:3)"), but it does need to be included for the prediction of the firing rate.

## 18.2.4 Predicting the PETH

In this case, the tuning curve obtained by fitting the exponential cosine tuning model directly to the data is very similar to that fit to the averaged firing rates (shown in Figure 18.2). That is because the predictor variable was identical across multiple trials, which meant that averaging first was not a completely unreasonable step. However, if the predictor variable is different across trials, averaging first may not be appropriate.

For example, the exact path taken by the neuron and the speed it travels will vary slightly from trial to trial, even for reaches to the same target. Thus far, we have just tried to predict the activity of the neuron over a large time window. However, in the last chapter, we visualized the peri-event time histograms (PETHs) of different neurons, and there was a clear temporal evolution of the neural activity not accounted for in the original cosine tuning model described by Georgopoulos and colleagues (1982).

This cosine-tuning model was extended by Moran and Schwartz (1999), who said that the current firing rate of the neuron is related to the sine and cosine of both the direction and the speed at a fixed time into the future. They used a linear model, but we will use a log-linear model and simplify their equation to state that the log of the current firing rate $D$ is a linear function of the $X$ and $Y$ velocity ($V_X$, $V_Y$) at a fixed time $\tau$ in the future. Moran and Schwartz also included a non-directional speed term, but that will be covered in the exercises. Because speed profiles are bell-shaped, this function can capture the gradual rise and fall in activity seen in the example PETH in Figure 17.7.

$$\log[D(t - \tau)] = b_0 + b_1 V_X(t) + b_2 V_Y(t) \tag{18.3}$$

In the dataset for this chapter, there are new variables which were not present in the Chapter 17 dataset. First, **binned** contains the spike count of all 158 neurons in a sequence of 50 ms bins. The variable **time** contains the timestamp in seconds at the end of each bin. Finally, **kin** contains a variety of kinematic variables, partitioned in the same 50 ms bins. For example, **kin.x** contains the $X$ hand position as a function of time, and **kin.xvel** contains the $X$ hand velocity.

To fit an encoding model, we just need to pull out a given neuron's spike count from the variable **binned**, and then pull out the kinematics we are interested in from **kin.** For now, we will assume our neuron leads velocity by 100 ms, or 2 time bins, so we need to align the data so neural firing and future velocity are matched.

```
lag = 2;                                          %Assume neuron leads velocity by 2 bins
fr = binned(neuronNum,1:end-lag)';                %Pull out spike counts
numBins = length(fr);                             %Number bins, accounting for lag
                                                  %Bundle future predictor variables
predictor = [ones(numBins,1) kin.xvel(lag + 1:end) kin.yvel(lag + 1:end)];
coeff = glmfit(predictor(:,2:end),fr,'poisson');  %Fit log-linear model
frFit = exp(predictor*coeff);                     %Predict spike count
```

You can now pull out the predicted spike counts centered on the go time, and average them across trials in the same direction to create a predicted PETH. We can then compare this to the empirical PETH we constructed using the code from Chapter 17. If you do this for neuron #2, you should come up with something like Figure 18.4.

## 18.3 EXERCISES

1. Implement the log-linear encoding model described in Equation 18.3, and plot the predicted PETH as shown in Figure 18.4. The Moran and Schwartz (1999) paper also included a non-directional speed term. You can access that in the variable **kin.speed**. Add speed as an extra term to your encoding model. Can you find a neuron where the predicted PETH is better if speed is included? Compare the correlation coefficient between the predicted and actual spike counts across all neurons with and without the speed term. Does adding speed seem to significantly improve the fit across neurons as a population?

2. We assumed a constant lag between neural firing. Instead, try several different lags for each neuron (from $+300$ ms to $-100$ ms, in 50 ms steps), and pick the best lag. What is the distribution of best lags across neurons? Compare the correlation coefficient between the predicted and actual spike counts across all neurons with a constant and a variable time lag. Does allowing a variable lag significantly improve the fit across neurons as a population?

## 18.4 PROJECT

Experiment with alternative encoding models. For example, another paper proposed that neurons encode a movement "pathlet," meaning that neurons encode velocity at several different time lags with different preferred directions (Hatsopoulos et al., 2007). Is

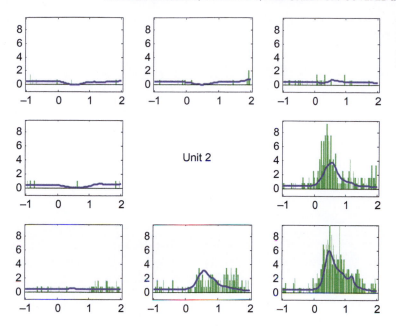

**FIGURE 18.4** The empirical PETH (green) for neuron #2, along with the velocity-encoding model's prediction (blue).

there evidence in this dataset that neurons encode velocity at different time lags? If so, how does the preferred direction change as a function of time?

## MATLAB FUNCTIONS, COMMANDS, AND OPERATORS COVERED IN THIS CHAPTER

**exp**
**poisspdf**
**glmfit**

# 19

# Principal Components Analysis

## 19.1 GOALS OF THIS CHAPTER

Previously, we explored how MATLAB® can be used to visualize neural data. This is a powerful tool. For example, a simple 2D tuning curve can demonstrate how a single neuron encodes a stimulus parameter in terms of a firing rate. However, it is not clear how this applies to multidimensional data. How do you represent stimulus encoding of a population of neurons or of a time-varying firing rate?

One solution is to try to compress data to make them easier to work with. If you can reduce the dimensionality to two or three dimensions, you can then use your visualization tools. In this chapter you will see how principal components analysis can be used to perform dimensionality reduction. You will also explore an application of this technique for spike-sorting neuronal waveforms. This will prepare you for the next chapter, where you will use principal components to capture the temporal aspects of a peri-event time histogram.

## 19.2 BACKGROUND

*Principal components analysis (PCA)* performs a linear transformation on data and can be used to reduce multidimensional data down to a few dimensions for easier analysis. The idea is that many large datasets contain correlations between the dimensions, so that a portion of the data is redundant. PCA will transform the data so that as much variation as possible will be crammed into the fewest possible dimensions. This allows you to compress your data by ignoring other dimensions. To apply PCA, you first need to understand how the correlations between dimensions can be described by a covariance matrix.

*MATLAB® for Neuroscientists.*
DOI: http://dx.doi.org/10.1016/B978-0-12-383836-0.00017-5

## 19.2.1 Covariance Matrices

You will start by analyzing some simulated 2D data. The MATLAB function **normrnd** can create zero-mean, Gaussian noise ("white noise"). You will use this to create two variables containing two columns, one for each dimension. In the first variable (*a*), the two dimensions will be uncorrelated, but in the second (*b*) there will be a significant correlation between the two dimensions. Use the following code to generate *a* and *b*:

```
n = 500;                        %n = number of datapoints
a(:,1) = normrnd(0,1,n,1);      %n random Gaussian values with mean 0, std. dev. 1
a(:,2) = normrnd(0,1,n,1);      %Repeat for the 2nd dimension.
b(:,1) = normrnd(0,1,n,1);      %n random Gaussian values with mean 0, std. dev. 1
%For b, the 2nd dimension is correlated with the 1st
b(:,2) = b(:,1)*0.5 + 0.5*normrnd(0,1,n,1);
```

If you plot the columns of these variables against one another (using **plot((a(:,1),a(:,2),'.')** for *a*), the data should look something like Figure 19.1.

You should already be familiar with the concepts of *mean, variance*, and *standard deviation*. If the sample data consist of *n* observations stored in a vector *x*, the sample mean is defined as follows:

$$\bar{x} = \frac{1}{n}\sum_{i=1}^{n} x_i \tag{19.1}$$

The variance ($\sigma^2$) of the sample is simply the expected value (mean) of the squared deviations from the sample mean:

$$\sigma^2 = \frac{1}{n}\sum_{i=1}^{n} (x_i - \bar{x})^2 \tag{19.2}$$

The standard deviation ($\sigma$) is just the square root of the variance. Unfortunately, it turns out that using the preceding expression as an estimate of the sample variance might be a

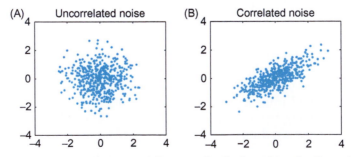

**FIGURE 19.1** Samples from a two-dimensional Gaussian distribution where the dimensions are uncorrelated (A) and correlated (B).

bad idea because this estimate is biased: it systematically underestimates the variance. However, it can be shown that the unbiased estimator is formed by replacing $n$ by $n-1$ in the first term. Thus, the *sample variance* ($s^2$) is usually defined as follows:

$$s^2 = \frac{1}{n-1} \sum_{i=1}^{n} (x_i - \bar{x})^2 \tag{19.3}$$

This is how the function **var** in MATLAB is defined. You can more compactly express the sample variance using matrix notation as follows:

$$s^2 = \frac{1}{n-1} (x - \bar{x})^T (x - \bar{x}) \tag{19.4}$$

The superscript $T$ signifies a transpose, whereby matrix columns are changed to rows and vice versa: an $m$ by $n$ matrix becomes an $n$ by $m$ matrix. In MATLAB, a transpose is designated with an apostrophe placed after the variable.

Let's compare the preceding formula with the function **var**. In the following code, do the two expressions give the same result?

```
var(a(:,1))                     %Compute sample variance of 1st dim of "a"
c = a(:,1)-mean(a(:,1));        %Subtract mean from 1st dim of "a"
c'*c/(n-1)                      %Compute sample variance of 1st dim of "a"
                                %Note the apostrophe denoting transpose(c)
```

The covariance is analogous to the variance, except that it is computed between two vectors, not a vector and itself. If you have a second data vector $y$ with $n$ independent values, then the *sample covariance* is expressed as follows:

$$\text{cov}(x, y) = \frac{1}{n-1} \sum_{i=1}^{n} (x_i - \bar{x})(y_i - \bar{y}) \tag{19.5}$$

You can see that if $x$ and $y$ are the same, the sample covariance is the same as the sample variance. Also, if $x$ and $y$ are uncorrelated, the covariance should be zero. A positive covariance means that when $x$ is large, so is $y$; while a negative covariance means that when $x$ is large, $y$ is small. The last thing you need to define is the covariance matrix. If the data have $m$ dimensions, then the covariance matrix is an $m$ by $m$ matrix where the diagonal terms are the variances of each dimension and the off-diagonal terms are the covariances between dimensions. If the additional dimensions are stored as extra columns in variable $x$ (so $x$ becomes an $n$ by $m$ matrix), then the sample covariance can be computed the same way as the sample variance:

$$\text{cov}(x) = \frac{1}{n-1} (x - \bar{x})^T (x - \bar{x}) \tag{19.6}$$

This is computed by the function **cov** in MATLAB. Compare the two methods by using the following code (the function **repmat** is used to create multiple copies of a vector):

```
cov(a)                  %Compute the covariance matrix for "a"
c = a-repmat(mean(a),n,1);   %Subtract the mean from "a"
c'*c/(n-1)              %Compute the covariance matrix for "a"
```

The covariance of the correlated noise should have large off-diagonal terms. One reason to compute the covariance is that it plays the same role in the multivariate Gaussian distribution as the variance plays in the univariate Gaussian. You can use the covariance and the function **mvnrand** (*mvn* stands for *multivariate normal*) in MATLAB to generate new multivariate correlated noise (*b2*). Plot *b2* on top of the correlated noise generated earlier (*b*). Did the covariance matrix adequately capture the structure of the data?

```
sigma = cov(b)                 %Compute the covariance matrix of b
%Generate new zero-mean noise with the same covariance matrix
b2 = mvnrnd([0 0],sigma,n);
```

## 19.2.2 Principal Components

Principal components analysis is essentially just a coordinate transformation. The original data are plotted on an $X$-axis and a $Y$-axis. For two-dimensional data, PCA seeks to rotate these two axes so that the new axis $X'$ lies along the direction of maximum variation in the data. PCA requires that the axes be perpendicular, so in two dimensions the choice of $X'$ will determine $Y'$. You obtain the transformed data by reading the $x$ and $y$ values off this new set of axes, $X'$ and $Y'$. For more than two dimensions, the first axis is in the direction of most variation; the second, in direction of the next-most variation; and so on.

How do you get your new set of axes? It turns out they are related to the eigenvalues and eigenvectors of the covariance matrix you just calculated (consult the mathematics tutorial in Chapter 3 for a review of what eigenvalues and eigenvectors are). In PCA, each eigenvector is a unit vector pointing in the direction of a new coordinate axis, and the axis with the highest eigenvalue is the axis that explains the most variation.

This concept may seem confusing, so start by looking at the correlated noise data (*b*) shown in Figure 19.1. You could make a decent guess at the principal components just by looking at the data: the first principal component line should fall on the long axis of the ellipse-shaped cluster. You can use the function **eig** in MATLAB to compute the eigenvectors and eigenvalues of the covariance matrix (**sigma**) you computed previously:

**[V, D] = eig(sigma)  %V = eigenvectors, D = eigenvalues for covariance matrix sigma**

This will output something like the following (because the noise was generated randomly, the exact values will vary):

```
V =
     0.5387    − 0.8425
   − 0.8425    − 0.5387
```

**D =**

| | |
|---|---|
| 0.2048 | 0 |
| 0 | 1.3341 |

The eigenvalues are stored on the diagonal of *D*, while the corresponding eigenvectors are the columns stored in *V*. Because the second eigenvalue is bigger, the second eigenvector is the first principal component. This means that a vector pointing from the origin to (−0.8445,−0.5387) lies along the axis of maximum variation in the data. Type the following to plot the new coordinate axes on the original data:

**plot(b(:,1),b(:,2),'b.'); hold on           %Plot correlated noise**
**plot(3*[-V(1,1) V(1,1)],3*[-V(1,2) V(1,2)],'k')  %Plot axis in direction of 1st eigenvector**
**plot(3*[-V(2,1) V(2,1)],3*[-V(2,2) V(2,2)],'k')  %Plot axis in direction of 2nd eigenvector**

This will produce a graph like the one in Figure 19.2.

Now you use these new coordinate axes to reassign the (*X,Y*) values to all your datapoints. First, you want to reorder the eigenvectors so that the first principal component is in the first row. Then you can simply multiply the data by this reordered matrix to obtain the new, transformed data. For example:

**V2(:,1) = V(:,2);  %Place the 1st principal component in the 1st row**
**V2(:,2) = V(:,1);  %Place the 2nd principal component in the 2nd row**
**newB = b*V2;    %Project data on PC coordinates**

If you plot these transformed data, it is clear that you just rotated the data so that most of the variation lies along the *X*-axis, as shown in Figure 19.3.

If you stop here, you haven't gained much, since the transformed data have just as many dimensions as the original data. However, if you wanted to compress the data, you can now just throw away the second column (the data plotted on the *Y*-axis in Figure 19.3). The whole

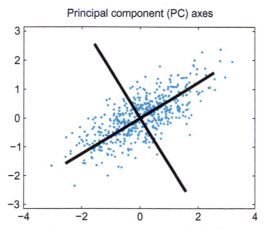

**FIGURE 19.2**   The first two principal component axes plotted with the original correlated data *b*.

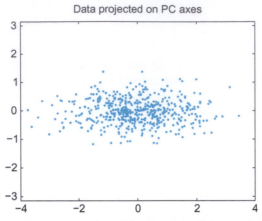

FIGURE 19.3    The correlated data *b* projected on the first two principal components.

point of PCA is that if you force as much of the variation as possible into a few dimensions, you can throw away the rest without losing much information.

How much variation can you capture by doing this? It turns out that the fraction of variation captured by each principal component is the ratio of its eigenvalue to the sum of all the eigenvalues. For the example, the first principal component has an eigenvalue of around 1.33, and the second principal component has an eigenvalue of around 0.20. That means if you keep just the first column of the transformed data, you still keep 87% of the variation in the data ($0.87 = 1.33/[1.33 + 0.20]$). That is, you can compress the size of the data by 50% but lose only 13% of the variation.

Conveniently, MATLAB already has a function that performs all these calculations in one fell swoop: **princomp**. Type the following to compute principal components for the correlated data:

**[coeff,score,latent] = princomp(b);   %Compute principal components of data in b**

The eigenvectors are stored in the variable **coeff**, the eigenvalues are stored in **latent**, and the transformed data (the old data projected onto the new PC axes) are stored in **score**. MATLAB even orders the eigenvectors so that the one with highest eigenvalue is first.

### 19.2.3  Spike Sorting

One common application of PCA is the spike sorting of neural data. Typically, a data acquisition system monitors a raw voltage trace. Every time the voltage crosses some threshold, the raw voltage is sampled during a time window surrounding this crossing to produce the recorded spike waveform. For example, in one commercially available data acquisition system (Cerebus system, Cyberkinetics Neurotechnology Systems, Inc.), each spike waveform consists of a 1600 μs section of the voltage trace sampled 30 times per millisecond for a total of 48 data points.

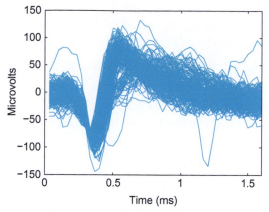

FIGURE 19.4 A plot of 200 extracellular action potential waveforms recorded from one electrode of a micro-electrode array implanted in the primary motor cortex of a macaque monkey.

Because any experimental system contains noise, the threshold crossing is often triggered by a chance deviation from the mean and not an action potential. Thus, after these recordings are made, the noise must be differentiated from the real spikes. You must also determine if the real spikes came from one or many neurons and then sort them accordingly. How do you compare waveforms? You can start by plotting them all on the same graph.

If you load the dataset for this chapter, you can use the code below to plot the first 200 waveforms from one electrode, as shown in Figure 19.4:

```
wf=session(2).chan48;  %Load waveforms
plot(wf(:,1:200));        %Plot 200 waveforms.
```

Since this is an extracellular recording, the sign of the voltage trace of the action potential is reversed, and the amplitude is much smaller than for intracellular recordings (microvolts instead of millivolts). You can immediately see that there is a large amplitude unit on the electrode. There may also be a smaller amplitude unit with a larger trough-peak spike width. There is also some noise. It is not immediately clear how to separate these categories, and you are looking at only 200 spikes. How do you deal with all 80,000 spikes that were recorded during an hour-long session? You could represent each waveform as a single point, but then each point would be in a 48-dimensional space. How do you make this analysis easier?

The solution (as you may have guessed) is to compress the data using principal components. Then you can plot the first versus the second principal component and see whether the data fall into clusters. When you use the code below to display all the spikes in a graph like the one in Figure 19.5, it becomes clear that there are two major clusters.

```
%Princomp doesn't work on integers, so convert to double
[coeff,score,latent]=princomp(double(wf'));
plot(score(:,1),score(:,2),'.','MarkerSize',1)
```

Unfortunately, with so many spikes, it's not clear how densely packed the clusters are. You can create a 3D histogram using the function **hist3** and then visualize the histogram using the function **surface**. Use the following code to reproduce Figure 19.6:

```
edges{1} = [-300:25:300];        %Bins for the X-axis
edges{2} = [-250:25:250];        %Bins for the Y-axis
h = hist3(score(:,1:2),edges);   %Compute a 2-D histogram
s = surface(h');                 %Visualize the histogram
set(s,'XData',edges{1})          %Label the X-axis
set(s,'YData',edges{2})          %Label the Y-axis
```

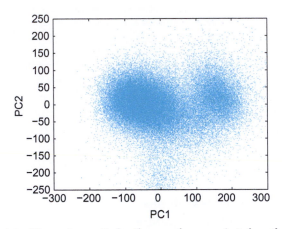

**FIGURE 19.5**    A scatterplot of the motor cortical spike waveforms projected on the first and second principal components.

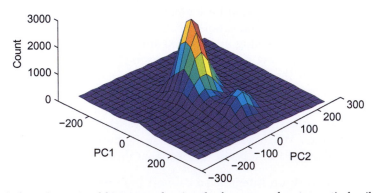

**FIGURE 19.6**    A three-dimensional histogram showing the frequency of motor cortical spike waveforms projected on the first (PC1) and second (PC2) principal components.

The default view is looking straight down on the surface. To make the figure prettier, click the Rotate 3D button (its icon is a counterclockwise arrow encircling a cube). Then click and drag on the figure to rotate it. Play with the function **colormap** to change the color scheme and type **help graph3D** for the list of color maps. For example, **colormap(white)** produces a black and white mesh grid.

Unfortunately, it is difficult to quickly spike sort these waveforms without a good graphical user interface. For example, you would like to be able to select a point in PC space and see what the corresponding waveform looks like. You want to be able to circle a group of points and then see both the average waveform and the interspike-interval histogram. This is the sort of functionality provided by commercial spike sorting packages, such as Offline Sorter by Plexon, Inc. You can implement something similar in MATLAB, but doing so is beyond our scope here.

In the project you will perform in this chapter, instead of using a graphical selection tool, you will select waveforms by looking at distances in PC space. Pick a point in PC1 versus PC2 space that you think is at the center of a cluster. Then calculate the Euclidean distance of every other point from this template point and pick all those that fall below a certain threshold. This is equivalent to drawing a circle on the PC graph and picking all the points that fall within the circle. Now you can calculate any statistic you want of the sorted waveforms, such as the average waveform or the interspike interval histogram. The function **find** may be useful. For example, if you store your Euclidean distances in *dist* and your distance threshold in *threshold*, you can find the indices of all the waveforms meeting this threshold with **ind = find(dist < threshold);**.

You can use this average waveform as the basis of a template sort, a common strategy in spike sorting. The average waveform is a template to which you compare all other waveforms. Calculate the mean-squared error for each waveform from this template waveform. Then keep all waveforms whose mean squared error falls below a certain threshold.

While this procedure would be easier with a commercial spike sorter, sometimes custom procedures in MATLAB can be useful. In the project for this chapter, you will also consider the problem of comparing waveforms from one day to the next. Because the principal components change depending on the data, instead of calculating the PCs of the second day, you will project the second day's data onto the first day's PCs. This isn't something that is usually possible with spike sorting software, so understanding how to implement this in MATLAB expands your analytical possibilities.

## 19.3 EXERCISES

### EXERCISE 19.1

When you computed the covariance matrix of the uncorrelated data *a*, why are the off-diagonal terms nonzero? Generate several new examples of uncorrelated noise. What do you think the average covariance matrix should be?

### EXERCISE 19.2

Use **princomp** to compute the principal components of the correlated noise you generated in *b*. Are they different from what you computed using the covariance matrix method? If they are, how would this affect the transformed data?

### EXERCISE 19.3

Use **princomp** to compute the principal components of uncorrelated noise. What are the PCs? What would you expect them to be?

## 19.4 PROJECT

In this project, you will build your own primitive spike sorter using principal components analysis to analyze extracellular data from recordings in the primary motor cortex of a nonhuman primate (data courtesy of the Hatsopoulos laboratory). The spike waveform and spike times data for this project are stored in a struct called *session*. You can access the data using the following code:

```
wf1 = session(1).wf;          %Waveforms from the 1st day
wf2 = session(2).wf;          %Waveforms from the 2nd day
stamps1 = session(1).stamps;  %Time stamps from the 1st day
stamps2 = session(2).stamps;  %Time stamps from the 2nd day
```

Specifically, you are asked to do the following:

1. Apply PCA to the first day's waveforms. What percent of variation is captured by the first two dimensions?
2. Spike sort the first day's waveforms using a template sort. First, select a region of interest in 2D PC space (a circle at the heart of a cluster) by finding all points in PC space within a certain Euclidean distance from a given point. Calculate the average waveform of the waveforms in this region. Use this average waveform as the template in a template sort. Plot the template and all the sorted waveforms for each neuron you think is present. Also plot the interspike interval histograms, which are just histograms of the times between sorted spikes. The function **diff** may be useful for this task.

**3.** Project the second day's data onto the first day's principal component's axes. How is this different from the second day's data projected on its own principal components? Repeat the sort you used for the first day's data. How do the neurons compare? Do you think they are the same neurons?

## MATLAB FUNCTIONS, COMMANDS, AND OPERATORS COVERED IN THIS CHAPTER

**cov**
**eig**
**hist3**
**mvnrand**
**normrnd**
**princomp**
**surface**
**transpose**
**repmat**
**find**
**diff**

# Information Theory

## 20.1  GOALS OF THIS CHAPTER

Thus far, we have assumed that a neuron encodes any relevant stimulus parameters by modifying its firing rate. We used this assumption to construct tuning curves describing this stimulus encoding. But a neuron could also encode a stimulus by changing the relative timing of its spikes. In this chapter we will introduce the methodology used in a series of papers by Richmond and Optican exploring temporal encoding in a primate visual area. They used principal components analysis and information theory to argue that a temporal code provided more information about the stimulus than a rate code did. You will apply similar methodology to data recorded from the primate motor cortex. Note that this chapter assumes familiarity with principal components analysis introduced in Chapter 19.

## 20.2  BACKGROUND

Richmond and Optican studied pattern discrimination in a primate visual area, the inferior temporal (IT) cortex. They addressed the question of temporal coding in IT in a well-known series of papers in the *Journal of Neurophysiology* (Optican and Richmond, 1987; Richmond and Optican, 1987; Richmond et al., 1987). They found that the firing rate of IT neurons modulated in response to the presentation of one of 64 two-dimensional visual stimuli. They also saw evidence of temporal modulation that was not captured by the firing rate.

To quantify the relevance of this temporal modulation, Richmond and Optican converted the raster plot for each trial into a spike density function (defined later in Section 20.2.2). They then computed principal components (PCs) of these functions. They computed the mutual information between the stimulus and either the firing rates (rate code)

*MATLAB® for Neuroscientists.*
DOI: http://dx.doi.org/10.1016/B978-0-12-383836-0.00018-7

or the first three PCs (temporal code). Their results indicated that the temporal code carried on average twice the information as the rate code.

### 20.2.1 Motor Cortical Data

In this chapter you will use a similar approach to examine encoding of movement direction in the primate motor cortex. You have already seen that motor neurons modulate their firing rate systematically depending on the direction of motion during a center-out task and that this modulation can be fit to a cosine-tuning curve. However, this analysis computed the firing rate over a coarse time bin (1 second). A tuning curve might predict a firing rate of 30 spikes per second for a preferred stimulus, but there are a lot of ways to arrange 30 spikes (each lasting 1–2 ms) over a 1-second time period.

When you use a rate code, you are implicitly assuming that there is no additional information contained in the relative timing of the spikes. This does not mean you assume there is no temporal variation. Instead, a rate code assumes this temporal variation is uninformative about the stimulus. For example, when you calculated a peri-stimulus time histogram (PSTH) centered on movement time in the preceding chapter, you saw the firing rate ramp up slowly 500 ms before movement initiation and ramp back down to baseline by 500 ms later. If each direction elicits this same temporal response scaled up or down, then the coarse rate code is appropriate. However, if the temporal response varies systematically across movement directions, then the rate code will ignore potentially useful information.

Figure 20.1 contains the raster plots (left) and PETHs (right) for unit #16 in the center-out dataset from Chapter 17 (available on the companion web site) recorded on electrode #19. The spike times are relative to the beginning of the instructed delay, where the peripheral

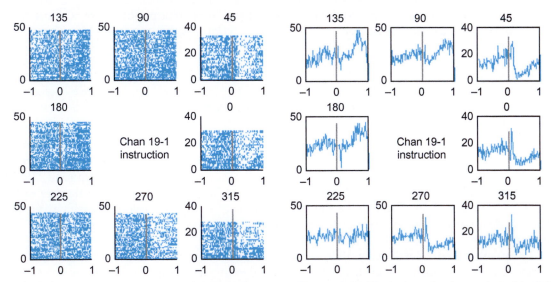

**FIGURE 20.1** *Left*: A raster plot of the sample neuron. The x-axis is time in seconds, the y-axis is the trial number, and the title reflects the movement direction. *Right:* A peri-event time histogram for the same neuron. Here, the y-axis reflects the number of spikes in a 10 ms bin across all trials.

target is visible but the animal is still holding on the center target. The responses to a preferred stimulus (135° or 180°) are similar; both show an increase in the firing rate about 500 ms after the target appears. However, the responses to an antipreferred stimulus (0° or 315°) are different; both show a transient increase in the firing rate in the first 100 ms followed by a marked depression for the following 900 ms. A rate code is unlikely to capture all the information in these responses.

How do you quantify such temporal information? Principal components analysis might work. However, you first need to think about how to format the spike times. Binning the data seems natural, but choosing the proper bin size can be tricky. If the bin is too small (1 ms), then you may make your potential response space huge ($2^{1000}$ possible responses) compared to the number of trials. If the bin is too large (1 second), then you lose potentially useful temporal variations within the time bin. The best bin size is close to the order of the temporal dynamics you are interested in. If you observed consistent variations on a 50–100 ms time scale, a 50–100 ms time bin would probably work.

## 20.2.2 Spike Density Functions

Another problem with binning spike times is that binning can introduce artifacts into the data. Suppose a response to a stimulus always consists of a single spike around time $t$, and the variability from trial to trial follows a Gaussian distribution around $t$. If $t$ sits right on an edge between two bins, then sometimes it will be counted in the first bin, and other times it will be counted in the second bin. This produces the illusion of a bimodal response when, in fact, there is only a single response.

An elegant solution to this problem is to convert each raster plot into a continuous spike density function. You first bin the spike times at a fine time resolution (1 ms) so that each bin has a 0 or 1. You then convolve this data with another function, called the *kernel*. The kernel captures how precise you think the spike times are: a wide kernel implies high variability, whereas a narrow kernel implies high precision.

You explored 2D convolution in Chapter 16; the 1D convolution function in the MATLAB® software is **conv**. Pay attention to the length of the resulting vector because **conv(*a*,*b*)** results in a function whose length is **length(*a*) + length(*b*)-1**. Suppose the kernel is a Gaussian function with a standard deviation of 15 ms. This is equivalent to putting a confidence interval on the spike times. This kernel means you believe that a neuron that wants to fire at 0 ms will actually fire at between −30 and 30 ms 95% of the time. To use this kernel in MATLAB, you need to evaluate the Gaussian over a range of time values (every 1 ms from −45 ms to 45 ms). If you convolve this function with 1 second of spike data (binned every 1 ms), the resulting vector will contain 1090 values. The corresponding time axis is −44 ms to 1045 ms. If you don't want values outside the 1-second time period of interest, you just select the middle section. Load the dataset for this chapter, which contains data from the sample neuron in the variable *trial*. In the following code, binned spike data from the first trial are stored in *binned* and the spike density function is stored in *s*:

```
trialNum=1;                     %Bin data from the first trial.
binned=hist(trial(trialNum).spikeTimes,[0:.001:1]);
sigma = .015;                   %Standard deviation of the kernel = 15 ms
```

```
edges = [-3*sigma:.001:3*sigma];          %Time ranges form -3*st. dev. to 3*st. dev.
kernel = normpdf(edges,0,sigma);          %Evaluate the Gaussian kernel
kernel = kernel*.001;                     %Multiply by bin width
s = conv(binned,kernel);                  %Convolve spike data with the kernel
center = ceil(length(edges)/2);           %Find the index of the kernel center
s = s(center:1000 + center-1);            %Trim out middle portion
```

An example of a spike density function is shown in Figure 20.2, along with the original raster plot.

Once you compute spike density functions for each neuron, you can compute the principal components of the spike density functions. However, to avoid computing a $1000 \times 1000$ dimensional covariance matrix, you should first sample the spike density functions every 10 or 20 ms and then apply principal components analysis.

### 20.2.3 Joint, Marginal, and Conditional Distributions

The goal here is to compute the amount of information contained in a firing rate code compared to a temporal code (as captured by principal components). However, before defining "information" precisely, we need to review the concept of joint, marginal, and conditional probability distributions. For a discrete variable (which is all we will consider here), a *probability distribution* is simply a function that assigns a probability between 0 and 1 to all possible outcomes such that all the probabilities sum to 1. A *joint probability distribution* is the same, except it involves more than one variable and thus assigns probabilities to combinations of variables.

In this example, you have a stimulus $S$, which can take on one of eight discrete values. Suppose you divide the firing rates $R$ of the same sample neuron from earlier into three

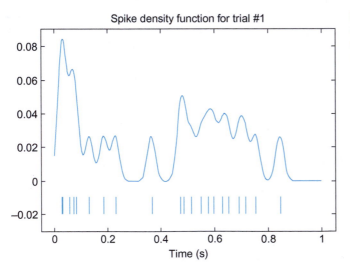

FIGURE 20.2  *Top:* An example of a spike density function using a Gaussian kernel with a standard deviation of 15 ms. *Bottom:* the original raster plot.

bins (low, medium, and high firing rate). If you use the code below to count how many times each of 24 possible combinations shows up in the data, you end up with something like Table 20.1.

```
for ii=1:8                    % Loop over all directions.
    ind=find(direction==ii);  % Find trials in same direction
    % Bin firing rate R into low, medium, high
    table(:,ii)=histc(R(ind),[0 19.5 30 100]);
end
```

From Table 20.1, it is clear that there is a relationship between the firing rate and stimulus direction. A high firing rate corresponds to a stimulus of 3, 4, or 5, while a low firing rate corresponds to a stimulus of 7, 8, 1, or 2. Remember that $S = 1$ corresponds to a movement direction of $0°$, $S = 2$ to $45°$, and so on. You can determine the observed joint probability distribution $P(S,R)$ simply by dividing each count by the total number of counts (here, this is 315). In addition, you can compute the marginal probability distributions, which are the probability distributions of one variable computed by summing the joint probability distribution over the other variable:

$$P(S) = \sum_{R} P(s,r) \quad \text{and} \quad P(R) = \sum_{S} P(s,r) \tag{20.1}$$

Note that in these equations, $S$ and $R$ refer to all possible stimuli or responses, respectively, and that $s$ and $r$ refer to a particular stimulus or response.

Table 20.2 shows the values of $P(S,R)$. The marginal distributions are listed in the last row and far right column because they are the sum taken across rows and across columns of the joint distribution.

TABLE 20.1    Observed Counts of Stimulus-Response Pairs for the Sample Neuron

|  | $S = 1$ | $S = 2$ | $S = 3$ | $S = 4$ | $S = 5$ | $S = 6$ | $S = 7$ | $S = 8$ |
|---|---|---|---|---|---|---|---|---|
| $R<20$ | 21 | 28 | 2 | 1 | 1 | 8 | 33 | 17 |
| $20 \le R<30$ | 7 | 5 | 23 | 20 | 17 | 29 | 9 | 6 |
| $R \ge 30$ | 1 | 0 | 21 | 26 | 27 | 7 | 1 | 5 |

TABLE 20.2    The Joint Probability Distribution $P(S,R)$ and the Marginal Distributions $P(R)$ and $P(S)$ for the Sample Neuron

| $P(S,R)$ | $S = 1$ | $S = 2$ | $S = 3$ | $S = 4$ | $S = 5$ | $S = 6$ | $S = 7$ | $S = 8$ | $P(R)$ |
|---|---|---|---|---|---|---|---|---|---|
| $R<20$ | 0.067 | 0.089 | 0.006 | 0.003 | 0.003 | 0.025 | 0.105 | 0.054 | 0.352 |
| $20 \le R<30$ | 0.022 | 0.016 | 0.073 | 0.063 | 0.054 | 0.092 | 0.029 | 0.019 | 0.368 |
| $R \ge 30$ | 0.003 | 0.000 | 0.067 | 0.083 | 0.086 | 0.022 | 0.003 | 0.016 | 0.279 |
| $P(S)$ | 0.092 | 0.105 | 0.146 | 0.149 | 0.143 | 0.140 | 0.137 | 0.089 | |

The last concept we need to address is the *conditional distribution*. The distribution of the response $R$ given knowledge of the stimulus $S$ is written $P(R|S)$ and is defined as $P(R|S) = P(S,R)/P(S)$. Likewise, the conditional distribution of $S$ given $R$ is $P(S|R) = P(S,R)/P(R)$. As an example, what is the probability distribution of firing rates given a movement direction of $0°$ ($S = 1$)? To find the answer, simply take the first column of values in Table 20.2, $P(S = 1, R)$, and divide by $P(S = 1)$ to get $P(R|S = 1)$. Hence, the probably of a low firing rate given $S = 1$ is $0.067/0.092 = 0.73$ and so on.

## 20.2.4 Information Theory

The foundation of information theory was laid in a 1948 paper by Shannon titled, "A Mathematical Theory of Communication." Shannon was interested in how much information a given communication channel could transmit. In neuroscience, you are interested in how much information the neuron's response can communicate about the experimental stimulus.

Information theory is based on a measure of uncertainty known as *entropy* (designated "H"). For example, the entropy of the stimulus $S$ is written H(S) and is defined as follows:

$$H(S) = - \sum_S P(s)\log_2 P(s) \tag{20.2}$$

The subscript $S$ underneath the summation simply means to sum over all possible stimuli $S = [1, 2 \ldots 8]$. This expression is called "entropy" because it is similar to the definition of entropy in thermodynamics. Thus, the preceding expression is sometimes referred to as "Shannon entropy." The entropy of the stimulus can be intuitively understood as "how long of a message (in bits) do I need to convey the value of the stimulus?" For example, suppose the center-out task had only two peripheral targets ("left" and "right"), which appeared with an equal probability. It would take only one bit (a 0 or a 1) to convey which target appeared; hence, you would expect the entropy of this stimulus to be 1 bit. That is what the preceding expression gives you, as $P(S) = 0.5$ and $\log_2(0.5) = -1$. The center-out stimulus in the dataset can take on eight possible values with equal probability, so you expect its entropy to be 3 bits. However, the entropy of the observed stimuli will actually be slightly less than 3 bits because the observed probabilities are not exactly uniform.

Next, you want to measure the entropy of the stimulus given the response, $H(S|R)$. For one particular stimulus, the entropy is defined similarly to the previous equation:

$$H(S|r) = - \sum_S P(s|r)\log_2 P(s|r) \tag{20.3}$$

To get the entropy $H(S|R)$, you just average over all possible responses:

$$H(S|R) = - \sum_R \sum_S P(r)P(s|r)\log_2 P(s|r) \tag{20.4}$$

Now you can define the information that the response contains about the stimulus. This is known as *mutual information* (designated $I$), and it is the difference between the two entropy values just defined:

$$I(R; S) = H(S) - H(S|R) = \sum_R \sum_S P(r)P(s|r)\log_2\left(\frac{P(s|r)}{P(s)}\right) \tag{20.5}$$

Why does this make sense? Imagine you divide the response into eight bins and that each stimulus is perfectly paired with one response. In this case, the entropy $H(S|R)$ would be 0 bits, because given the response, there is no uncertainty about what the stimulus was. You already decided the $H(S)$ was theoretically 3 bits, so the mutual information $I(R;S)$ would be 3 bits − 0 bits = 3 bits. This confirms that the response has perfect information about the stimulus.

Suppose instead that you divide the response into two bins, and that one bin corresponds to stimuli 1−4 and the other bin corresponds to stimuli 5−8. Each bin has four equally likely choices, so the entropy $H(R|S)$ will be 2 bits. Now the mutual information is $I(R;S)$ = 3 bits − 2 bits = 1 bit. This means that response allows you to reduce the uncertainty about the stimulus by a factor of 2. This makes sense because the response divides the stimuli into two equally likely groups. This also emphasizes that the choice of bins affects the value of the mutual information.

Note that you can use the definition of conditional probability to rearrange the expression for mutual information. The following version is easier to use with the table of joint and marginal probabilities computed earlier. Mutual information can also be defined as follows:

$$I(R; S) = \sum_R \sum_S P(s, r)\log_2\left(\frac{P(s, r)}{P(s)P(r)}\right) \tag{20.6}$$

Applying this equation to the joint distribution of the sample neuron gives a mutual information of 0.50 bits for a rate code.

## 20.2.5 Understanding Bias

Now we will try estimating the mutual information of an uninformative neuron. "Uninformative" means that the firing rate probabilities are independent of the stimulus probabilities. By the definition of independence, the joint probability distribution of two independent variables is the product of their marginal distributions $P(R,S) = P(S)P(R)$. If you substitute this into the previous expression for mutual information, you will see the quantity inside the logarithm is always 1. Since $\log_2(1) = 0$, this means the mutual information of two independent variables is also 0.

To make things easy, assume that each of the three responses is equally likely and that each of the eight stimuli is equally likely. Thus, $P(R) = 1/3$ and $P(S) = 1/8$, and each value of the joint probability distribution is the same: $P(S,R) = 1/24$. The mutual information of this distribution is still 0 bits.

However, even if this is the true probability distribution, the observed probabilities would likely be different. You can simulate the values of observed probabilities with the following code. Here, you will simulate 24 random trials, so the expected count for each cell is 1:

```
edges = [0:24]/24;          %Bin edges for each table entry
data = rand(24,1);          %Generate 24 random values between 0 and 1
count = histc(data,edges);  %Count how many fall in the bin edges
count = count(1:24);        %Ignore the last value (counts values equal to 1).
count = reshape(count,3,8); %Reformat the table.
```

This might lead to the counts shown in Table 20.3.

If you calculate the mutual information of the associated joint distribution (divide each count by 24 for the joint distribution), you get 0.50 bits, which is much higher than the 0.00 bits you expect. In fact, this is the same information as the sample neuron found previously. How can you now trust this earlier calculation?

Calculating mutual information directly from the observed probability distribution (as done here) leads to a biased estimate. A *biased* estimate is one that will not equal the true value, even if the estimate is averaged over many repetitions. The estimate of mutual information becomes unbiased only when you have infinite data. Such datasets are hard to come by.

However, all hope is not lost, because you can reduce the size of this bias. To start with, note that the number of trials is the same as the number of bins in the joint distribution. The sample data had 315 trials, which should reduce the chance of spurious counts. Repeating the previous exercise with 315 random trials might give the counts shown in Table 20.4.

TABLE 20.3    The Observed Counts of the Stimulus-Response Pairs for 24 Random Trials of an Uninformative Neuron

|                  | $S=1$ | $S=2$ | $S=3$ | $S=4$ | $S=5$ | $S=6$ | $S=7$ | $S=8$ |
|------------------|-------|-------|-------|-------|-------|-------|-------|-------|
| $R<20$           | 0     | 2     | 1     | 1     | 1     | 2     | 1     | 0     |
| $20 \leq R<30$   | 1     | 0     | 1     | 0     | 0     | 1     | 0     | 2     |
| $R \geq 30$      | 1     | 2     | 1     | 2     | 0     | 2     | 3     | 0     |

TABLE 20.4    The Observed Counts of the Stimulus-Response Pairs for 315 Random Trials of an Uninformative Neuron

|                  | $S=1$ | $S=2$ | $S=3$ | $S=4$ | $S=5$ | $S=6$ | $S=7$ | $S=8$ |
|------------------|-------|-------|-------|-------|-------|-------|-------|-------|
| $R<20$           | 14    | 25    | 11    | 16    | 15    | 21    | 15    | 14    |
| $20 \leq R<30$   | 12    | 11    | 11    | 16    | 12    | 14    | 9     | 12    |
| $R \geq 30$      | 11    | 11    | 16    | 6     | 8     | 14    | 7     | 14    |

The mutual information calculating from this table's joint distribution is now just 0.03 bits, which is much closer to the expectation (0 bits). This gives you more confidence in the 0.50 bits you estimated for the sample neuron. However, these numbers are each generated from a single random experiment. In the exercises you will see that repeated experiments confirm this trend: increasing the number of trials does decrease the bias of the estimate of mutual information.

Note that the relevant parameter is actually the number of trials per bin. This means that if you have 315 trials but you also have 315 bins, you will still have a significant bias. Therefore, you must choose the number of bins carefully. It seems that large bins throw away information (shouldn't a spike count of 4 be treated differently than 16?), but smaller bins introduce a larger bias.

### 20.2.6 Shuffle Correction

These simulations are similar to a simple method (called *shuffle correction*) that corrects for the bias in the mutual information estimate. Suppose you store the stimulus values in a vector **direction**. You are interested in determining what the estimated mutual information would be if the firing rates were independent of the stimulus values. To do this, you randomly rearrange (or "shuffle") the stimulus vector and then compute a new joint distribution and estimate mutual information from that. If you repeat this operation many times, you can derive a "null distribution" of mutual information estimates of a firing rate which carries no information. Thus, you can conclude that any neuron whose mutual information value that is significantly different from this null distribution is informative. You can also calculate a "shuffle corrected" mutual information estimate by subtracting the mean of this null distribution. The following code shows how you can shuffle the stimulus values in MATLAB:

```
x = rand(315,1);         %Vector of 315 random numbers between 0 and 1
[temp ind] = sort(x);    %Sort random numbers and keep the indices in vector "ind"
dirSh = direction(ind);  %Use "ind" to randomly shuffle the vector of stimuli
```

Now you can compute a table counting combinations of the original firing rate and the shuffled stimulus. One example might be the one shown in Table 20.5.

The mutual information of the associated joint distribution is 0.03 bits. Repeating the shuffling 30 times gives a mean information of 0.03 bits and a standard deviation of 0.01

TABLE 20.5    The Observed Counts of the Stimulus-Response Pairs for the "Shuffled" Version of the Sample Neuron

|             | S = 1 | S = 2 | S = 3 | S = 4 | S = 5 | S = 6 | S = 7 | S = 8 |
|-------------|-------|-------|-------|-------|-------|-------|-------|-------|
| R<20        | 8     | 10    | 12    | 20    | 18    | 16    | 16    | 11    |
| 20 ≤ R<30   | 13    | 12    | 22    | 14    | 17    | 15    | 11    | 12    |
| R ≥ 30      | 8     | 11    | 12    | 13    | 10    | 13    | 16    | 5     |

bits for the null distribution. Hence, the "shuffle corrected" mutual information of the original neuron is 0.50 bits − 0.03 bits = 0.47 bits. Thus, you can be confident that a rate code contains significant information about the stimulus direction.

Bias correction is particularly important when you are comparing mutual information of joint distributions that have different numbers of bins. In the project for this chapter, you will compare a rate code to a temporal code. Comparing a rate code to the first principal component is straightforward because you could use the same number of bins for each variable. However, if you use the $n$ bins to look at the first principal components, you would have to use $n^2$ bins to look at the first two principal components together using the same bin widths. As you know, increasing the number of bins increases the bias. If you compared uncorrected estimates, it would by easy to assume the second principal component provides additional information when it actually doesn't.

It should also be noted that accurate estimation of information measures is an active research field and that shuffle correction is perhaps the simplest of available techniques. Refer to Panzeri et al. (2007) for a more recent review of bias correction techniques and to Hatsopoulos et al. (1998) for another example of the use of shuffle correction.

## 20.3 EXERCISES

### EXERCISE 20.1

Compute the entropy of the observed values of the stimulus.

### EXERCISE 20.2

Run 30 simulations each of the observed joint distribution of the uninformative neuron (8 stimuli, 3 responses) with $n$ trials, where $n = 25, 50, 100, 200, 400, 800$. Plot the mean and standard deviation of the bias as a function of the number of trials.

### EXERCISE 20.3

Repeat Exercise 20.2, but with 6 and 12 response bins. Plot the mean and standard deviation of the bias as a function of the number of trials.

**EXERCISE 20.4**

Combine data from Exercises 20.2 and 20.3, and plot the mean bias as a function of the number of trials per bin.

## 20.4  PROJECT

Choose at least five active neurons from the dataset from Chapter 17, "Neural Data Analysis I: Encoding" (the sample neuron here is unit #16) available at the companion web site to analyze. Convert each raster plot into a spike density function. Report the details of the kernel you used. Calculate principal components (PCs) of the spike density functions. Calculate the shuffle-correction mutual information between the stimuli and, in turn, the firing rate, first PC score, and first and second PC score considered together.

In addition, answer the following questions:

1. Is there evidence for a temporal code?
2. How similar are the first few PCs (not the scores) calculated for the different neurons? What do you think they represent?
3. How does a temporal code that depends only on the first PC differ from one which depends on two or more PCs?

## MATLAB FUNCTIONS, COMMANDS, AND OPERATORS COVERED IN THIS CHAPTER

**conv**
**reshape**

# Neural Decoding I: Discrete Variables

## 21.1 GOALS OF THIS CHAPTER

In the previous chapters, we saw how neural firing can be expressed as a function of some behavioral variable. But what about the inverse of that problem? This chapter will introduce an open-ended approach toward solving the problem of neural decoding, whereby an estimate of behavior is generated from observed neural activity. Specifically, this chapter will address how to predict the upcoming direction of movement from a population of neuronal signals recorded from motor areas of a macaque monkey. Note that this chapter assumes completion of Chapter 17, "Neural Data Analysis I: Encoding," as it makes use of the preferred directions calculated in the exercises.

## 21.2 BACKGROUND

What is *neural decoding*? Simply put, it is a mathematical mapping from the brain activity to the outside world. In the sensory domain, the outside world consists of the received visual, auditory, or other sensory information. In the motor domain, the outside world consists of the state of the skeletomuscular system. This is the inverse of *neural encoding*, which maps the outside world to brain activity. For example, in Chapter 17, you looked at how a cosine tuning curve specifies how a neuron modulates its firing rate depending on the upcoming direction of movement. In contrast, estimating this movement direction from one (or many) observed firing rate(s) is an example of neural decoding. Because signals about motor intention precede movement, decoding can be thought of as "mind reading." Neuroscientists seek to predict an action as soon as it is intended, before it ever takes place.

Neural decoding can also be thought of as pattern recognition. A set of neuronal spike times represents a pattern, and the goal of the decoder is to figure out which stimuli or movements are associated with which patterns. This is a common problem in science.

Doctors perform pattern recognition when they produce a diagnosis from a collection of physical and physiological findings. For example, an electrocardiogram (EKG) trace contains a repeated, stereotypical pattern that corresponds to a single heart beat. Each part of the trace corresponds to de- and repolarization of a different part of the heart. Therefore, doctors can use deviations from the normal EKG as clues about underlying abnormalities. An elevation in one part of the trace (ST elevation) is used to help diagnose heart attacks. This is pattern recognition.

Interpreting a raster plot is not quite that straightforward (neither is interpreting EKGs, but that's another matter). Figure 21.1 shows 10 raster plots and a peri-event time histogram for a motor neuron. Each neuron's spike times is centered on the start time of repeated movements in the same direction. There is a clear pattern. On every trial, there is a transient increase in the neuron firing rate starting a few hundred milliseconds before the movement starts. However, this is only an approximate relationship. If you look at each raster individually, it is not clear exactly when the movement begins. In addition, each raster plot is different. This means you can't simply perform pattern recognition by using a "look-up table" because it is unlikely that a neuron's response will ever exactly repeat itself.

In this chapter we will implement different strategies for predicting the direction of an upcoming movement based on the firing rates of a population of neurons. This is relevant to two distinct goals of neuroscience. First, neuroscientists would like to understand the brain on a functional level. Neuroscientists ask, "What is the brain doing and how is it doing that?" Second, neuroscientists are interested in using neuronal signals to do something useful. Neural prosthetics seek to do this. For example, cochlear implants convert sound into digital signals used to stimulate the auditory nerve, thus restoring speech perception in the deaf (Papsin and Gordon, 2007). A decoding strategy introduced in the next

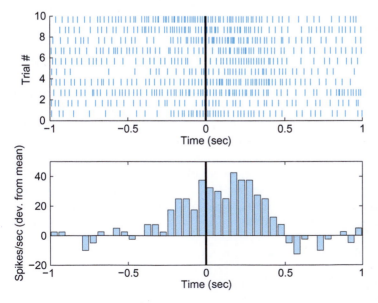

FIGURE 21.1  *Top:* A raster plot of 10 trials of a center-out task. *Bottom:* A peri-event time histogram of the same data. The start of movement occurs at *time* = 0 seconds. The firing rate is expressed as the deviation from the mean firing rate.

chapter (the linear filter) was used in a neuroprosthetic that allowed a human with tetra-plegia to control a computer cursor and other devices (Hochberg et al., 2006).

It is important to note, however, that decoding to understand how the brain works is different from decoding for control. You have seen that neuronal activity in the motor cortex is directionally tuned, but that is not the same as saying these neurons encode direction. Properly interpreting what is being encoded requires more experiments than what we have described thus far. In the canonical center-out experiment, the posture is the same for all eight directions. Thus, instead of encoding direction, the neuron might be encoding the specific sequence of muscle activations, the desired end posture, or the spatial location of the target (see Kakei et al., 1999, 2001, for experiments which varied forearm posture during a center-out task). The adage "correlation does not imply causation" applies here.

One area of debate in neuroscience is, "At what time scale should we look for information?" *Rate encoding* holds that the firing rate calculated over some broad time span (usually hundreds of milliseconds) contains all the necessary information. *Temporal coding* holds that additional information is available at smaller time scales. At the extreme, you could use a 1 ms time bin where each bin either has a 0 (no spike) or a 1 (spike). This approach likely contains more information about the stimulus than a coarse firing rate. However, it also greatly increases the dimensionality of the potential responses. Instead of one discrete variable (the firing rate over 1 second), which might reasonably take a few dozen values, a 1 ms time bin gives a $1000 \times 1$ vector of binary variables with as many as $2^{1000}$ possible values, which is a number larger than the estimated number of atoms in the universe. Thus, for computational simplicity, you can start by assuming a coarse rate code. The idea of temporal encoding is explored in more detail in Chapter 20.

## 21.2.1 Population Vector

In Chapter 17, we introduced the concept of directional tuning of motor cortical neurons. This is an encoding model that translates an upcoming direction of movement into a neuronal firing rate. If you now want to decode direction from a firing rate, you are faced with two problems. First, since a cosine-tuning curve is symmetric, most firing rates are ambiguous because they could be associated with two movement directions. Second, the firing rate signal is very noisy. This is due to both intrinsic neuronal noise as well as measurement noise introduced by the equipment. How, then, can you decode movement direction?

The solution is averaging over a population of neurons, which decreases the effects of noise and allows disambiguation of the movement direction. The "population vector" algorithm introduced by Georgopoulos and colleagues is an intuitive way to use cosine-tuning information from a population of neurons to decode movement direction (1986). In Chapter 17, you determined the preferred direction of each neuron using information from all trials.

Having done this, proceed as follows:

1. Assume that each neuron "votes" for its preferred movement direction. Specifically, each neuron is going to contribute a "response vector" that is aligned with its preferred direction.
2. The magnitude (or length) of each neuron's response vector is determined by the neural activity of the neuron during each trial. This is the weight given to each neuron's vote. For now, assume the weight is simply the firing rate during the hold period.
3. Sum all the response vectors from all neurons to arrive at the population vector for this trial. The direction of this population vector corresponds to the predicted direction.

This can be expressed as a formula,

$$\vec{P} = \sum_{i=1}^{n} w_i \vec{C}_i \tag{21.1}$$

where for $n$ neurons, $P$ is the population vector, $w_i$ is the weight given to each vote, and $C_i$ is a unit vector pointing in the $i$th neuron's preferred direction. The arrows above $P$ and $C_i$ indicate that these quantities are vectors. Recall that if you represent vectors with Cartesian coordinates, you can sum the $X$ and $Y$ components separately. Thus, if $\theta_i$ is the $i$th neuron's preferred direction in radians, then the population vector can be broken down into its $X$ and $Y$ components:

$$P_X = \sum_{i=1}^{n} w_i \cos(\theta_i) \quad P_Y = \sum_{i=1}^{n} w_i \sin(\theta_i) \tag{21.2}$$

Our perspective has changed. Previously, you considered all trials to determine the preferred direction of a given neuron. Now, you consider the neural activity of all cells in a given trial to determine the combined response of the population of neurons: the population vector. The population vector points toward the upcoming movement direction.

The population vector is useful because it is easy to implement and intuitive to understand. It is based on the theory that motor cortex neurons fire to produce muscle forces, which, given a certain posture, act in the neuron's preferred direction. However, the population vector is limited because a number of conditions must be met for the method to perform well (Georgopoulos, 1986). For example, the neuron's tuning curves must actually follow a cosine or at least be radially symmetric around the preferred direction. Also, the preferred directions must be uniformly distributed.

## 21.2.2 Maximum Likelihood

You can develop a more general decoding algorithm by relaxing some of the assumptions made by the population vector. One way to do this is to use statistical methods. You assume that a neuron modulates its firing rate in response to the upcoming movement direction. However, you do not assume exactly how (which a cosine tuning curve does). You assume that the neuron's target firing rate will be corrupted by noise, so that for a given direction you will observe a distribution of firing rates rather than a single firing rate. If you assume the form of this distribution, you can use standard statistical methods to estimate its parameters. For example, if you thought the distribution of firing rates was Gaussian, you could characterize it fully by computing the mean and standard deviation of all the firing rates for trials moving in one direction.

Once you have estimated the parameters, you can calculate the probability of any firing rate giving a certain direction. This is a conditional probability. We reviewed this in Chapter 20, "Information Theory." Briefly, the conditional probability of event $A$ given event $B$ is denoted as $P(A|B)$. It is defined as the joint probability $A$ and $B$ divided by the probability of $B$:

$$P(A|B) = \frac{P(A, B)}{P(B)} \tag{21.3}$$

Intuitively, this can be thought of as the probability of $A$ taking into account some piece of information (that $B$ has happened). Here, you are interested in the conditional probability of a firing rate $R$ given the direction $d$: $P(R|d)$. For one neuron, the maximum likelihood approach is to look at the firing rate and select the direction for which this firing rate is most likely. As an example, consider a simplified center-out experiment in which targets are presented at equal probabilities to the left or right, and you are recording from a neuron that prefers movement to the right. Say this neuron fires 25 spikes/s with a standard deviation of 5 spikes/s before right targets and fires 10 spikes/s with a standard deviation of 3 spikes/s before left targets. If you assume these firing rates follow Gaussian distributions, then the maximum likelihood algorithm would predict a right target for a firing rate $\geq 17$ spike/s and a left target for firing rates $\leq 16$ spike/s (see Figure 21.2).

If there is more than one neuron, the situation is more complicated. You make the simplifying assumption that the neuron's firing rates are independent. At first, this seems at odds with our understanding of the brain: aren't connections between neurons the whole point? However, most neurons in the sample are separated by large distances (relative to the size of the neurons), and you may attempt to relax this assumption in the exercises. Independence means you can express the probability of a set of firing rates as the product of the probabilities for each individual firing rate. This calculation is performed when you

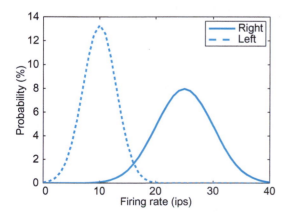

FIGURE 21.2 Firing rate distributions of a hypothetical, rightward-preferring neuron.

say the chance of flipping four heads in a row is $1/16 = (1/2)^4$. The probability of a set of firing rates $R = [r_1, r_2, \ldots, r_n]$ for a given direction $d$ can be expressed as follows:

$$P(R|d) = \prod_{i=1}^{n} P(r_i|d) \tag{21.4}$$

The maximum likelihood approach predicts the direction associated with the highest likelihood $P(R|d)$. However, when you're implementing this in MATLAB, it is more convenient to pick the direction that maximizes the log-likelihood: $\log[P(R|d)]$. Because the natural logarithm is a monotonically increasing function, the choice of direction that maximizes the log-likelihood will also maximize the likelihood. This approach avoids calculating the product of very small numbers in MATLAB, which, due to numerical precision constraints, quickly becomes inaccurate. Instead, you sum negative numbers (since the probabilities are less than 1). The log-likelihood can be expressed as follows:

$$\log[P(R|d)] = \sum_{i=1}^{n} \log[P(R_i|d)] \tag{21.5}$$

The algorithm we have described here is more accurately called the "maximum a posteriori" estimation, which will be discussed in more detail in Chapter 22. In this formulation, we are relying on an experimental trick: the prior probability of each direction is equal. This is important, because if you are decoding direction, you want to maximize the conditional probability of the direction given the firing rates, $P(d|r)$. This algorithm is maximizing the reverse: the conditional probability of the firing rates given the direction, $P(r|d)$. This simplification is valid only if all directions are equally probable. If they are not, you will need to calculate $P(d|r)$ using Bayes' rule, which will be discussed in Chapter 22.

### 21.2.3 Data

Here, you will use the dataset from Chapter 17, "Neural Data Analysis I: Encoding." However, there is a second dataset for this chapter, available on the companion web site. The first dataset will be used to train the decoding algorithms, meaning these data are used to estimate the parameters of the model (such as preferred direction). The second dataset will be used to test the algorithm built using the first dataset. It is important not to test a prediction algorithm using the same data you trained with. Otherwise, the optimal prediction would be "the exact same thing is going to happen." Testing on novel data helps ensure that the model does not overfit the data.

To compare the population vector method to the maximum likelihood method, you will need to bin the population vector to force it to assume one of the eight discrete directions. You can use the function **histc** in MATLAB to do this, though you should remember that the 0-degree direction will correspond to two bins: 0 to 22.5 degrees and 337.5 to 360 degrees.

You also need to define an accuracy metric for the predictions made. The percentage correct is the easiest to compute. However, the mean squared error may be more appropriate, as it penalizes large errors more than small errors.

# 21.3 EXERCISES

## EXERCISE 21.1

### IMPLEMENTATION

1. Implement a population vector decoder. Use the preferred directions determined for the training dataset. Predict the movement direction of the test dataset by using the firing rates during the hold period as weights. The hold period start times are stored in the variable *instruction*. The hold period ends 1 second after the start time. Bin the predicted direction to convert it to one of the eight discrete movement directions.

2. Implement a maximum likelihood decoder. Assume a Poisson firing rate model and independent firing rates.

Determine the mean firing rate for each neuron and each direction for the training dataset. Use the function **poisspdf** to determine the likelihood of each direction. You might want to threshold the observed mean firing rates if they are below 1 spike/s or so to avoid unrealistically small probabilities. Pick the direction that maximizes the log-likelihood of all firing rates for a given trial.

3. Compare the accuracy of these two decoding methods, using a percent accuracy or mean squared error.

## EXERCISE 21.2

### VARIATIONS

1. The population vector methods make assumptions about the data: that neurons were cosine tuned and that preferred directions are uniformly distributed. Are these assumptions valid? Provide evidence for your answer.

2. Instead of weighting the response vectors using the firing rate, weight using the change in firing rate from baseline. Use the mean firing rate across all directions to determine the baseline firing rate. Does this affect the decoding accuracy? Why might this be?

3. Use a Gaussian firing rate model for the maximum likelihood decoder. The likelihood can be determined with the function **normpdf** and the mean and standard deviation for each neuron and each direction. Does this affect the decoding accuracy? Why might this be?

## 21.4 PROJECT

Create a new decoder by either modifying the algorithms introduced here or by developing your own ideas. Report the accuracy of the new decoder as compared to population vector or maximum likelihood methods. Here are some suggested approaches:

*Easy*: Change the specific implementation of one of the decoders. For the population vector, create a new function to determine the weights from the firing rates. For maximum likelihood, use a different probability distribution as a firing rate model. Alternatively, try to include temporal coding by using principal components analysis or using smaller time bins.

*Medium*: Transform the data to make it conform to the assumptions made by decoders. For the population vector, change the weighting scheme of the population vector algorithm to compensate for a nonuniform distribution of preferred directions. For maximum likelihood, try to compensate for correlations between neurons, or between a neuron's current firing rate and its past firing rates.

*Hard*: Create a new decoding technique. For example, the algorithms introduced treat firing rates as independent. More information might be available if firing rates are treated has a single vector. You could then use a distance metric to classify a novel vector of firing rates that falls into one of eight clusters (for each direction of movement). Using dimensionality reduction techniques (such as principal components analysis, covered in Chapter 19) might allow you plot the clusters in a low dimensional space.

## MATLAB FUNCTIONS, COMMANDS, AND OPERATORS COVERED IN THIS CHAPTER

**poisspdf**
**normpdf**

# 22

# Neural Decoding II: Continuous Variables

## 22.1 GOALS OF THIS CHAPTER

The preceding chapter explored methods of decoding movement direction from neuronal signals. The movement direction was a discrete stimulus, taking on one of eight possible values. In this chapter, you will look at how to decode a continuous, time-varying stimulus from neuronal signals. Specifically, this chapter will address how to decode the instantaneous hand position from a population of motor cortical neurons recorded from a macaque monkey. You will also see how information about how the hand position changes over time can be used to improve your decoding.

Note that this chapter assumes completion of Chapter 17, "Neural Data Analysis I: Encoding," Chapter 18, "Neural Data Analysis II: Binned Spike Data," and Chapter 21, "Neural Decoding I: Discrete Variables."

## 22.2 BACKGROUND

We previously discussed how neurons in the motor cortex carry information about upcoming movements. In the last chapter, you were able to use this information to decode the direction of a movement made to one of eight targets. But what do you do if movement isn't so constrained as it is in the center-out task? Another common experimental paradigm in motor control literature can be described as the "pinball" task: a target appears somewhere in a 2D playing field, and as soon as the participant moves the cursor within this target, a new target appears at a different, randomly selected position. There are no hold times in this task, unlike the center-out task, so the hand is constantly moving.

You are interested in decoding the trajectory of the hand, meaning you want to know X- and Y-positions of the hand at each time point. X- and Y-positions are examples of kinematic variables, meaning they describe the motion of the object (the hand), but not the forces that generated the motion. Considering just the limb kinematics is easier than considering the full limb dynamics, which requires a model of how muscle forces and the physical properties of the limb interact to produce motion.

What method can you use to decode these kinematic variables? You could just modify the algorithms you have already seen. Recall that the population vector has a magnitude as well as direction. If you assume this magnitude is proportional to the instantaneous speed, you could simply compute a population vector for each time bin and then add them tail to tip to create an estimate of the trajectory. This was, in fact, an early approach to the problem (Georgopoulos, 1988).

In this chapter we will take a couple of different approaches. We mentioned previously that a simple neuronal encoding model assumes the firing rate varies linearly with stimulus intensity. You can apply this to motor cortical neurons and assume they fire linearly with the instantaneous hand position. For decoding, you simply invert this equation and derive an estimate of hand position using a linear function of firing rates. Another approach is to divide the X- and Y-axes into a grid and then use the same maximum likelihood method we used for center-out data to determine which grid square the hand is located in. As you will see, the fact that the hand position varies smoothly as a function of time introduces some new wrinkles.

Load the dataset for this chapter, available on the companion web site (data courtesy of the Hatsopoulos laboratory). There are two main variables: *kin* stores the X- and Y-positions (sampled every 70 ms), and *rate* stores the number of spikes in each 70 ms bin. Now look at the relationship between just one kinematic variable (Y-position) and one neuron's spike count. Notice how the indices are offset to create a vector of spike counts that lead the kinematics by two time bins (or 140 ms):

```
lag = 2;                  %Lag between neural firing and hand position
y = kin(lag + 1:end,2);   %Y-position of the hand
s = rate(1:end-lag,36);   %Corresponding spike count of one neuron
```

We are interested in how the spike count varies as a function of hand position (for simplicity, right now we only focus on position in the Y-direction). However, if you were to plot the position versus the raw spike count as a scatter plot, the result would be confusing, because the spike counts are highly variable. It is better to look at how the mean spike count varies as a function of position. To do this, we will divide the position variable into 15 1-cm wide bins, and average the corresponding spike counts. We will also keep track of the standard deviation, so we can compute a standard error: this is equal to the standard deviation divided by the square root of the number of data points. We can then plot a confidence interval for the empirical mean, as the "true" mean should fall within two standard errors of empirical mean 95% of the time. The error bars corresponding to a confidence interval can be plotted using the MATLAB® function **errorbar.**

The following code uses that function to plot the mean spike count as a function of the position of the hand:

```
yEdge = [0:15];                                     %Bin edges
for ii = 1:length(yEdge)-1
    ind = find(y > yEdge(ii) & y < = yEdge(ii + 1));   %Find positions within bin
    meanS(ii) = mean(s(ind));                       %Mean of spike counts for this bin
    stdS(ii) = std(s(ind));                         %Standard deviation
    errS(ii) = stdS(ii)/sqrt(length(ind));          %Standard error
end
yCenter = yEdge(1:end-1) + .5;                      %Center of each bin
errorbar(yCenter,meanS,2*errS,'.')                  %Plot mean with error bars
```

In this case, the mean spike count seems to vary roughly linearly with the position. If you fit the spike counts as a linear function of position using **polyfit** (see Chapter 17) and plot the fit, you will end with something like Figure 22.1.

### 22.2.1 Linear Filter

You now have a decoder relating one neuron's spike counts $S$ to the instantaneous $y$-position $Y$, which takes the familiar linear form, $Y = mS + b$, where m and b are the coefficients returned by polyfit. But what exactly is the MATLAB software doing when you run polyfit? It is determining the optimal linear regression that minimizes the squared residuals (which are the differences between the fitted data and the actual data). The optimal linear regression can be expressed analytically. If you express the linear relationship in matrix form as $Y = Sf$, where $Y$ is the kinematics, $S$ is the spike counts, and $f$ is the decoding filter (the $y$-intercept b has disappeared, but you will add a column of 1s to $S$ to account for this), then minimizing the squared residuals is the same as solving the following equation:

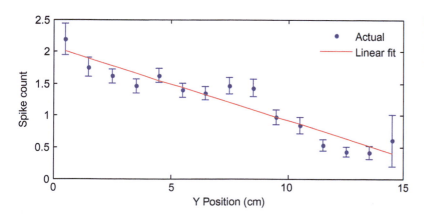

FIGURE 22.1 A linear fit of the relationship between $Y$-position and the spike count.

$$(S^T S)f = S^T Y \tag{22.1}$$

You can use this to solve for the desired decoder:

$$f = (S^T S)^{-1} S^T Y \tag{22.2}$$

A matrix inverse can be computed in MATLAB with the function **inv**, and a transpose is denoted with an apostrophe. Compare this solution to the values you derived with **polyfit**:

```
p = polyfit(s,y,1)          %Linear regression
s = [s ones(length(s),1)];  %Add a row of 1's to allow y-intercept
f = inv(s'*s)*s'*y          %Analytical linear regression
```

The advantage of this approach is that you can easily add more neurons or more kinematics to the model: you just add more columns to $S$ and $Y$. The following code shows how to fit a linear filter to the training data for this dataset, using all of the kinematics and all the neuronal firing rates lagged two time bins (140 ms):

```
numBin = length(kin);                   %Number of datapoints
yTrain = kin(lag + 1:numBin,:);         %Kinematic training data
sTrain = rate(1:numBin-lag,:);          %Neural training data
sTrain = [sTrain ones(numBin-lag,1)];   %Add vector of ones for baseline
f = inv(sTrain'*sTrain)*sTrain'*yTrain; %Create linear filter
```

Once you fit the linear filter to the training data, load the test data and use the following code to see how well the decoder performs (remember, you always want to test on different data than you trained with to prevent over-fitting):

```
numBin = length(kin);
sTest = [rate(1:numBin-lag,:) ones(numBin-lag,1)]; %Test neural data
yActual = kin(lag + 1:numBin,:);                   %Actual test kinematics
yFit =  sTest*f;                                   %Predicted test kinematics
```

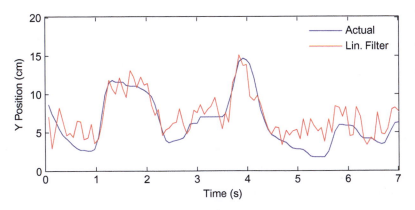

**FIGURE 22.2** The actual *Y*-position of the hand (blue) compared to the reconstruction using a linear filter (red). Only the first 100 datapoints are shown here.

If you plot the actual Y-position versus the estimated position, the linear filter seems to be doing a pretty good job (Figure 22.2). This approach of using a linear filter as a decoder was used in 2002 to show that a macaque monkey could successfully control a computer cursor with neuronal signals (Serruya et al., 2002) and again in 2006 to show that a human with tetraplegia could control a variety of neural prosthetics with neuronal signals (Hochberg et al., 2006). Of course, the linear filter can be used on data recorded outside the motor cortex. For example, it has been used to decode visual information from retinal ganglion cells (Warland et al., 1997).

## 22.2.2 Maximum a Posteriori (MAP) Estimation

We just saw that a linear filter can be used to generate a continuous estimate of the position of the hand from neural data. This is different, though, than what we did in Chapter 21, where we computed the probability of the neural data given each movement direction and picked the direction which corresponded to the most likely data. Can we use that approach here? Yes; however we will simplify the problem by approximating the continuous estimate with a discrete one, by binning the hand position. Since the targets were 1 cm square, we can try to decode the position at a 1 cm resolution. How do we do that?

First, we need a prediction of the neural firing rate. In Chapter 18, we introduced log-linear models, which can be fit with generalized linear model techniques using the function **glmfit**. This has the advantage of always predicting a non-negative firing rate. We will use that here to fit an encoding model relating the Y-position to neural firing (recall that we assumed fixed lag of 2 time bins, or 140 ms). We can then use this model to predict the firing rate of each neuron at one of 15 evenly spaced Y-positions which cover the playing field.

```
yCenter = [0.5:1:14.5];                               %Discrete set of positions
numNeuron = size(rate,2);                             %Number of neurons
for n = 1:numNeuron                                   %Loop over all neurons
   coeff(n,:) = glmfit(yTrain(:,2),sTrain(:,n),'poisson');  %Fit the encoding model
   sFit(n,:) = exp(coeff(n,1) + coeff(n,2)*yCenter);  %Predict firing rate
end
```

We already assumed that the neuronal firing rates are Poisson distributed for the encoding model. If we assume all neurons are conditionally independent, then the probability of a vector of spike counts is just the product of the probabilities of each spike count given our prediction (see Equation 21.4). We mentioned in Chapter 21 that it is sometimes advantageous to work with the sum of the logs of probabilities instead of the product of the probabilities, but it doesn't turn out to be an issue with this data. What this gives us is the conditional probability (discussed in Chapter 20) of observing a vector of spike counts $S$ given we are at position $Y$: $P(S|Y)$.

```
for t = 1:length(sTest)                               %Loop over all trials
   frTemp = sTest(t,1:numNeuron);                     %Select spike counts for one time bin
   prob = poisspdf(repmat(frTemp',1,15),sFit);        %Prob. of each count given position
   probSgivenY(t,:) = prod(prob);                     %Prob. of all counts given position
end
```

We aren't done yet, though, because this isn't actually the probability we are interested in. Rather, we want the conditional probability of the position $Y$ given the observed spike counts $S$: $P(Y|S)$. This is more useful, because to decode position, we just select the position $Y$ with the highest probability. But how do we convert from one to the other? Recall from Chapter 20 that conditional distributions can be written in terms of a joint distribution: $P(S|Y) = P(S,Y)/P(Y)$ and $P(Y|S) = P(S,Y)/P(S)$. We can make a simple substitution to come up with an expression known as Bayes' rule:

$$P(Y|S) = \frac{P(S|Y)P(Y)}{P(S)} \tag{22.3}$$

When people talk about Bayes' rule or Bayesian statistics, they often talk about prior and posterior distributions. In the context of this decoding problem, the prior distribution is $P(Y)$. That is, the prior represents whatever prior knowledge we have about the position of the hand, before we see the neural spike counts. We then use Bayes' rule to incorporate this prior with the information we get from neural data. The result of this combination is $P(Y|S)$, called the posterior distribution. This represents all of our knowledge of the position of the hand after we get to see the neural firing rates.

What prior information do we have? To start with, we can just look at how long the cursor dwells near one of the 15 discrete positions, and compute a histogram. If we then normalize the histogram to sum to 1, that gives us $P(Y)$. Now, the expression above also contains $P(S)$, but it turns out we don't actually have to calculate this, because we don't care about what the probability of a given position actually is, we just want to know which position is most probable. This strategy is known as *maximum a posteriori* (MAP) estimation. "A posteriori" is just the Latin name for knowledge possessed after making an observation (in this case, of neural spike counts), whereas "a priori" refers to knowledge possessed before making an observation. The MAP estimate of the position is shown below (the hat over the $Y$ indicates it is an estimate of position and not the actual value $Y$):

$$\hat{Y}_{MAP} = \max_Y P(S|Y)P(Y) \tag{22.4}$$

The code below determines the prior distribution $P(Y)$ and the MAP estimate of the position:

```
probY = histc(yTrain(:,2),[0:15]);           %Histrogram of position values
probY = probY(1:15)/sum(probY(1:15));        %Convert to a probability
for t = 1:length(sTest)
   probYgivenS(t,:) = probSgivenY(t,:).*probY';   %P(Y|S) is proportional to P(S|Y)P(Y)
   [temp maxInd] = max(probYgivenS(t,:));          %Find index of maximum
   mapS(t) = yCenter(maxInd);                       %Find position with highest P(S|Y)
end
```

The results of the MAP decoding are shown in Figure 22.3 for the first few seconds.

While the MAP estimate does okay, it is a little disappointing that it appears to do worse than a linear filter, but takes more effort to compute. How can we improve on this algorithm?

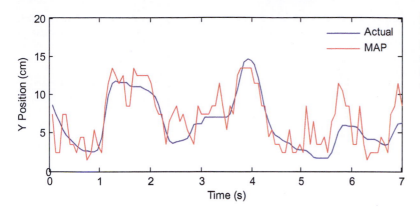

FIGURE 22.3 The actual Y-position of the hand (blue) compared to the reconstruction using maximum a posteriori (MAP) estimation (red).

### 22.2.3 Recursive Bayesian Estimation

Our MAP estimate contained two components: $P(S|Y)$ and $P(Y)$. The first expression, $P(S|Y)$, is called the *observation equation*, because it details the probability of observing a given set of spike counts $S$ given the position at $Y$. This is the same as the encoding models we developed in Chapter 18. But what about the prior distribution, $P(Y)$? If you actually look at values for the variable "probY," computed above, you'll see that all it says is that the hand is more likely to be in the middle of the playing field than at an edge. That's not a lot in terms of prior information. An alternative way to determine the prior distribution for Bayes' rule is to use a *state equation*, which describes how the position of the hand evolves as a function of time. The hand can only move so much during a 70 time bin, so we will make the assumption that the position of the hand in the next time bin $y_{t+1}$ is a function of the hand in the current time bin $Y_t$, plus some error term $e_{t+1}$ (which in this case is the hand's velocity).

$$Y_{t+1} = Y_t + E_{t+1} \tag{22.5}$$

So what kind of distribution can we define for this error term? We can look at this by simply differentiating the position for our training data, and looking at what kind of values we see. Since the $Y$-position is more or less between 0 and 15 cm, the error term should be between $-15$ and $+15$ cm/bin. So we will compute a histogram of the observed errors between the position terms for each time bin.

```
yDiff = diff(yTrain(:,2));           %Compute error term
yDiffEdge = [-15.5:15.5];            %Define edges
yHist = histc(yDiff,yDiffEdge);      %Compute histogram
probDiffY = yHist(1:31)/sum(yHist);  %Convert to probability
bar([-15:15],probDiffY);             %Plot probability
```

Your output should look like Figure 22.4.

What is interesting about this figure is how tightly clustered it is around zero. In fact, over 97% of the bins have an absolute change in position of less than 2.5 cm. So the position doesn't change that quickly as a function of time. This means that if we have a good

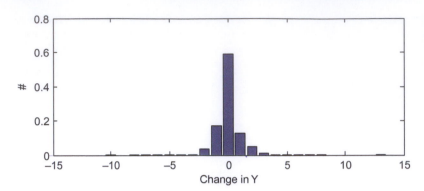

FIGURE 22.4 A histogram of the changes in Y-position, in units of cm per time bin.

estimate of where the position is in the current time bin, we also have a pretty good estimate of where it will be in the next time bin. How do we actually calculate the probability of cursor position of the next time bin if we know the probability of the current time bin? We just have to add up all the different combinations which could lead to a given position. For example, if I know the probability of being at each position now, what is the probability of being at 5 cm one time step from now? We just need to sum the probabilities of all combinations which add up to 5: start at 5 and stay there, start at 6 and move down one, start at 4 and move up one, etc. This kind of procedure is known as *convolution*, which we touched on in Chapters 16 and 20. Basically, we just need to take the probability distribution of the position at the last time point, and "blur" it by how much we think the position can shift in one time step. The probability of the current position at location $j$ can be written:

$$P(Y_t = j) = \sum_{k=1}^{15} P(E_t = j - k)P(Y_{t-1} = k) \tag{22.6}$$

In MATLAB, we could use a loop to evaluate this expression, but it is faster to use the built-in function **conv**. We can use this probability distribution as an alternative form of the prior information about position at the current time point $t$: $P(Y_t)$. We generate this prior information from our previous best estimate of the position at the last time point $t - 1$: $P(Y_{t-1})$. This is what is meant by recursive Bayesian estimation; the estimate of the current time point is based on an estimate of the last time point to create prior information, which is incorporated with neural data using Bayes' rule. Given Bayes' rule, we can describe the probability of the current position $Y_t$ given all the neural observations up to the current time point $S_{1:t}$ as the following:

$$P(Y_t|S_{1:t}) = \frac{P(S_{1:t}|Y_t)P(Y_t)}{P(S_{1:t})} \tag{22.7}$$

We're going to make a few assumptions to simplify things. First, we assume that the current position only affects the current neural observation (or rather, the neural observation at a specific lag), which means we can replace $P(S_{1:t}|Y_t)$ with $P(S_t|Y_t)$. Second, for the expression $P(Y_t)$ we will make use of the state equation, which assumes that the position at the current time is a function of the position of the last time point. However, we don't know the exact position at the last time point, but we will assume that we have probability distribution of the position given past neural observations: $P(Y_{t-1}|S_{1:t-1})$. We can convolve this expression with the velocity term in our state equation, $P(E_t)$, to come up with a prior distribution of position for the current time point: $P(Y_t|S_{1:t-1})$. Finally, we will ignore the normalizing factor $P(S_{1:t})$, and look at a proportionality (designated with an "$\propto$") instead of an equality. Because we are dealing with a discrete set of positions, we can normalize by forcing the probabilities to sum to 1. Thus, we can combine these assumptions to come up with the following expression for our a posteriori (after we see the neural observations) estimate of position:

$$P(Y_t|S_{1:t}) \propto P(S_t|Y_t)P(Y_t|S_{1:t-1}) \tag{22.8}$$

The expression for our *a priori* (before we see neural data) knowledge, $P(Y_t|S_{1:t-1})$, is found by convolving the last estimate, $P(Y_{t-1}|S_{1:t-1})$, with the error term ($E_t$) in our state equation, as we did earlier:

$$P(Y_t = j|S_{1:t-1}) = \sum_{k=1}^{15} P(E_t = j - k)P(Y_{t-1} = j|S_{1:t-1}) \tag{22.9}$$

There is one curious thing about this expression—we always reference the last estimate $P(Y_{t-1}|S_{1:t-1})$. But what do we do on the first time step? Well, it turns out we just need to make a guess, and the estimate will converge to a better estimate over a few time steps. For simplicity, we can assume a uniform distribution over our 15 possible positions for the first time step: $P(Y_0) = 1/15$. We then convolve this with the error distribution we determined earlier. In MATLAB, we can do this with the following code:

```
probTemp = conv(probDiffY,ones(1,15)/15);  %Convolve last estimate with error term
%Trim out the middle
probPriorY(1,:) = probTemp(16:30)/sum(probTemp(16:30));
```

However, when we convolve, this actual gives use a distribution of positions from $-14$ cm to 30 cm. We are assuming the position can only be between 0.5 and 14.5 cm, so we trim out the middle portion of the distribution, and re-normalize so the probabilities sum to one.

This gives us the prior estimate of position. To get the posterior estimate, we just need to multiply (element by element) this prior distribution with the probability of the neural data $S$ given the position $Y$, which we already computed for the MAP decoder. Again, we will normalize to make the probabilities sum to 1. Finally, instead of picking the most likely position, our estimate will be a probability-weighted sum of the possible positions.

```
probPostY(1,:) = probPriorY(1,:).*probYgivenS(1,:);    %Combine prior with neural data
probPostY(1,:) = probPostY(1,:)/sum(probPostY(1,:)); %Normalize probabilities
bayesY(1) = sum(probPostY(1,:).*yCenter);              %Bayesian estimate of position
```

That's just the first time point, though. For the rest of the time points, we do the exact same thing, but we proceed recursively, always building off of our previous estimate.

```
%Recursive Bayesian decoder
for t = 2:length(sTest);
    %Convolve last estimate with error term for prior
    probTemp = conv(probPostY(t-1,:),probDiffY);
    probPriorY(t,:) = probTemp(16:30)/sum(probTemp(16:30));


    %Combine prior with neural data for the posterior
    probPostY(t,:) = probPriorY(t,:).*probYgivenS(t,:);
    probPostY(t,:) = probPostY(t,:)/sum(probPostY(t,:));


    %Convert distribution to a single estimate of position
    bayesY(t) = sum(probPostY(t,:).*yCenter);
end
```

If you compare the recursive Bayesian estimate to the actual position, your results should look similar to those shown in Figure 22.5. Notice how the estimate is much less jerky than the MAP estimate or the linear filter, because we enforce continuity of the position trace with our state equation.

There are some caveats to this comparison. First, the linear filter doesn't traditionally make use of just one time bin's worth of data. Instead, the linear filter can make use of all neural data going back as much as a second. This averages out some of the noise inherent in measuring spike counts, so the position estimate becomes much smoother. You will look at the effect of adding more neural data to the linear filter in the exercises. Second, our recursive Bayesian decoder only looked at position tuning in the neural data, even though we know that neurons also encode velocity. You will also add velocity to the Bayesian decoder in the exercises.

One drawback of the discrete Bayesian decoder we present here is that the computational requirements increase quickly as you add more degrees of freedom. Here, we have 15 possible states for the $Y$-position. If we wanted to track $X$ and $Y$ concurrently, we

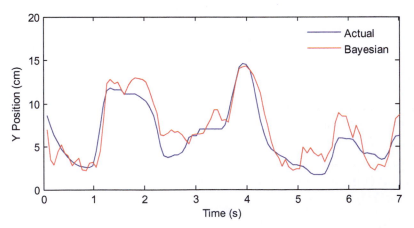

**FIGURE 22.5** The actual $Y$-position of the hand (blue) compared to the reconstruction using recursive Bayesian estimation (red).

would have 15^2 = 225 possible states, and tracking X and Y velocity would increase the state space even further. For this reason, real-time Bayesian decoding for brain machine interfaces often makes use of a Kalman filter (Wu et al., 2004), which provides a computationally efficient, closed-form solution to the decoding problem, if certain assumptions are made (linear observation and state equation; Gaussian noise). The advantage of discrete Bayesian decoding is that the observation and state equation can be changed to anything you want, so they can handle nonlinearities. You can explore that in the project for this chapter, where you will design and implement your own decoder.

## 22.3 EXERCISES

---

### EXERCISE 22.1

**Linear Filter**

1. Train a linear filter on the first dataset for this chapter using a two-time-bin (140 ms) lag, and test it on the second dataset. Report both the mean-squared error (MSE) and the correlation coefficient between the actual Y-position and the decoded output.

2. Test lags from 1 to 10 time bins, and see how this affects the MSE and the correlation coefficient. Is two time bins the optimal lag?

3. You can use more than one lag by appending the lagged rates as extra columns of X. For example, if you used lags of 1 and 2 time bins, you would have 84 columns instead of 42 (not including the column of 1s). This corresponds to a filter length (which is the total number of lags) of 2. Test a variety of filter lengths. What do you think the best filter length is?

---

### EXERCISE 22.2

**Recursive Bayesian Decoder**

Try adding a Y-velocity component to the state and observation equation. This means tracking both the position and the velocity as a function of time, and making the position at the current time a function of previous estimates of position Y and velocity V, while the velocity would be a function of its previous estimate. You will also need an encoding model (or observation equation) that is a function of position and velocity. How does adding velocity affect performance?

$$Y_{t+1} = Y_t + V_t + E_Y$$

$$V_{t+1} = V_t + E_V$$

(22.10)

## 22.4 PROJECT

Create your own decoding algorithm. The simplest scenario would involve modifying the state or observation equations for the recursive Bayesian decoder. For example, perhaps neural firing is actually a non-linear function of the hand position or velocity (the function **nlinfit** might be useful). Or perhaps the current position is a function of more than just the previous position. You could also try your own techniques, as long as they are optimized solely from the training data. Comment on the assumptions that you are changing, and why you think this is an improvement. Report both the mean-squared error and the correlation coefficient of your fit for Y-position of the test data.

## MATLAB FUNCTIONS, COMMANDS, AND OPERATORS COVERED IN THIS CHAPTER

**diff**
**inv**
**nlinfit**
**polyfit**

# Local Field Potentials

## 23.1 GOALS OF THIS CHAPTER

Thus far, we have analyzed data in the form of neural spike times, in the context of how neurons encode the state of the outside world, and in the context of how the state of the world can be decoded from neuronal spiking. Here, we will examine a different neuronal signal, the local field potential. We will look at encoding and decoding by making use of frequency analysis (covered in Chapters 11 and 12). Finally, we will introduce Chronux, an open-source software package written in MATLAB®, which can be used to analyze neural data.

## 23.2 BACKGROUND

An extra-cellular electrode placed in the brain registers the activity of nearby neurons as a change in the voltage across the electrode. Action potentials are very fast, and so they cause very quick changes to the voltage trace. In an attempt to isolate the activity of just one neuron (or perhaps a few neurons), the voltage trace is high-pass filtered (for frequencies above 250 Hz) before spikes are detected by threshold crossings. However, the lower frequencies have information about neurons as well. When the voltage trace is low-pass filtered (for frequencies below 250 Hz), the result is known as the *local field potential*, or LFP. LFPs reflect the summed activity of thousands of neurons near the electrode. "Near" is an imprecise term, but one paper looking an orientation tuning in visual cortex concluded that LFPs reflected the activity of neurons within 250 microns of the electrode tip (Katzner et al., 2009). Because unsynchronized neural activity should cancel out, LFPs are thought to reflect synchronous or oscillatory firing of the neuronal population.

What does the LFP look like? In the dataset for this chapter, the variable **lfp** contains the raw LFP recorded from the dorsal premotor cortex, sampled at 1 kHz and in units

*MATLAB® for Neuroscientists.*
DOI: http://dx.doi.org/10.1016/B978-0-12-383836-0.00023-0

of microvolts ($\mu$V). However, instead of looking at the raw data, neuroscientists usually subdivide the LFP by band-pass filtering it. We will follow the methodology of a paper by O'Leary and Hatsopoulos (2006), which examined directional tuning in LFPs recorded in motor cortex. This paper divided the LFP into high (25–45 Hz), medium (10–25 Hz), and low (<10 Hz) frequency bands. The high and medium bands are sometimes referred to as gamma and beta oscillations, respectively, though the usage varies among authors.

Filtering is basically a way to smooth data by computing a weighted, moving average. For example, in Chapter 20 we smoothed binary spike data with a Gaussian kernel to produce a spike density function. MATLAB has several built-in functions to help with filtering, including **filter** and **filtfilt**. If we have a vector $x$ and want to create a filtered vector $y$, we can do this by calling **y = filter(B,A,x)**. Here, the vector $B$ specifies the coefficients that are applied to raw signal $x$. The vector $A$ specifies coefficients applied to the filtered signal, but we will ignore that for now, and set $A$ equal to 1. Let's take an example where the raw signal $x$ is a ramp from 1 to 10 over 10 seconds, plus some amount of corruption from noise. We can create a filtered signal by taking a three-point moving average. We do this with **filter** by setting the vector of coefficients $B$ to be [1/3 1/3 1/3]. The function **filter** only uses current and past data, so this means the third element of $y$ is the average of the first three elements of $x$. The function will ignore the missing paste values in the beginning, so here the first element of $y$ will just be one-third the first element of $x$.

```
x = [1:10] + rand(1,10)*2;    %Raw signal X, plus some noise
y = filter([1/3 1/3 1/3],1,x);  %Filter x with a three point moving average
```

However, there is one issue with the function **filter.** If we ignore the noise, the third element of $y$ will be the mean of the vector [1 2 3], or 2. But the original value (before the noise) was 3, so our filtered value is one off. To put it another way, the filtered signal lags the raw signal by 1 second because it won't reach the value of 3 until 4 seconds in, not at 3 seconds. Thus filtering in this manner introduces a time-delay. One way to fix this is to filter the data again, but to do the process in reverse order so that the two time-delays cancel. That is what the function **filtfilt** does (see code below). If you plot the raw signal $x$ and the two filtered versions $y$ and $z$ from the two different functions, you will end up with something like Figure 23.1. This figure makes it clear that while the output of **filter** lags the raw data, the output of **filtfilt** does not.

```
z = filtfilt([1/3 1/3 1/3],1,x);  %Filter x forwards and backwards, to avoid a time lag.
```

To filter our LFP signal, we will use a more complicated filter called the Butterworth filter. The mathematics of filter design is beyond our scope here, but a relevant discussion can be found in the book *Signal Processing for Neuroscientists* by Wim van Drongelen. In short, we can specify the frequency range we want to focus on, and then use the MATLAB command **butter** to create the appropriate coefficients of a Butterworth filter that we can then apply to our data using **filtfilt**. There are many other choices for a band-pass filter, such as the Elliptic filter (MATLAB command **ellip)** or the Chebyshev filter

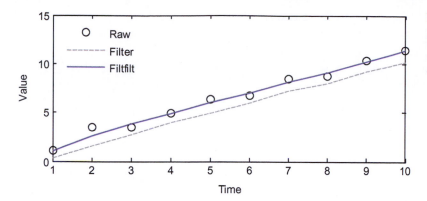

FIGURE 23.1 A comparison of a raw signal and the filtered version using the **filter** and **filtfilt** commands.

(**cheby1**), but the Butterworth filter is commonly used and will serve our purposes here. The Butterworth is an auto-regressive filter, meaning it acts on past filtered values as well, so the coefficient *A* won't be 1 like in the previous example. For function **butter**, you need to specify the order, which is how many past samples the filter will look at. For example, the earlier three-point moving average is order two, because it looked at the current and two past values of the raw signal. You can use the code ahead to filter the raw LFP into the three frequency bands we want:

```
Fs = 1000;                                  %Sampling frequency
n = 4;                                      %Controls the order of the filter
fpass = [0 10; 10 25; 25 45];              %Frequency bands
[b,a] = butter(n,2*fpass(1,2)*2/Fs,'low');  %Low-pass filter
lfp(:,2) = filtfilt(b,a,lfp(:,1));
for ii = 2:3
   [b,a] = butter(n,fpass(ii,:)*2/Fs);     %Band-pass filter
   lfp(:,ii + 1) = filtfilt(b,a,lfp(:,1));  %Run forwards and backwards
end
```

In this code, we also specified the sampling rate *Fs*, which is 1000 times per second for these data. Also, the frequency range is given as a number between 0 and 1, where 1 is the Nyquist limit (*Fs*/2). Recall that the Nyquist limit (see Chapter 11) specifies that we cannot faithfully capture a frequency greater than the sampling rate divided by 2. Thus it doesn't make sense to target a frequency above the Nyquist limit.

The previous code filters the LFP recorded over the entire experiment. However, we will want to examine the LFP when something behaviorally interesting is going on. These data were collected during a center-out reaching task with an instructed delay, just like the neural spiking data from Chapter 17. The dataset contains the time (in seconds) that the target first appeared (*instruction*), the time the go cue was triggered (*go*), and the time the research participant actually left the center target (*startMove*). To convert the time of a cue to the indices of the LFP, just multiply it by the sampling rate, and round off. The code ahead plots

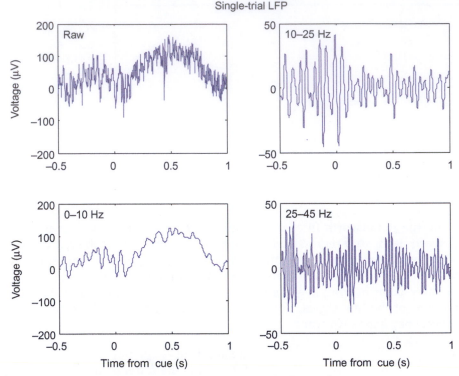

**FIGURE 23.2**    A comparison of the single-trial raw and filtered LFP signals.

the raw LFP in a window centered on the first instruction cue. If you plot the filtered LFPs as well, you will get something like Figure 23.2.

```
cueInd = round(instruction(1)*Fs);   %First instruction cue
indTrial = cueInd-Fs/2:cueInd + Fs;  %-500 ms to +1000 ms around cue
time = [-500:1000]/Fs;
plot(time,lfp(indTrial, 1));
```

## 23.2.1 Evoked Potentials

While it is clear from Figure 23.2 that the amplitude of the signals are changing as a function of time, it is not clear how much of that change is related to the onset of the instruction cue. This is because, like any physiological signal, the LFP is partially corrupted by noise. We can help eliminate the noise by averaging across trials to create an *evoked potential*. The code ahead will average the LFPs across trials in a window centered on the instruction time. If you plot these averaged evoked potentials, you will get something like Figure 23.3.

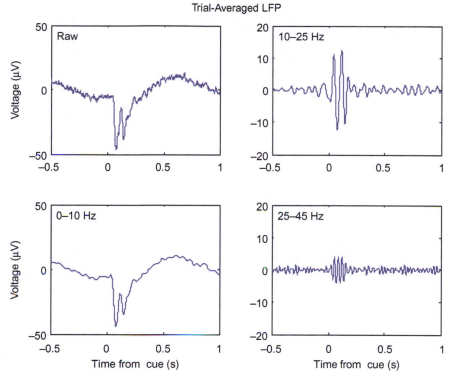

**FIGURE 23.3** A comparison of the trial-averaged, or evoked, LFP signals.

```
lfpEvoked = zeros(1501,4);                                    %Initialize evoked lFP
for ii = 1:length(instruction);                              %Loop over trials
   cueInd = round(instruction(ii)*Fs);                       %Find cue for this trial
   for jj = 1:4
                                                             %Sum LFPs across trials
      lfpEvoked(:,jj) = lfpEvoked(:,jj) + lfp(cueInd-Fs/2:cueInd + Fs,jj);
   end
end
```

Now it is clear that something is happening around the instruction cue. First, both the raw and the low-pass filtered LFP show a sharp downward deflection after the instruction appears, before returning to baseline. Second, both the high and medium frequency bands show an increase in the amplitude (peak-trough distance of the oscillations), which starts at the instruction cue and lasts for around 200 ms.

## 23.2.2 Directional tuning

This LFP was recorded from motor cortex during a center-out task, and we saw in Chapter 17 that single units in motor cortex encode the direction of the upcoming reach. What about LFPs? The simplest way to examine directional tuning is to look at evoked potentials averaged across trials in the same direction. The direction of each trial is stored in the variable *direction*, where 1 is 0 degrees (movement to the right), 2 is 45 degrees, and so on up to 8, which is 315 degrees. If you average the LFPs on the instruction cue for the 10–25 Hz frequency band (using the **find** command to isolate trials in a given direction), and plot the evoked LFPs in subplots corresponding to the direction of movement, you will get Figure 23.4.

The evoked LFP for each direction looks like a scaled version of the evoked LFP across all trials. However, the LFP response is strongest for movement down and to the right. How do we quantify this response? Our signal here consists of oscillations, thus both the positive and negative deviations indicate a response. The simplest thing to do is to square the measured voltage, and then average it over a time window of interest, similar to O'Leary and Hatsopoulos (2006). Taking the square root of this number is helpful, as it yields a measure in the original units (μV). The following code will compute the LFP response in a 200 ms window following the instruction cue:

```
for ii = 1:length(instruction)
    cueInd = round(instruction(ii)*Fs);        %Find index of instruction cue
    temp = lfp(cueInd:cueInd + 199,3);         %Take 200 ms window after cue
    lfpResponse(ii) = sqrt(mean(temp.^2));     %Root-mean square of voltage
end
```

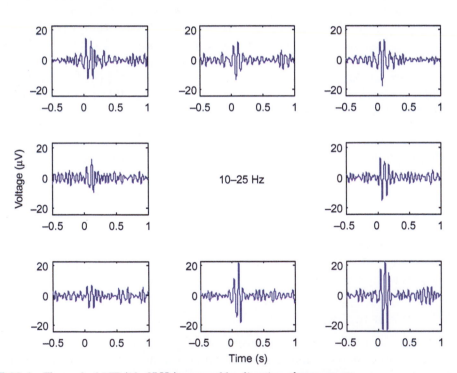

FIGURE 23.4   The evoked LFP (10–25 Hz) arranged by direction of movement.

Now we want to look at how this LFP response varies as a function of direction. We can take the mean of the response for trials of each direction, but we want to have a sense of the variability of the data as well, so we will also compute the standard deviation of the data. However, directional tuning implies that the mean of the response will vary as a function of the direction of movement. We therefore want to know how good our estimate of the mean is. This is captured by the *standard error of the mean*, which is the standard deviation of the expected values of our sample mean. The standard error can be estimated by dividing the sample standard deviation by the square root of the number of samples. The following MATLAB code computes the mean, standard deviation, and standard error of the LFP response we just computed.

```
for jj = 1:8
  ind = find(direction == jj);          %Find trials in given direction
  lfpMean(jj) = mean(lfpResponse(ind));  %Mean
  lfpStd(jj) = std(lfpResponse(ind));    %Standard deviation
  lfpErr(jj) = lfpStd(jj)/sqrt(length(ind)); %Standard error
end
```

If the data are assumed to follow a normal distribution, then the sample mean should follow a normal distribution with the standard error as its standard deviation. Thus we can use the standard error to calculate a *confidence interval*. A random variable drawn from a normal distribution has a 95% chance of being within 1.96 standard deviations of the mean (you might verify that with a simulation using the function **randn**). A 95% confidence interval computed on our data implies that if we were to repeat our experiment 100 times, in 95 of them the true mean would fall within the confidence interval we compute. It is not quite correct to say that there is a 95% chance that the confidence interval contains the true mean, but the precise meaning of a confidence interval is tricky. For our purposes, it is sufficient to know that it is a measure that indicates the reliability of an estimate (in this case, the mean).

For practical purposes, confidence intervals are used to visually display how much variability there is in the data, and to give an approximation as to whether two variables are significantly different (their confidence intervals should not overlap). In MATLAB, we can plot data with error bars corresponding to plus or minus one standard error using the function **errorbar.** Plotting our data this way confirms our suspicion that the data vary by direction (Figure 23.5).

```
deg = [0:45:315]; %Direction of movement in degrees
errorbar(deg,lfpMean,lfpErr,lfpErr,'o')
```

It does not appear that the data are cosine tuned, as there are two local maxima in the figure, and a sharp drop-off from the maximal value. However, we can perform a test of whether the means are significantly different by performing a one-way analysis of variance, as covered in Chapter 7, by using the following code:

**anova1(lfpResponse,direction)  %Are means significant different?**

Another way to explore whether the LFP responses really are related to direction is to compare the response vector to a randomized vector of directions, instead of the actual

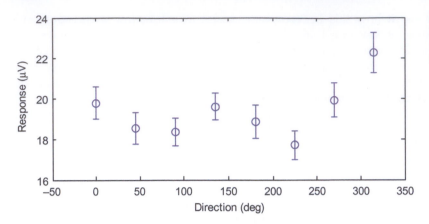

FIGURE 23.5 The mean LFP response (in the 10–25 Hz band) versus the direction of movement.

directions. In a sense, this is a negative control; if you repeat the analysis and it still shows the relationship with the random data, there is likely a mistake somewhere. The code below stores the random vector in the variable "dirRand." You can then go back to the last for-loop, and replace the variable "direction" with variable "dirRand." When you plot the response as a function of direction, the responses should be flat, and the analysis of variance should show that they are not significantly different.

```
dirRand = ceil(rand(length(go),1)*8);   %A random vector of 8 directions
```

### 23.2.3 Spectrograms

We introduced the concept of a spectrogram in Chapter 12, which shows the power of a time-varying signal at different frequencies. Looking at a spectrogram is useful here, because it might allow us to make a better choice of frequency bands and time windows used to examine directional tuning. We will analyze the data in a similar manner as the evoked potential, by computing a spectrogram on the single trials and then averaging all trials in a given direction.

We will start as we did in Chapter 12, using the built-in function **spectrogram.** We will compute the spectrogram of the first trial that we plotted in Figure 23.2. We also need to decide on how many data points to use in our window to compute the spectrum (we will use 128, or 0.128 seconds of data) and what frequencies we are interested in (we will look between 6 Hz and 60 Hz). Then you can run the code ahead to plot the spectrogram (shown in the left-hand panel of Figure 23.6).

```
n = 128;                                 %Number of points in moving window
freq = [6:3:60];                         %Frequencies we are interested in
[S,F,T,P] = spectrogram(lfp(indTrial, 1),n,[],freq,Fs,'yaxis');
maxDb = 15;                              %Maximum on scale for decibels.
imagesc(T-.5,F,10*log10(P),[0 maxDb])   %Plot spectrogrm
axis xy; xlabel('Time(s)'); ylabel('Freq (Hz)')
```

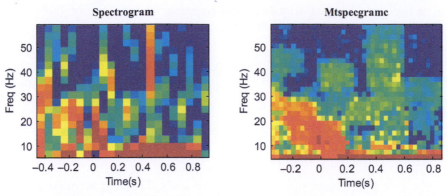

**FIGURE 23.6** A comparison of a single-trial spectrogram using the built-in function **spectrogram** and the Chronux function **mtspecgramc.**

Note that what we are actually plotting is the power of the spectrogram (stored in "P"), and that we are plotting power in units of decibels (dB). We converted to dB by taking the base-10 logarithm and multiplying it by 10. This means that something valued at 10 dB is 10 times greater than a reference value at 0 dB, while 20 dB is 100 times greater than the reference value at 0 dB. Such a logarithmic scale is useful for things that vary widely in scale (such as sound, where decibels are most often used). We also adjusted the time axis ("T-.5") to reflect that fact that time starts 0.5 s before the cue.

To compute an average spectrogram, we could repeat this code for all trials, store the power spectrogram, and average them all before plotting them. That is the procedure we used before for evoked potentials. However, this is a bit cumbersome here, and it turns out we can compute spectrograms faster and more accurately by using an open-source software package for neural data analysis known as *Chronux*, developed for MATLAB by the laboratory of Partha Mitra at Cold Spring Harbor. It can be downloaded for free from the web site http://chronux.org/. This software package can be used for a variety of purposes, such as generating raster plots and peri-event time histograms as covered in Chapter 17, but we will focus here on the frequency analysis code. After downloading the package you will need to add two sub-directories to your current path (under File then Set Path): "chronux\spectral_analysis\continuous" and "chronux\spectral_analysis\helper."

This software package makes use of a mathematical technique for computing the power spectrum known as multitaper spectral analysis. Discussion of multitaper analysis can be found in the book *Observed Brain Dynamics*, and an example of its application to LFP data can be found in Pesaran et al. (2002). In short, the "multi" in multitaper refers to the fact that it is possible to get several different, statistically independent estimates of the spectrum of a signal. These can then be averaged together, which gives a more precise estimate (and error bars, if desired) of the spectrum than is possible using the function **spectrogram**. Using the frequency analysis code in Chronux involves setting a number of parameters, including the number and resolution of the tapers. However, for now you can just use the parameters ahead, which work well for these data, and plot the spectrogram for a single trial using the Chronux function **mtspecgramc**. The output of the two functions (**spectrogram** and **mtspecgramc**) are compared in Figure 23.6.

```
params.Fs = 1000;                           %Sampling rate
params.tapers = [2 3];                      %Taper parameters
params.fpass = [5 60];                      %Look at 5-60 Hz band
movingwin = [0.3 0.03];                     %Window to compute spectrum
maxDb = 15;                                 %Limit power range to 15 Db
[P2,T2,F2] = mtspecgramc(lfp(indTrial, 1),movingwin,params);
imagesc(T2-.5,F2,10*log10(P2'),[0 maxDb])   %Plot power in dB
axis xy; xlabel('Time(s)'); ylabel('Freq (Hz)')
```

There are a few parameters here you can play with. The first element of the moving window (0.3) specifies how much data (in seconds) you use to compute the spectrum, analogous to the number of data points used earlier. The first element of "params.tapers" specifies the time-bandwidth product *TW*. This is a constant, so if you shrink the moving window, you will increase the corresponding frequency window. The second element of "params.tapers" specifies the number of tapers, which must be less than or equal to 2*TW*-1. See *Observed Brain Dynamics* for further discussion of the interpretation of these parameters.

You can see in Figure 23.6 that the output of the built-in function is noisier than that of the Chronux function, though they do show the same overall pattern. However, what we want to compute is an average spectrum over many trials, not just one. We can do this using the Chronux function **mtspecgramtrigc**, as demonstrated in the code ahead. This code will compute the average spectrum for just one direction, but the average spectrograms for each of the eight directions of movement are shown in Figure 23.7.

```
ind = find(direction == 1);                 %Find trials in given dir.
cueTime = instruction(ind);                 %Find instruction times
params.trialave = 1;                        %Average across trials
win = [1 1];                                %1 sec before to 1 sec after cue
[S,t,f] = mtspecgramtrigc(lfp(:,1),cueTime,win,movingwin,params);
figure
imagesc(t-win(1),f,10*log10(S'),[0 maxDb])  %Plot power in dB
axis xy                                     %Flip Y-axis
```

In Figure 23.7, high power is indicated by the color red, and low power is indicated by the color blue, with yellow in between (use the function **colorbar** for details on the scale). You can see that there is an increase in power between 5 and 30 Hz from the instruction cue to 200 ms afterwards. This increase is directionally tuned, and it is maximal for movement to the right and down. This is consistent with what we saw in the evoked LFP in the 10–25 Hz range (Figure 23.4).

Unlike the evoked potential, there is a second increase in power from 500 ms to 700 ms after the instruction cue. This increase is also directionally tuned, but now it is maximal for movement to the left. This did not show up in the evoked potential of the same data previously shown. This is because the evoked potential will only pick up an increase in

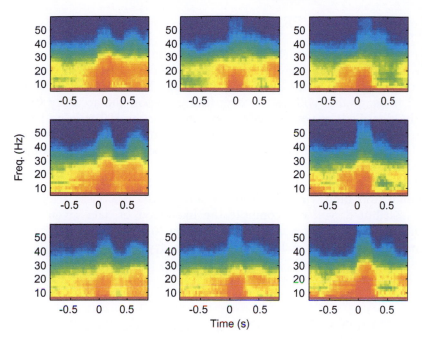

FIGURE 23.7 The spectrograms of the LFP averaged across all trials in a given direction.

amplitude where the responses across trials have a similar phase, which they do immediately after the instruction cue (see O'Leary and Hatsopoulos, 2006). If the responses are not phase-locked, they will cancel out even if the amplitude is increased if one averages across trials. Thus, for these data, the spectrogram has more information than the evoked potential.

## 23.3 EXERCISES

1. Compute evoked potentials for each direction in both the 10–25 Hz and 25–45 Hz bands for all three behavioral cues: the instruction cue, the go cue, and the start of movement. Compute an LFP response for the window from 0 to 200 ms after each event. Does the LFP response show directional tuning in all conditions? If so, what are the preferred directions for each condition? Do any conditions show cosine tuning (see Chapter 17)?

2. Compute an LFP response in the time window from 500 ms to 700 ms for the 10–25 Hz and 25–45 Hz bands. Is the direction tuning different from the time window from 0 to 200 ms? If so, what are the preferred directions for each band?

3. Compute evoked potentials for the <10 Hz band. We saw that this band shows a downward deflection after the instruction cue (Figure 23.2). Compute a baseline for each single trial by averaging the voltage for the 500 ms period preceding the instruction cue. Then integrate the filtered LFP from 0 to 200 ms after each behavioral cue. Does the low-pass filtered LFP show directional tuning for each condition? If so, what are the preferred directions?

## 23.4 PROJECT

Devise a decoder to map the single-trial LFP data to an estimate of the direction of movement for that trial. Train the decoder on the first 300 trials of data, and when you are done tweaking your decoder, apply it to the remaining 163 trials. Try not to test multiple decoders on the last 163 trials, as that will lead to over-fitting. Report the accuracy (percentage of trials correctly identified) of your decoder separately for the 300 training trials and 163 test trials. Since there are eight discrete movement directions, the baseline percentage correct should be 12.5%. You should be able to build a decoder which gets around 30% of test trials correct.

One approach is to record the mean and standard deviation of the LFP response for a handful of frequency bands and time windows. Then, assuming the responses are independent and normally distributed, you can compute the likelihood of data given each movement direction, and then pick the direction corresponding to the highest likelihood (see Chapter 21 for more on this approach). However, you may try a different decoder if you wish.

## MATLAB FUNCTIONS, COMMANDS, AND OPERATORS COVERED IN THIS CHAPTER

**filter**
**filtfilt**
**butter**
**mtspecgramc** (requires Chronux)
**mtspecgramtrigc** (requires Chronux)

# Functional Magnetic Resonance Imaging

## 24.1 GOALS OF THIS CHAPTER

This chapter will introduce you to functional magnetic resonance imaging (fMRI) as a fundamental noninvasive tool in understanding brain functioning in human and non-human primates. We will describe the basic physics behind both structural and functional magnetic resonance imaging. We will then describe the major experimental paradigms used in fMRI research and the kinds of data that are collected in an fMRI experiment. Finally, using existing fMRI data from a visual attention experiment and face recognition study, we will show you how to analyze and visualize the data to come up with a statistical parametric map of activation in the brain. After completing this project, you should expect to understand how researchers take fMRI data to infer activation associated with a behavioral task in different parts of the brain.

## 24.2 BACKGROUND

Functional magnetic resonance imaging has emerged as the dominant form of noninvasive functional imaging in humans. Although it is a relatively young technology that began in the early 1990s, it now plays a major role in many subfields of psychology, cognitive science, and neuroscience. It is even creeping up in other disciplines such as sociology and economics. As of the middle of 2013, a PubMed search revealed over 338,055 papers that reference the use of fMRI.

The development of fMRI began more than 70 years ago when Linus Pauling discovered that oxyhemoglobin and deoxyhemoglobin have different magnetic properties (Pauling and Coryell, 1936). He noted that deoxyhemoglobin was paramagnetic. However, it wasn't until the 1990s that researchers used these properties of hemoglobin to measure what would be

*MATLAB® for Neuroscientists.*
DOI: http://dx.doi.org/10.1016/B978-0-12-383836-0.00024-2

later termed the blood oxygen level-dependent (BOLD) signal (Bandettini et al., 1992; Kwong et al., 1992; Ogawa et al., 1992). Since that time, thousands of studies have been completed that have dramatically increased our understanding of the human brain. In 2001, Logothetis and colleagues demonstrated that the BOLD response correlates with local field potentials in monkeys viewing a flickering checkerboard, leading to the conclusion that the BOLD response is a post-synaptic phenomenon (Logothetis et al., 2001). Thus, the BOLD response is a reflection of the post-synaptic response to the local neural activity.

Functional magnetic resonance imaging using BOLD imaging provides a rapid *in vivo* glance into human brain function. These experiments are fundamentally based on the principle that the brain oversupplies blood and oxygen to regions that increase their activity relative to the state before a stimulus is presented; yet this indirect measurement must be interpreted cautiously and has led to some criticism of the field. Specifically, the term "BOLD response" is not interchangeable with "neural activity." While the BOLD response is a reflection of the neural activity, it is also dependent on neurovascular coupling including the calcium dynamics in astrocytes and the vasodilation of arterioles. Thus, group differences indicate physiological differences, but they do not necessarily indicate differences in neural activity. Additionally, negative results cannot be interpreted as not having a physiological difference. Finally, BOLD is based on the relative signal between two conditions, usually an experimental state and fixation or two experimental conditions. Since BOLD is relative and indirect, Logothetis and colleagues concluded that fMRI is limited to "conditionally confirm" prior knowledge from other studies (Logothetis et al., 2001). Therefore, inferences of brain function derived from fMRI should draw upon other psychophysical and physiological measurements.

## 24.2.1 Basic Physics of the MRI Signal

We will describe the basic physical principles that create the MRI signal. Although the physics behind MRI is inherently quantum mechanical, most of the ideas can be expressed in classical terms that you would learn in a high school physics course. There are two phenomena that must be understood: precession and relaxation. Atoms with an odd number of protons (or neutrons) in their nucleus such as hydrogen act like tiny magnetic dipoles because they possess a quantum mechanical spin. As you may remember from your high school physics, an electrical charge that is rotating will generate a magnetic field perpendicular to its rotational plane according to the right-hand rule. In the presence of an external magnetic field, $B_o$, these proton magnetic dipoles will tend to align with it by precessing around the $B_o$ axis like a top precesses about the gravitational field. The precession frequency, $F_o$, of the protons in the nucleus is proportional to the strength of $B_o$ with a proportionality constant that depends on the type of atomic nucleus:

$$F_o = \gamma B_o \tag{24.1}$$

This is called the *Lamour frequency*, and it characterizes the resonant frequency of the atomic material that is being imaged. In the presence of the static magnetic field, $B_o$, the spinning protons will eventually settle and align their spins with the external magnetic field and by doing so will create their own internal magnetization, $M_o$. The time constant that characterizes this settling or relaxation time is called the $T_1$ *time*.

To create an MRI signal, an external oscillating magnetic pulse, $B_1$, is applied in the transverse direction perpendicular to $B_o$. This pulse is called an *RF pulse* because the magnetic field frequency is in the radio frequency range (i.e., megahertz range) and typically lasts for a millisecond or so. This pulse is generated by a wire coil that lies in a plane parallel to $B_o$. If the oscillating frequency of $B_1$ is close to the resonant frequency (i.e., the precession frequency of the protons), the internal magnetization will be perturbed and shift its orientation toward the transverse direction. This is very much like a forced harmonic oscillator that will begin to oscillate with a very large amplitude if a forcing frequency matches the resonant frequency. The shifted internal magnetization of the protons will precess at its resonant frequency, and will inductively generate an electrical signal in the same coil that generated the RF pulse. Again from high school physics, you know that a changing magnetic field generates an electrical field, and will create an electric current if a wire is nearby. This inductive current will decay in time with a relaxation time constant $T_2$ after the RF pulse is turned off because the precessing protons that initially were in phase with each other will no longer be phase locked with each other. The time between pulses is referred to as the repetition time, *TR*.

To create an image from the MRI signal, additional gradient coils create a gradient in the static magnetic field, $B_o$, such that the strength of the static field varies linearly along different spatial axes. According to the Lamour frequency equation, the resonant frequency is proportional to the magnitude of the static field. In one dimension, a gradient will shift the resonant frequency of the atomic material. If an RF pulse is applied at a particular frequency, this will predominantly excite only one point along the spatial dimension. Imagine a set of harmonic oscillators with linearly varying resonant frequencies placed along one axis. A forcing oscillation at a particular frequency (the RF pulse in MRI) will excite those harmonic oscillators whose resonant frequencies are close to the forcing frequency. More importantly, the relative phase between the oscillator and the forcing oscillation will vary linearly along the axis. This is the essence of MRI imaging.

## 24.2.2 BOLD Signal (fMRI)

We will now describe the basic principles of the blood oxygen level-dependent (BOLD) signal. The oxygenation concentration of blood was discovered to alter the MRI signal (Ogawa et al., 1992). In particular, as the ratio of oxygenated to deoxygenated hemoglobin increased, the MRI signal increased. It was soon found that brain activation in the human also affected the MRI signal, presumably due to changes in blood oxygenation levels surrounding the brain tissue (Kwong et al., 1992). The time course of the BOLD signal initially shows a weak decrease, due to a relative increase in deoxyhemoglobin concentration associated with increased oxygen utilization, followed by a much stronger increase, due to relative decrease in deoxyhemoglobin concentration associated with the oversupply of oxygenated blood without a change in oxygen utilization, that peaks several seconds ($\sim$5 seconds) after a stimulus is presented to the subject. The mismatch between supply and utilization is critical to the BOLD phenomenon (Raichle and Mintun, 2006). Vascular physiology suggests that the source of this BOLD signal is primarily the venous and capillary blood as opposed to arterial blood and more specifically the dynamic relationship between cerebral blood flow, cerebral blood volume, and the cerebral metabolic rate of oxygen or oxygen utilization. Most studies have focused on the later, robust increase in the MRI signal. You should keep the relatively slow dynamics of the signal in mind when interpreting fMRI data because this places a limit to the temporal resolution of tracking neural activity and the ability to separate neural events close in time.

Functional magnetic resonance imaging initially began by simply subtracting the mean value of two states akin to computing the correlation between a simple box car function and the data; however, more recent studies use more complex modeling methods by incorporating knowledge of the hemodynamic response to identify active regions (Friston et al., 1995; Worsley and Friston, 1995). The hemodynamic response is the shape and amplitude of the BOLD signal in response to a stimulus. As the stimulus approaches an instantaneous impulse, the response becomes the hemodynamic response function (HRF). This is used in the convolution with the stimulus to create knowledge of the predicted BOLD response, which is then used to identify regions that are active. Initial studies of the HRF were completed in visual cortex in the mid-1990s (Boynton et al., 1996). More recently, the hemodynamic response has been shown to be variable across participants, regions, and trials due to the timing of neural activity and differences in neurovascular coupling (Aguirre et al., 1998; Handwerker et al., 2004; Steffener et al., 2010). To account for more of the variability, researchers have begun using the temporal and dispersion derivatives of the HRF (Calhoun et al., 2004; Steffener et al., 2010). In 2001, Henson and colleagues statistically demonstrated that the inclusion of these terms accounted for most of the variance in the hemodynamic response (Henson et al., 2001).

## 24.2.3 Preprocessing of the BOLD Signal

We will describe the data structure and the basic preprocessing steps for functional magnetic resonance data. The data that are acquired in a functional magnetic resonance imaging (fMRI) experiment require a number of preprocessing steps before formal statistical data analyses can be performed. The data we will give you, available on the companion web site (courtesy of Christian Buechel, Karl Friston, and Rik Henson), have already been preprocessed. However, it is helpful to understand what has been done to the data before you work with it (Smith, 2003a, b).

Functional magnetic resonance imaging consists of a series of whole brain images that are usually collected at uniform intervals typically between 1–5 seconds. Each whole brain volume is made up of a series of 2D brain slices, although newer structural imaging is collected in 3D rather than as a series of 2D slices. The signal is initially represented in the Fourier domain as a set of complex numbers at different frequencies and phases (k-space representation). This needs to be transformed into a set of real numbers in the time domain using a 2D Fourier transformation for each slice and each time point. The output of the Fourier transformation is an image for each slice in a proprietary format of the company that produced the scanner. The Fourier transforms allow the data to be viewed as shown in Figure 24.1. For illustrative purposes, a T1-weighted slice is shown in Figure 24.1 (top left). Each 2D slice is made up of many subunits known as voxels, or discrete volume elements (Figure 24.1, top right). The middle image in Figure 24.1 shows a

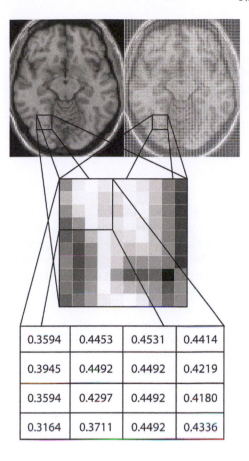

FIGURE 24.1 An illustrative example of voxels using a T1-weighted image. *Top left*: the view of the entire slice. *Top right*: the view of all subunits known as voxels, or discrete volume elements, within a single slice. *Middle*: a magnified view of a subset of voxels. *Bottom*: the numeric value of each voxel.

| 0.3594 | 0.4453 | 0.4531 | 0.4414 |
|--------|--------|--------|--------|
| 0.3945 | 0.4492 | 0.4492 | 0.4219 |
| 0.3594 | 0.4297 | 0.4492 | 0.4180 |
| 0.3164 | 0.3711 | 0.4492 | 0.4336 |

zoomed view of the voxels. Each of these voxels is actually a numeric value that can be viewed in color or grayscale as shown in Figure 24.1 (bottom). These images are then collated into whole brain images as either an ANALYZE format file (.img/.hdr pair) or a NIFTI-1 format file (.img/.hdr pair or .nii). The latter is more common with newer software packages. The difference between the image formats is the information stored in the header of the files. One key difference is that NIFTI-1 files now store the origin and rotation of the image, which allows linear transformations to be stored in the image headers.

Once you have collated the data into whole brain images, preprocessing can begin. In fMRI, we are generally analyzing each voxel individually. Each voxel has a 3D location associated with it and is measured across time to produce a time series that can be represented as a 4D matrix. Because each brain volume image is acquired as a series of two-dimensional slices, each slice is acquired at a slightly different time. Therefore, slice-timing correction is performed, although this is less important in block designs. Following slice-time correction, motion correction is then performed such that each whole brain volume is spatially aligned. Because motion and slice acquisition time are interdependent and processes for simultaneously correcting for both do not exist, some have argued that slice-time correction should be applied after motion correction. Importantly, if slice-time correction is performed first, then the motion correction will not require reslicing of the data. Next, data are spatially normalized to an atlas space, typically MNI space for fMRI (Evans et al., 1993), using both affine and non-linear transformations. Following this transformation, the data are resampled to a common grid to produce isotropic voxels that are generally 2 or 3 cubic millimeters in size. Using 2 cubic millimeter voxels will require about 4 times as much disk space as using 3 cubic millimeters. After the data have been resampled, it is spatially smoothed using a Gaussian kernel with a full-width half maximum of 1−2 times the voxel size, typically 8 mm. With the exception of resting state data, temporal filtering is usually included as part of the statistical analysis.

### 24.2.4 Experimental Designs

We will describe the four major types of experimental design used in fMRI: (1) the block design, (2) the event-related design, (3) the mixed block event-related design, and (4) the resting state design (Donaldson and Buckner, 2003). The block design was the first approach used in early fMRI experiments and is quite easy to implement. Blocks of time (typically tens of seconds long) are defined in which subjects are either presented with multiple stimuli or perform a task repeatedly (experimental block), or are presented with nothing or asked to rest (control block). These blocks are typically presented in an alternating fashion. The BOLD signal is then compared between the experimental and control blocks. Block designs are useful in clinical studies where the main question is where and to what extent a region is activated. In contrast, the event-related design uses a series of short, 0−10 second, rapidly presented stimuli. The event-related design is used when you want to examine the relationship between a behavioral event and the dynamics of the BOLD signal. Additionally, several different stimuli can be used and presented in a pseudo randomized fashion expanding the number of comparisons that can be made from a single study. However, each stimulus condition or experimental manipulation should have at least 30 events (Huettel and McCarthy, 2001). The mixed block event-related design is simply a combination of the block design and event-related designs. Short stimuli are grouped into separate blocks with the idea that the design will elucidate the brain state of the task as well as the responses to individual trials. This is the least commonly utilized design. Finally, the resting state design is the easiest design to implement. Researchers using a resting state design instruct the participant to lie still and either: (1) keep their eyes open; (2) keep their eyes closed; or (3) fixate on a crosshair in the middle of a screen (Biswal et al., 1995; Fox et al., 2005). This type of design is advantageous in clinical studies where patients might not be able to perform behavioral tasks or provide a response to the researcher.

### 24.2.5 Analysis Methods

We will briefly describe several analysis techniques that are used in fMRI analyses as well as different software packages. In this chapter, we will focus on the general linear model and apply it to each voxel in a mass univariate approach (Friston et al., 1995). The mass univariate approach utilizes the general linear model because of the ease of finding the solution using matrix inversion:

$$Y = X\beta + \varepsilon \tag{24.2}$$

$Y$ is a column vector and represents the fMRI time series data from a single voxel with $N$ data points. $X$ is an $N \times M$ matrix representing the predicted brain responses to the experimental stimulus and is formed by convolving each condition in the experimental design with the hemodynamic response function. $M$ is the number of experimental conditions. $\beta$ is a column vector of the unknown amplitudes of the BOLD response for each condition that is being estimated by the model and $\varepsilon$ is normally distributed noise. The solution to the general linear model is:

$$\beta = (X'X)^{-1}X'Y \tag{24.3}$$

where $X'$ is the transpose of $X$. Inferences are then based on whether $\beta$ is significantly greater than or less than 0 or different between two conditions.

While the examples in this chapter use the mass univariate approach, it is useful to know about other approaches. One commonly utilized approach is independent component analysis (ICA; Calhoun et al., 2001). Two advantages of ICA are that it: (1) does not require an *a priori* model of the expected BOLD response; and (2) does not suffer from the multiple-comparisons problem of mass univariate analyses. Another approach is partial least squares (PLS; McIntosh et al., 1996). The advantage of PLS is that it is a multivariate technique that can separate the effects of different conditions. Additionally, it also does not suffer from the multiple-comparisons problem.

An alternative to investigating evoked brain activity is to investigate brain connectivity during different tasks. Two main methods have been developed and used for this purpose. The first is psychophysiological interactions (PPI), which investigate whether the connectivity between two brain regions is modulated by the experimental condition (Friston et al., 1997; Gitelman et al., 2003; McLaren et al., 2012). PPI analyses are implemented using the general linear model with additional predictors that are added to $X$. Another approach that has been developed is called dynamic causal models (DCM), which investigates how connectivity between brain regions is modulated by experimental contexts using generative models that better capture the dynamic and nonlinear nature of the signal (Friston et al., 2003).

Finally, it is useful to know that there are several free programs, in addition to MATLAB for neuroimaging, that include: (1) Statistical Parametric Mapping 8 (SPM8; Wellcome Department of Imaging Neuroscience, University College London, UK; http://www.fil.ion.ucl.ac.uk/spm/) is written for MATLAB; (2) FMRIB Software Library (FSL; Functional MRI of the Brain Centre, University of Oxford, UK; http://www.fmrib.ox.ac.uk/fsl/); (3) Analysis for Functional NeuroImages (AFNI; Medical College of Wisconsin, USA; http://afni.nimh.nih.gov/afni/); (4) CARET (Washington University School of Medicine, USA; http://brainvis.wustl.edu/wiki/index.php/Caret:About); and (5) Freesurfer (Athinoula A. Martinos Center for Biomedical Imaging, Massachusetts General Hospital, USA; http://surfer.nmr.mgh.harvard.edu/). Often times, studies involving more advanced methodology will utilize components from each of these programs. Additionally, each of these programs provides an email list for imaging questions and technical support.

## 24.2.6 Multiple Comparisons

We will briefly describe several ways to correct for the problem of multiple comparisons. As mentioned previously, this chapter focuses on the mass univariate analysis of neuroimaging data, which can involve over 100,000 statistical tests. Applying a Bonferroni correction to the data is too conservative to find significant effects. Thus, several other methods have been developed and/or applied to neuroimaging data.

The first method is called the family-wise error (FWE) correction. Instead of correcting for the number of voxels, the FWE correction is a correction for the number of independent tests, or resels (*resolution elements*). The number of resels in an image is a function of the number of voxels and the spatial correlation as measured in full-width at half-maximum (FWHM). From the number of resels, one can compute the statistic required to achieve an expected Euler characteristic of 0 or 1. This will be the height threshold to determine if a voxel is significant or not.

The second method is called the false discovery rate (FDR) correction. The FDR correction leads to the inference that less than $q$ of the significant voxels are false positives. Briefly, after sorting the $p$-values (P) from smallest to largest, find the largest $k$ that satisfies the following equation:

$$P_{(k)} \leq \frac{k}{m \bullet C(m)} q \tag{24.4}$$

where $m$ is the number of voxels and $C(m)$ is a constant describing the relationship between voxels. If the voxels are independent or positively correlated, then $C(m) = 1$; otherwise:

$$C(m) = \sum_{i-1}^{m} 1/i = \ln(m) + .5772 \tag{24.5}$$

All voxels with a $p$-value less than $P_{(k)}$ are significant at an FDR of less than $q$ (Genovese et al., 2002). As with FWE, this is a voxel-wise correction. Recent work has implemented FDR at the cluster level (Chumbley and Friston, 2009).

The third method uses both a voxel-wise threshold and an extent threshold. This correction leads to the inference that a cluster is significant, which is fundamentally different from the first two methods that usually correct the data at the voxel level. To find the extent of a cluster, a Monte Carlo simulation needs to be performed (Forman et al., 1995). One program that will perform the simulation is 3dClustSim (AFNI; Cox, 1996). The cluster extent is based on the voxel threshold, the smoothness of the image, and the spatial distribution of voxels being investigated. One caveat of using this method is that one should be cognizant of using a higher voxel threshold in a smaller cluster compared to using a lower voxel threshold in a larger cluster.

Multiple-comparison correction procedures that correct inferences at the voxel level penalize studies that have higher number of voxels. As the number of voxels increases, the number of individuals in the study needs to increase to detect the same area. While there are several possible solutions, all of them have their drawbacks. Decreasing the resolution of the image will result in fewer voxels and, assuming the statistics are the same, potentially reveal more regions that are truly significant. As the voxel size increases, the ability to detect an effect will be decreased. Given that functional changes are likely to occur in only parts of any anatomical region, the cluster size must be significantly smaller than the region. Thus, as imaging improves more robust results will be needed to reduce the cluster extent threshold. An alternative to voxel-wise procedures is multivariate approaches (e.g., ICA and PLS).

## 24.2.7 Caveats and Limitations

We will briefly describe several of the caveats and limitations of imaging. First, the small sample sizes and low trial numbers used in fMRI studies typically result in lower power. Thus, negative results cannot be interpreted as the absence of the effect of interest. Rather, interpretation should be focused on the positive findings. Furthermore, group comparisons rely on the assumption that neurovascular coupling is the same across all participants and groups. If the neurovascular coupling changes, then the hemodynamic response function used in the analysis will not be correct. However, the goal of most studies is to show differences in the neural correlates, which encompass neurovascular coupling differences. Second, the interpretation of the data is only as good as the cognitive construct or theory that is being tested. If the manipulation does not test the theory or test it properly, then the interpretation will also be invalid. Thus, researchers should make sure the cognitive construct is valid and that the experiment is manipulating the cognitive theory of interest, before conducting the imaging portion of the study. Finally, in terms of the analysis, the most critical thing is the proper model. Functional MRI suffers from its own hemodynamic inverse problem. Specifically, neural activity is linearly related to the hemodynamic response, but we use the hemodynamic response to infer neural activity by specifying the onset and duration of the expected neural activity (Buckner, 2003). However, neither of these parameters are known and must be estimated from the task design. For example, in a visual search task, the stimulus may be present for 5 seconds, but the individual finds the target after 2 seconds. The neural activity in some regions may have only lasted 2 seconds and this timing needs to be taken into account in the model (Shulman et al., 2003). Thus, it is critical to know when the neurons increase and decrease firing during your task and not simply when the stimulus comes on and goes off. Despite the caveats and limitations, fMRI is a very powerful non-invasive tool for understanding brain function.

## 24.3 EXERCISES

These exercises are presented to introduce some of the techniques used to analyze fMRI data. Before you begin, you will need to download SPM8 (http://www.fil.ion.ucl.ac.uk/spm/software/) and the scripts available on the companion site (M-files directory) and then add them to your matlab path using **addpath** and **genpath**:

```
>> addpath(genpath('...\spm8'))
```

The example data are from two studies: (1) an experiment involving visual motion from one human subject (data courtesy of Karl Friston and Christian Buchel at the Functional Imaging Laboratory at University College London; Buchel and Friston, 1997); and (2) an experiment involving the repeated presentations of famous and non-famous faces from one human subject (data courtesy of Rik Henson at the Wellcome Department of Cognitive Neurology, University College London; Henson et al., 2002) and are available on the companion web site.

The first example dataset uses a block design (attention folder) with four conditions: viewing of stationary dots, attending to moving dots, viewing moving dots with no attention, and viewing a fixation crosshair. A complete image of the brain was acquired every 3.22 s (i.e., $TR = 3.22$ seconds) and there are 360 time samples for the whole experiment. The data are stored in image files (*.img) and a timing file (multi_condition.mat). Often times imaging data are stored in a series of 3D images that are hard to manipulate in MATLAB, so conversion to a 4D file is helpful. To concatenate the 3D images in the example data set, type **concatimages('fMRIdata.nii', 'snff*img')**. To load the images, type **img = openIMG('fMRIdata.nii')**. The variable **img** is a four-dimensional matrix corresponding to the $x$ (medial-lateral; sagittal plane), $y$ (anterior-posterior; coronal plane), $z$ (inferior-superior; axial plane) spatial dimensions and time. Each element of the variable **img** represents the BOLD signal from one voxel or volume element from the brain at one time point. One way researchers examine activation at a particular voxel is to cross-correlate the signal with the expected hemodynamic response. Begin with a simple boxcar hemodynamic response that is on (i.e., 1) during the active blocks (stationary dots, moving dots without attention, and attending to moving dots) and off (i.e., 0) during the control blocks (e.g., when the subject is viewing a fixation crosshair). The experiment begins with a rest (control) block of 10 TRs:

```
hemo = [repmat([repmat(0,10,1);repmat(1,10,1);repmat(0,10,1);repmat(1,10,1);repmat
(0,10,1);repmat(1,10,1);repmat(0,10,1);repmat(1,10,1);repmat(1,10,1)],4,1)];
```

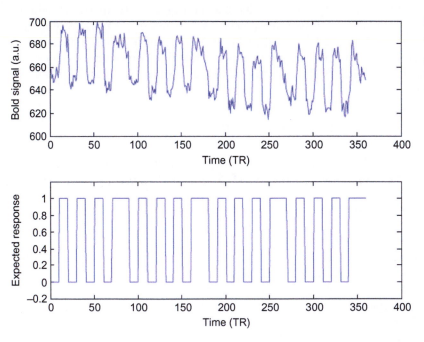

FIGURE 24.2 The BOLD signal from one sample voxel (top) along with the expected hemodynamic response (bottom).

Begin by looking at one voxel. Plot **img(31,6,25,:)** along with the expected hemodynamic response in two subplots of the same figure using the **subplot** function. You will first need to use the **reshape** function to convert **img(31,6,25,:)** into a column vector using **voxel = reshape(img(31,6,25,:), 360, 1)**. Adjust the y-axis of the hemodynamic plot to range from −0.2 to 1.2 using the **axis** function. The result should look like Figure 24.2.

Notice how the BOLD signal oscillates at the same frequency as the expected hemodynamic response.

---

**EXERCISE 24.1**

Compute a power spectrum of that voxel using **pwelch**.

---

**EXERCISE 24.2**

Compute the cross-covariance between the voxel activation and the expected hemodynamic response using **xcov**:

>> [b a] = xcov(voxel,hemo,'coeff');

The result is shown in Figure 24.3 (top panel).

The **xcov** function generates the cross-covariance instead of the cross-correlation because you want to examine how the two signals covary with respect to their respective means. The function **xcorr** would consider their covariation ignoring their mean values despite the fact the two signals have completely different units and magnitudes. The **'coeff'** flag makes sure that the output represents the normalized correlation coefficient ranging from −1 to 1. If you zoom in the figure (Figure 24.3, bottom panel), you will notice that the peak in the cross-covariance occurs at a lag time of 2. This is the biophysical delay between the performance-based neural activation and the hemodynamic response.

To quantitatively determine which voxels are significantly activated, you will apply a regression model or general linear model (GLM) to find a linear relationship between the expected hemodynamic response and the actual BOLD signals. Specifically, you will find the optimal (in the least-squares sense) offset and gain parameters that relate the expected hemodynamic response and the voxel's BOLD signal such that

$$voxel = offset + gain \times hemo + \varepsilon \qquad (24.6)$$

where $\varepsilon$ is normally distributed noise. However, before computing the parameters, you need to introduce the biophysical delay into the expected hemodynamic response using **spm_hrf** and **conv**.

>> hrf = spm_hrf(3.22);
>> canonical_ hemo = conv(hemo,hrf);
>> canonical_hemo = canonical_hemo (1:360)

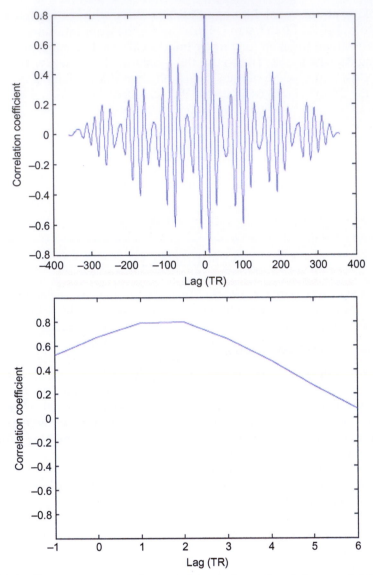

**FIGURE 24.3** The cross-covariance between the sample voxel and the expected hemodynamic response at a broad time scale (top) and a narrow time scale (bottom).

## EXERCISE 24.3

Run statistics on the sample voxel using the canonical hemodynamic response. What are the constant/offset and gain parameters that are computed and their $p$-values?

```
>> B = pinv([ones(360,1)
canonical_hemo])*voxel;
```

**B(1)** is the constant/offset and **B(2)** is the gain parameter relating to the active blocks.

```
>> residuals = voxel-([ones(360,1)
canonical_hemo])*B;
>> ResSS = sum(residuals.^2);
```

*T*-statistics and their associated *p*-values are based on contrasts. In this example, the contrast ($\lambda$) for the offset, the first parameter, is [1; 0] since we want to test if the offset is greater than 0 and the contrast for the gain, the second parameter, is [0; 1] since we want to test if the gain is greater than 0. The equation for generating the *T*-statistic is:

$$T = \frac{\lambda'\beta}{sqrt(\sigma^2 \lambda'(X'X)^{-1}\lambda)} \qquad (24.7)$$

where $\lambda$ is the contrast and $X$ is the predictors. $\sigma^2$ is residual sums of squares (**ResSS**) divided by the degrees of freedom in the model.

```
>> poffset = 2 * tcdf(-abs(B(1)/ sqrt
(ResSS/(360-2)*[1 0]*([ones(360,1)
canonical_hemo]'*[ones(360,1)
canonical_hemo])^-1*[1 0]')), 360-2);
>> pgain = 2 * tcdf(-abs(B(2)/ sqrt
(ResSS/(360-2)*[0 1]*([ones(360,1)
canonical_hemo]' *[ones(360,1)
canonical_hemo])^-1*[0 1]')), 360-2);
```

## EXERCISE 24.4

Run statistics on the sample voxel for each condition (stationary dots, s; moving dots with no attention, natt; attending to moving dots, att) using the canonical hemodynamic response. What is the difference in the gain parameters for each condition and their *p*-values?

```
>> s = [repmat(0,80,1);repmat(1,10,1);
repmat(0,80,1);repmat(1,10,1);repmat
(0,80,1); repmat(1,10,1);repmat(0,80,1);
repmat(1,10,1)];
>> natt = [repmat(0,30,1);repmat(1,10,1);
repmat(0,30,1);repmat(1,10,1); repmat
(0,40,1);repmat(1,10,1);repmat(0,30,1);
repmat(1,10,1);repmat(0,20,1);repmat
(1,10,1);repmat(0,30,1);repmat(1,10,1);
repmat(0,40,1);repmat(1,10,1);repmat
(0,30,1);repmat(1,10,1); repmat(0,30,1)];
>> att = [repmat(0,10,1);repmat(1,10,1);
repmat(0,30,1);repmat(1,10,1); repmat
```

```
(0,40,1);repmat(1,10,1);repmat(0,30,1);
repmat(1,10,1);repmat(0,60,1);repmat
(1,10,1);repmat(0,30,1);repmat(1,10,1);
repmat(0,40,1);repmat(1,10,1);repmat
(0,30,1);repmat(1,10,1); repmat(0,10,1)];
>> canonical_hemo_s = conv(s,hrf);
>> canonical_hemo_s = canonical_
hemo_s(1:1:360);
>> canonical_hemo_natt = conv(natt,
hrf);
>> canonical_hemo_natt = canonical_
hemo_natt(1:1:360);
>> canonical_hemo_att = conv(att,hrf);
>> canonical_hemo_att = canonical_
hemo_att(1:1:360);
>> B = pinv([ones(360,1)
canonical_hemo_s canonical_hemo_natt
canonical_hemo_att])*voxel
>> residuals = voxel-([ones(360,1)
canonical_hemo_s canonical_hemo_natt
canonical_hemo_att])*B;
```

```
>> ResSS = sum(residuals.^2);
```

Examples of the *p*-value computations for the constant/offset parameter and comparing the gain of moving dots without attention to attending to moving dots:

```
>> poffset = 2 * tcdf(-abs(B(1)/ sqrt
(ResSS/(360-4)*[1 0 0 0]*([ones(360,1)
canonical_hemo_s canonical_hemo_natt
canonical_hemo_att]' *[ones(360,1)
```

```
canonical_hemo_s canonical_hemo_natt
canonical_hemo_att])^-1*[1 0 0 0]')),
360-4);
>> pnattgtatt = 2 * tcdf(-abs(B(3)-B(4)/
sqrt(ResSS/(360-4)*[0 0 1 -1]*([ones(360,1)
canonical_hemo_s canonical_hemo_natt
canonical_hemo_att]' *[ones(360,1)
canonical_hemo_s canonical_hemo_natt
canonical_hemo_att])^-1*[0 0 1 -1]')),
360-4);
```

---

## EXERCISE 24.5

Run **spm_firstlevel** on the entire dataset. The M-file **spm_firstlevel** does the same computations as in the previous exercise. The input is a structure variable or a mat-file containing a structure variable described below. What are the constant and gain parameters that are computed and their *t*-statistics?

```
>> SPMin.nscan = 360;
>> SPMin.TR = 3.22;
>> SPMin.Units = 'scans';
>> SPMin.xVi = 'AR(1)';
>> SPMin.HP = 128;
>> SPMin.timingfiles =
{'multi_condition.mat'};
```

The names of the three conditions are: NoAtten, Attention, and Stationary. Here is an example of the **Contrasts** field structure. Make this for all 26 combinations of conditions including compared against fixation (use **'none'**).

```
>> SPMin.Contrasts(1).left =
{'NoAtten'};
>> SPMin.Contrasts(1).right =
{'Attention'};
>> SPMin.Contrasts(1).STAT = 'T';
```

```
>> SPMin.Contrasts(1).
name = 'NoAtten_minus_Attention';
>> SPMin.Contrasts(1).MinEvents = 1;
```

In the M-files directory, the file **Allfields_firstlevel.mat** has an example structure with all fields. Additionally, **AllContrasts_firstlevel.mat** has an example structure identical to above, but with all 26 contrasts included. Definitions of each of the fields can be found in the help for **spm_firstlevel**.

```
>> spm_firstlevel(SPMin)
```

In the file selection window, change the file filter from **.*** to **fMRIdata.***. Then select all entries with the filename **fMRIdata.nii**. Now view the design matrix.

```
>> load SPM.mat; figure; imagesc
(SPM.xX.X); colormap('gray'); shg
```

The design matrix should look like Figure 24.4.

The **beta_*img** stores the gain parameters for each condition, and the constant/offset parameter. There will be one image for each column of the design matrix. The **con_*img** stores the difference in or averages of the

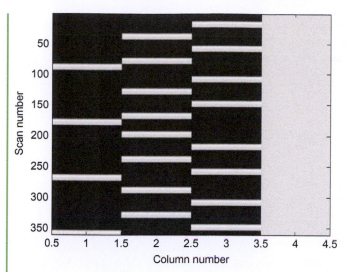

FIGURE 24.4 The design matrix for the general linear model. The first three columns are the expected hemodynamic responses for the three conditions: stationary dots, moving dots without attention, and attending to moving dots. The fourth column is the constant/offset. Each row represents one fMRI volume.

gain parameters and/or averages across runs. The **spmT_*img** stores the *t*-statistics for each contrast. These images can be found in the results directory (e.g., attention/results) and opened with **openIMG**:

```
>> openIMG
('spmT_NoAtten_minus_Attention.img')
```

## EXERCISE 24.6

Use **peak_nii** to identify the significant clusters and peak voxels in the comparison of moving dots without attention to attending to moving dots. What are the regions that show a significant difference?

```
>> mapparameters.out = [];
>> mapparameters.sign = 'pos';
>> mapparameters.type = 'T';
>> mapparameters.voxlimit = 1000;
>> mapparameters.separation = 8;
>> mapparameters.SPM = 1;
>> mapparameters.conn = 18;
>> mapparameters.cluster = 10;
>> mapparameters.mask = [];
>> mapparameters.df1 = 356;
>> mapparameters.nearest = 1;
```

```
>> mapparameters.label = 'aal_
MNI_V4';
>> mapparameters.thresh = .001;
>> [peaks regions] = peak_nii
('spmT_NoAtten_minus_Attention.img',
mapparameters);
```

For a full description of these fields, see the help for **peak_nii**. The results are in two variables: (1) **peaks** contains two cells, the first lists the cluster size, the *t*-statistic and x,y,z coordinates, number of peaks averaged if finding the center of gravity, and the cluster number containing the peak and the second lists the region of each peak; and (2) **regions** contains a list of the regions corresponding to each peak.

Additionally, two images, which can be used for visualizing the results, are created: (1) *_clusters.nii stores the clusters with each cluster labeled with a different number; and (2) (image)_peaks_date_thresh*_extent*.nii stores the thresholded data.

---

### EXERCISE 24.7

Use **slover** to overlay the significant clusters from the contrast of moving dots without attention versus attending to moving dots on every other axial slice.

>> **slover('basic_ui')**

In the image selection window, select **single_subj_T1.nii**, which can be found in the canonical folder inside the spm8 directory and then select **spmT_NoAtten_ minus_Attention_peaks_ < date > _thresh3.113_extent10.nii** from the results directory. Click Done. Next, specify the **single_subj_T1.nii as a *Structural image** and the spmT image as a **Blobs**. Set the Colormap to **gray**. Set the range to **0.5 4**. Since the input image was thresholded at $p<0.001$, the voxels being viewed will exceed that threshold. Click **Axial**. Set the slices to be **0:2:30** representing $Z = 0$ to $Z = 30$. The resulting figure should look like Figure 24.5.

---

## 24.4 PROJECT

This project involves analyzing the second example fMRI dataset (repetition folder) from an experiment involving the repeated presentations of famous and non-famous faces using the general linear model and then inspecting the results. The pre-processed images follow the filename structure **swr*img** and the timings file is **all-conditions.mat**. Specifically, you should do the following:

Apply the **spm_firstlevel** function to determine significant activation due to the task as in Exercise 18.4.

Use a $p$-value of 0.001 for the threshold for activation and a cluster extent of 20 using **peak_nii** as in Exercise 24.5 to determine significant areas.

Generate a grayscale plot for every other $x$-$y$ slice of the brain ($Z = -60$ to $Z = 80$) for the contrast famous faces greater than non-famous faces ($F1 + F2 - N1 - N2$) using **slover**. Your result should look like Figure 24.6.

### 24.4.1 Methods Used to Collect fMRI Data

For the repetition fMRI dataset, the paradigm is a $2 \times 2$ factorial event-related design. Twenty-six famous and 26 non-famous grayscale faces were each presented twice for 500 ms. An oval checkerboard was present throughout the interstimulus interval and is the implicit

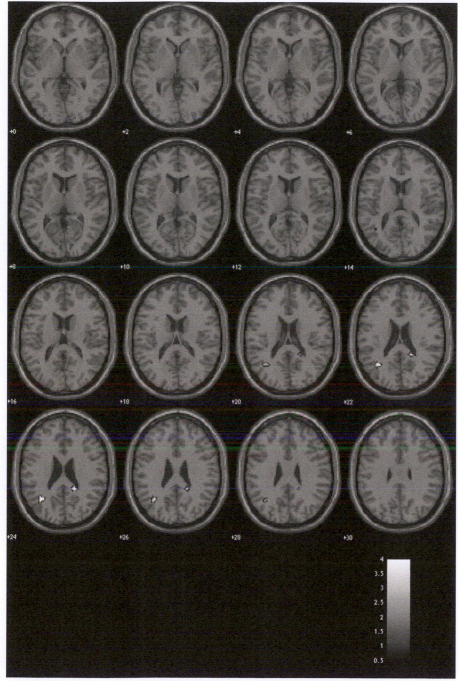

FIGURE 24.5    Axial slices from the contrast no attention greater than attention at $p<0.001$ in at least 10 contiguous voxels overlaid onto a T1-weighted image from a single subject. White represents significant activation. Numbers are the slice plane in millimeters from the origin.

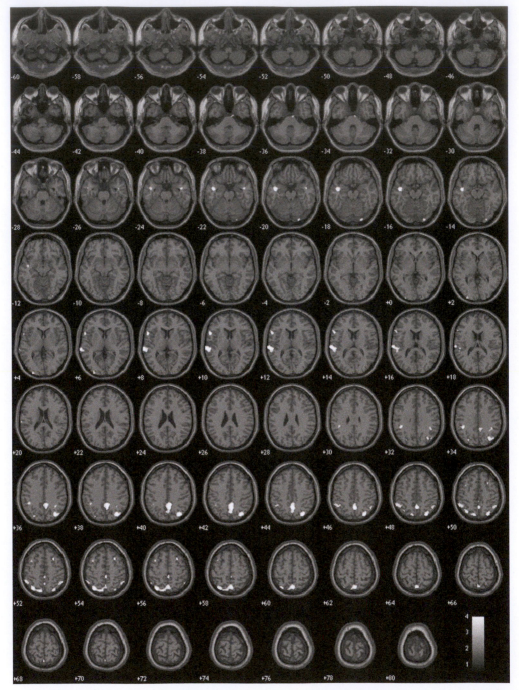

**FIGURE 24.6**    Axial slices from the contrast famous faces greater than non-famous faces at $p<0.001$ in at least 20 contiguous voxels overlaid onto a T1-weighted image from a single subject. White represents significant activation. Numbers are the slice plane in millimeters from the origin. Negative numbers indicate the slice is inferior to the origin.

baseline in the analysis. Stimulus onset asynchrony (SOA) was at least 4.5 s and null events were randomly intermixed to increase the variability in the timing of stimuli and increase subject vigilance. Subjects used buttons to indicate whether a face was famous or non-famous.

BOLD data were collected on a 2T VISION system (Siemens, Erlangen, Germany) using a sequential transverse echo planar imaging (EPI) sequence with the following parameters: repetition time (TR) = 2000 ms, echo time (TE) = 40 ms, matrix = 64 × 64, field of view (FOV) = 192 × 192 mm, flip angle was not reported, and 24 transverse 3 mm thick slices with a 1.5 mm gap. A total of 356 volumes were collected with the first five volumes discarded to allow for T1 equilibration effects.

## 24.4.2 Group Analysis

In an actual fMRI study, the same experimental paradigm would have been administered to multiple individuals in one or more groups. The contrast images formed by **spm_firstlevel** are used in a second general linear model to assess whether the BOLD activity related to a particular condition or between conditions is significantly different than 0 across subjects in a random effects analysis. If there is only one group, then $X$ is a column vector of 1 s and the test is synonymous with a one-sample $t$-test. However, second-level analyses are complicated by between-subject designs, within-subject designs, and mixed designs, which are outside the scope of this chapter. Thus, it is best to use existing programs (e.g., SPM) to conduct these tests.

## MATLAB FUNCTIONS, COMMANDS, AND OPERATORS COVERED IN THIS CHAPTER

**concatimages**
**conv**
**imagesc**
**openIMG**
**peak_nii**
**pinv**
**pwelch**
**repmat**
**reshape**
**shg**
**slover**
**spm_firstlevel**
**spm_hrf**
**tcdf**
**xcov**

# PART IV

# DATA MODELING WITH MATLAB

# Voltage-Gated Ion Channels

## 25.1 GOAL OF THIS CHAPTER

This chapter will explore the dynamics of ion channels using methods similar to those introduced by Hodgkin and Huxley in 1952. You will derive ordinary differential equations that approximate real ion channel behavior and solve these equations using a numerical integrator written in the MATLAB® software. Finally, you will visualize the dynamics by predicting current responses to single channel voltage-clamp experiments.

## 25.2 BACKGROUND

Ion channels are a class of multimeric transmembrane proteins with a hydrophilic pore that facilitates transport of ions across the cell membrane. The size of the ion channel pore and the charge of amino acids near the opening of the pore help exclude entry of some ions while promoting the entry of others. This confers upon the ion channel a selective permeability to different ions. Several factors can induce conformational changes in the ion channel, altering its quaternary structure and therefore its permeability. These factors are referred to as *gating variables* because they function as a gate between the ion channels' different conformational states.

Most ion channels are classified according to the nature of their gating and their selectivity of ions. The largest subclasses of ion channels classified by gating are the *ligand-gated* and *voltage-gated* ion channels. Ligand-gated ion channels change conformation when a ligand binds to them. The most common ligand-gated ion channels found at the postsynaptic membrane of neurons include NMDA, kainate, AMPA, and GABA$_A$ receptors.

The second class of ion channels, voltage-gated ion channels, undergo conformational changes corresponding to alterations in membrane potential. Voltage-gated ion channels can show selectivity for sodium, potassium, or calcium. In this chapter you will model voltage-gated potassium channels (K$_v$ channels) and voltage-gated sodium channels (Na$_v$

*MATLAB® for Neuroscientists.*
DOI: http://dx.doi.org/10.1016/B978-0-12-383836-0.00025-4

channels). $K_v$ channels generally gate between two conformations—an open conformation permeable to potassium and a closed conformation impermeable to potassium—while $Na_v$ channels often gate between three stable conformations—an open conformation permeable to sodium, a closed conformation impermeable to sodium, and an inactive conformation also impermeable to sodium. Although both the inactive and closed states of $Na_v$ channels are impermeable to sodium, they represent different conformational states of the channel and have different kinetics.

## 25.2.1 The Model

Suppose you are interested in building a model to predict the current response to a single ion channel voltage-clamp experiment. Since ionic currents pass through the ion channel to enter the cell, you can consider the ion channel as an electrical resistor whose resistance depends on the conformational state of the ion channel. The equation that relates the resistance of a resistor, $R$, to the current it passes, $I$, and the voltage drop across the resistor, $V$, is Ohm's law:

$$V = IR \tag{25.1}$$

If you divide this equation through by the resistance, then you can write this equation as

$$I = gV \tag{25.2}$$

where $g$ is the conductance of the resistor (note $g = 1/R$). Finally, if you suppose that the conductance is directly proportional to the probability that the channel is in the open conformation, then Equation 25.1 becomes

$$I = g_{max} * P_o * V, \tag{25.3}$$

where $g_{max}$ is the maximum conductance of the channel, and $P_o$ is the probability that the channel is in the open conformation. Therefore, determining the conductance of the channel is equivalent to determining the probability that the channel is open.

## 25.2.2 $K_v$ Channel

Let's begin with the simplest case, the $K_v$ channel. For the $K_v$ channel, Equation 25.3 will take the form

$$I_K = \bar{g}_K * n * V \tag{25.4}$$

where $\bar{g}_K = 36 \ \mu S/cm^2$ is the maximum conductance of the $K_v$ channel, and $n$ is the probability that the channel is in the open conformation. Next, suppose that the ion channel can exist only in an open or closed conformation as depicted by the reversible reaction in the equation

$$(K_v)_{closed} \underset{k_{-1}}{\overset{k_1}{\longleftrightarrow}} (K_v)_{open} \tag{25.5}$$

where $k_1$ is the rate the channel goes from closed to open, and $k_{-1}$ is the rate the channel goes from open to closed. Next, assume that the change in the probability of open channels over time is equal to the probability of the channel being closed, and then going from closed to open (at rate $k_1$), minus the probability of its being open, and then closing (at rate $k_{-1}$). This can be represented by the equation

$$\frac{dn}{dt} = (1 - n)k_1 - nk_{-1} = k_1 - (k_1 + k_{-1})n, \tag{25.6}$$

since all channels are either open or closed. Now further assume that at time $t = 0$ all the channels are closed so that $n(0) = 0$. Finally, recall that for a voltage-gated ion channel the gating between conformational states depends on the membrane potential, so the rates $k_1$ and $k_{-1}$ are both functions of voltage. If you were modeling ligand-gated ion channels, then these rates would depend on the concentration of the ligand. Hodgkin and Huxley used voltage clamp experiments to help determine these rates. In this chapter assume the following functional forms (see Hodgkin and Huxley, 1952) for the transition rates between conformational states of the $K_v$ channel:

$$k_1 = \frac{0.01 * (V + 10)}{\exp\left(\dfrac{V + 10}{10}\right) - 1}$$

$$\tag{25.7}$$

$$k_{-1} = 0.125 * \exp\left(\frac{V}{80}\right).$$

### 25.2.3 The $Na_v$ Channel

The $Na_v$ channel is slightly more complicated, since it has three stable confirmations and therefore a greater number of possible transitions between conformational states. For simplicity, we will ignore transitions between inactive and closed conformational states, and assume that the channel is governed by the following reversible reactions that act independently of each other:

$$(Na_v)_{\text{closed}} \underset{k_1}{\overset{k_{-1}}{\leftrightarrow}} (Na_v)_{\text{open}}$$

$$(Na_v)_{\text{inactive}} \underset{k_{-1}}{\overset{k_1}{\leftrightarrow}} (Na_v)_{\text{open}}. \tag{25.8}$$

If you let $m$ represent the probability of the channel being open given that it was closed previously, and you let $h$ represent the probability of the channel being open given that it was inactive previously, then Equation 25.3 takes on the form:

$$I_{Na} = \bar{g}_{Na} * m * h * V, \tag{25.9}$$

where $\bar{g}_{Na} = 120\,\mu\text{S/cm}^2$. The previous reversible reactions lead to the following differential equations:

$$\frac{dm}{dt} = (1-m)k_1 - mk_{-1} = k_1 - (k_1 + k_{-1})m$$

$$\frac{dh}{dt} = (1-h)k_1 - hk_{-1} = k_1 - (k_1 + k_{-1})h. \tag{25.10}$$

For the $Na_v$ open-close kinetics, assume:

$$k_1 = \frac{0.1*(V+25)}{\exp\left(\dfrac{V+25}{10}\right) - 1}$$

$$k_{-1} = 4*\exp\left(\frac{V}{18}\right), \tag{25.11}$$

and for $Na_v$ inactivation kinetics, assume:

$$k_1 = 0.07*\exp\left(\frac{V}{20}\right)$$

$$k_{-1} = \frac{1}{\exp\left(\dfrac{V+30}{10}\right) + 1}. \tag{25.12}$$

Simple intuition has led naturally to a model that expresses the current you expect to flow through a channel in terms of a differential equation for the probability of the channel being open.

Next, we will discuss a simple algorithm to numerically solve differential equations such as Equation 25.6.

### 25.2.4 Solving Differential Equations Numerically

To understand the current-voltage properties of an ion channel, you can solve an ordinary differential equation describing the probability of the channel being open given the initial condition that $P_o(0) = 0$. In general, solving differential equations can be very tricky (if not impossible), but some simple techniques for approximating solutions do exist. Perhaps the simplest method for approximating the solution to a differential equation is Euler's method. It is based on the definition of the derivative:

$$\frac{df}{dx} = \lim_{\Delta x \to 0} \frac{f(x+\Delta x) - f(x)}{\Delta x}, \tag{25.13}$$

which for small nonzero values of $\Delta x$ implies that:

$$\frac{df}{dx} \approx \frac{f(x + \Delta x) - f(x)}{\Delta x} \Rightarrow f(x + \Delta x) \approx f(x) + \Delta x * \frac{df}{dx}. \tag{25.14}$$

Equation 25.14 lets you determine an approximation for the value of a function $f$ at a point $(x + \Delta x)$ given information about the value of the function of $f$ at $x$ and the derivative of $f$. You can examine Euler's method by choosing a differential equation whose solution is known and compare it to the approximation obtained from Equation 25.14. Now try to apply Equation 25.14 to a simple differential equation:

$$\frac{df}{dx} = 2x, \text{ where } f(0) = 1. \tag{25.15}$$

This differential equation has the obvious solution $f(x) = x^2 + 1$, since it satisfies Equation 25.15 with the condition that $f(0) = 1$. To approximate the solution to this equation using Euler's method, you proceed by first plugging Equation 25.15 into Equation 25.14 to get:

$$f(x + \Delta x) \approx f(x) + \Delta x * 2x. \tag{25.16}$$

Next, you choose $\Delta x = 0.1$ (i.e., something small, since the approximation is most valid for small $\Delta x$) and $x = 0$ and plug into Equation 25.16 to obtain:

$$f(0 + 0.1) \approx f(0) + 0.1 * 2 * 0 \Rightarrow f(0.1) \approx 1. \tag{25.17}$$

This equation can be repeated to give an estimate of $f(0.2)$ such that:

$$f(0.1 + 0.1) \approx f(0.1) + 0.1 * 2 * 0.1 \Rightarrow f(0.2) \approx 1.02, \tag{25.18}$$

where the previous value $f(0.1)$ has been substituted into Equation 25.16 instead of the initial condition. Although you might be tempted to approximate $f(0.2)$ directly from the initial condition by letting $\Delta x = 0.2$, recall that the approximation is best when $\Delta x$ is as close to 0 as possible. This procedure can be repeated to approximate $f(x)$ over any range of $x$ desired. Of course, without the help of modern computers, this method would be too laborious to be practical for any moderately large range of $x$.

Another method for numerically solving differential equations is known as the Runge-Kutta method (RK method). We shall derive the second-order RK method for the general differential equation:

$$\frac{dy}{dx} = f(x, y), \tag{25.19}$$

with the initial condition $y(x_o) = y_o$. Note that the ion-channel gating Equations 25.6 and 25.10 are of this general form. Our first step in deriving the formulae for the RK method is to assume that the numerical solution $y(x)$ that we are looking for has a Taylor series expansion that converges on the interval $I$ for which we want to find the numerical solution. If it does, then:

$$y(x) = y(x_o) + \frac{y'(x_o)}{1!} * (x - x_o) + \frac{y''(x_o)}{2!} * (x - x_o)^2 + \cdots + \frac{y^{(n)}(x_o)}{n!}(x - x_o)^n \tag{25.20}$$

is the Taylor series expansion. Let us define $\Delta x = (x - x_o)$, and substitute this into Equation 25.20, which gives:

$$y(x_o + \Delta x) = y(x_o) + \frac{y'(x_o)}{1!} * (\Delta x) + \frac{y''(x_o)}{2!} * (\Delta x)^2 + \cdots + \frac{y^{(n)}(x_o)}{n!}(\Delta x)^n. \qquad (25.21)$$

We can make a polynomial approximation to the series in Equation 25.21. If we make a second-order approximation, then:

$$y(x_o + \Delta x) \approx y(x_o) + \frac{y'(x_o)}{1!} * (\Delta x) + \frac{y''(x_o)}{2!} * (\Delta x)^2. \qquad (25.22)$$

Notice that the first-order approximation reduces to the key equation in iterating Euler's method (see Equation 25.14). Next, we realize that Equation 25.19 gives us a substitution for $y'(x_o) = f(x_o, y_o)$. We can take the derivative of Equation 25.19 to obtain a substitution for $y''(x_o)$ as follows:

$$y''(x_o) = \frac{\partial f(x_o, y_o)}{\partial x} + \frac{\partial f(x_o, y_o)}{\partial y} y'(x_o) = \frac{\partial f(x_o, y_o)}{\partial x} + \frac{\partial f(x_o, y_o)}{\partial y} f(x_o, y_o). \qquad (25.23)$$

Making these two substitutions into Equation 25.22 gives:

$$y(x_o + \Delta x) \approx y(x_o) + f(x_o, y_o) * (\Delta x) + \left[ \frac{\partial f(x_o, y_o)}{\partial x} + \frac{\partial f(x_o, y_o)}{\partial y} * f(x_o, y_o) \right] * \frac{(\Delta x)^2}{2}. \qquad (25.24)$$

If $f$ were a function whose derivatives could easily be calculated, then we could simply end here, and use Equation 25.24 in much the same way as we used Equation 25.14 to iterate Euler's method. This does not happen often, however, so we will try to find a simplification for Equation 25.24 that does not involve partial derivatives of $f$. Since $y$ was assumed to have a Taylor series expansion, then its derivative does too, so by Equation 25.19 $f$ must also have a Taylor series expansion. The Taylor series expansion of $f$ is slightly complicated, however, since $f$ is a multivariable function. The Taylor expansion is as follows:

$$f(x_o + a, y_o + b) = f(x_o, y_o) + \frac{\partial f(x_o, y_o)}{\partial x} * a + \frac{\partial f(x_o, y_o)}{\partial y} * b + \cdots \qquad (25.25)$$

The terms shown in Equation 25.25 are through first order. If we let $a = \Delta x$ and $b = \Delta x * f(x_o, y_o)$, and we substitute these into Equation 25.25, keeping only up to the first-order terms of the series, then we get:

$$f[x_o + \Delta x, y_o + \Delta x * f(x_o, y_o)] = f(x_o, y_o) + \left[ \frac{\partial f(x_o, y_o)}{\partial x} + \frac{\partial f(x_o, y_o)}{\partial y} * f(x_o, y_o) \right] * \Delta x. \qquad (25.26)$$

If we subtract $f(x_o, y_o)$ from the right-hand side of Equation 25.26 and multiply through by $\Delta x/2$, we obtain:

$$\frac{\Delta x}{2}\{f[x_o + \Delta x, y_o + \Delta x * f(x_o, y_o)] - f(x_o, y_o)\} = \left[\frac{\partial f(x_o, y_o)}{\partial x} + \frac{\partial f(x_o, y_o)}{\partial y} * f(x_o, y_o)\right] * \frac{(\Delta x)^2}{2}.$$

(25.27)

The right-hand side of Equation 25.27 is the last term of Equation 25.24. If we now substitute Equation 25.27 into Equation 25.24, then we will have removed the partial derivatives of $f$. After slight simplification we achieve:

$$y(x_o + \Delta x) \approx y(x_o) + \{f(x_o, y_o) + f[x_o + \Delta x, y_o + \Delta x * f(x_o, y_o)]\} * \frac{(\Delta x)}{2}.$$

(25.28)

We now have an equation to give us the next $y$ value given the previous value that requires only the initial condition $y(x_o) = y_o$, the step size $\Delta x$, and the differential equation, which gives us $f$. Since Equation 25.28 is quite cumbersome looking, it is often presented as a set of equations in the following way:

$$y(x_o + \Delta x) = y(x_o) + \frac{1}{2}(u_1 + u_2)$$

where

$$u_1 = \Delta x * f(x_o, y_o)$$

and

(25.29)

$$u_2 = \Delta x * f(x_o + \Delta x, y_o + u_1)$$

Make the substitutions and convince yourself that this system is equivalent to Equation 25.28. In Equation 25.22 above we truncated the Taylor expansion of $y$ to the second order. For that reason, the set of equations shown above are called the second-order RK equations. We can build higher-order RK equations by keeping higherorders of the series expansion of $y$ and using higher-order expansions of the function $f$ to remove partial derivatives. The following set of equations is the result of truncating $y$ to the fourth order.

$$y(x_o + \Delta x) = y(x_o) + \frac{1}{6}(v_1 + 2v_2 + 2v_3 + v_4)$$

where

$$v_1 = \Delta x * f(x_o, y_o)$$

$$v_2 = \Delta x * f\left(x_o + \frac{\Delta x}{2}, y_o + \frac{v_1}{2}\right)$$

(25.30)

$$v_3 = \Delta x * f\left(x_o + \frac{\Delta x}{2}, y_o + \frac{v_2}{2}\right)$$

$$v_4 = \Delta x * f(x_o + \Delta x, y_o + v_3)$$

The fourth-order RK equations are the most popular numerical method for solving differential equations used today. Although a higher-order expansion in $y$ would give a more accurate solution, the increased processing time required by a computer to achieve the solution is often not worth the minor improvement.

## 25.3 EXERCISES

For these exercises, begin by writing a function called **ode_euler**, which will implement Euler's method to solve the sample differential equation of Equation 25.14 for $x = 0:0.1:10$:

```
function f = ode_euler(x, f_o)
%This function takes two arguments, x and f_o.
%x is a vector that specifies the time points that the function f should be
%approximated for.
%f_o is the initial condition.
%The function returns a vector, f, representing the approximate solution to the
%differential equation, df/dx = 2x with f(0) = f_o.
%Set delta_x as the difference between successive x values.
delta_x = x(2)-x(1);
%Determine how many points we need to approximate by finding the length of
%vector x.
l_x = length(x);
%Initialize f by creating a vector of the right length. We will reset the elements to
%the correct values in the for loop below.
f = zeros(1, l_x);
%Set the initial value of f to f_o.
f(1) = f_o;
%Use a for-loop to implementEq. 25.14
for ii = 1:(l_x-1)
f(ii + 1) = f(ii) + delta_x*2*x(ii); % line 24
end;
```

Now visualize the solution by plotting this approximation for $f$ alongside the exact solution, $f(x) = x^2 + 1$. See whether your solution looks like the one shown in Figure 25.1. Before proceeding, you should explore the relationship between the value of $\Delta x$ and the validity of the approximation of Euler's method. For example, if you plot the exact solution alongside several approximations, each with a different $\Delta x$, how quickly does the approximation cease to be reasonable? Similarly, at what point does decreasing $\Delta x$ fail to provide significant improvement in the approximation despite increased run time? Basic questions such as these are important to consider whenever a numerical method is employed to approximate the solution to a differential equation.

The function **ode_euler** can solve only the sample differential equation of Equation 25.14 because the derivative was plugged into Euler's method explicitly in line 24. You can generalize this function to solve any differential equation by introducing the **feval()** function. The **feval()** function evaluates functions by taking a functional handle that

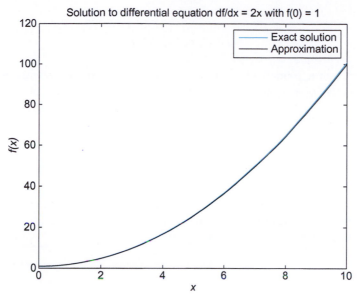

**FIGURE 25.1** Exact and approximate solution to differential equation.

references the function to be evaluated and a variable number of arguments depending on the number of input arguments required by the function referenced by the function handle. As an example, consider the following command:

>>f = feval(@ode_euler, 0:0.1:10, 1);

This line is equivalent to typing

>>f = ode_euler(0:0.1:10, 1);

except with the **feval()** function the name of the function is a variable argument, so it can be changed. To see how this can be applied to generalize the code for Euler's method, first create another function called **f_prime()**:

```
function df = f_prime(x)
% This function takes a point x and calculates the derivative of f at the point x.
df = 2*x;
```

Now modify the first line of **ode_euler** so that it takes an additional argument, a function handle to a differential equation, and modify line 24 to use **feval()** to determine the value of the derivative according to the function referred to by the function handle. If everything is done correctly, then the command

>>f = ode_euler(@f_prime, 0:0.1:10, 1);

should produce the same results as before **feval()** was introduced into the code, but now **ode_euler** makes no explicit reference to any particular differential equation.

---

### EXERCISE 25.1

Write a function **n_prime(t,V)** that calculates the derivative of $n$ at the point $t$ given
that the membrane potential is $V$. *Hint:* You
will need to use Equations 25.6 and 25.7.

---

### EXERCISE 25.2

Modify the first line of **ode_euler()** so
that it takes an additional input argument,
$V$, and line 24, so that **feval()** takes three
arguments, a function handle such as
**@n_prime** and two additional input arguments, $t$ and $V$, to the function referred to
by the function handle.

---

### EXERCISE 25.3

Write a function **RK4(fhandle, x, f_o)**
that uses the fourth-order Runge-Kutta
method to numerically solve the differential
equation referenced by the function handle
fhandle. Use RK4 to solve the differential
equation 25.14 for $\Delta x = 0.1$. Compare this to
the exact solution and the solution using
Euler's method (see Figure 25.1).

---

## 25.4 PROJECT

In this project, you will use the Euler method to derive the current kinetics for the
voltage-gated potassium and sodium channels:

1. Write a function called **K_v(t,V)** that takes a time interval $t$ and a holding potential $V$
   and returns the current response of a $K_v$ channel over the time range specified by $t$.
   *Hint:* **K_v** should call **ode_euler** or RK4 with the inputs **@n_prime**, $t$, $n\_o$, and $V$, and
   use the result along with Equation 25.4 to determine $I_K$.
2. Use **K_v** to plot the current response of a $K_v$ channel when the membrane potential is
   clamped to $-30$ mV. Repeat this for holding potentials from $-30$ mV to $50$ mV in
   $10$ mV increments, and plot the solutions on the same graph. *Hint:* See **hold on**
   command.
3. Write functions **m_prime(t,V)** and **h_prime(t,V)** that calculate the derivative of $m$ and $h$
   at the point $t$ given that the membrane potential is $V$. This will be completely analogous
   to the code for **n_prime**.

4. Write a function called **Na_v(t,V)** that takes a time interval $t$ and a holding potential $V$, and returns the current response of an $Na_v$ channel over the time range specified by $t$. *Hint:* **Na_v** should call **ode_euler** or RK4 twice, once with the inputs **@m_prime**, $t$, $m\_o$, and $V$ and another with the inputs **@h_prime**, $t$, $h\_o$, and $V$; use these results along with Equation 25.9 to determine $I_{Na}$.

5. Use **Na_v** to plot the current response of an $Na_v$ channel when the membrane potential is clamped to $-30$ mV. Repeat this for holding potentials from $-30$ mV to 50 mV in 10 mV increments, and plot the solutions on the same graph.

## MATLAB FUNCTIONS, COMMANDS, AND OPERATORS COVERED IN THIS CHAPTER

**feval**
**hold on**

# Synaptic Transmission

## 26.1 GOALS OF THIS CHAPTER

This chapter will use a number of methods to characterize the processes surrounding synaptic transmission, in particular the release and diffusion of neurotransmitters. This chapter will also introduce handles for graphic objects to update images dynamically. By the end of this chapter, you should have an understanding of different random variables, discrete distributions, finite difference approximations to partial differential equations, and the use of graphic handles for rudimentary animation.

## 26.2 BACKGROUND

Chemical synapses use the release of chemical neurotransmitters to propagate signals from one neuron (presynaptic) to another (postsynaptic). The two cells are separated by the *synaptic cleft*, a gap of approximately 40 nm between the presynaptic and postsynaptic membranes.

On the presynaptic side of the cleft, depolarization from the arrival of the action potential triggers the opening of voltage-sensitive $Ca^{+2}$ channels. The subsequent influx of calcium ions causes the fusion of neurotransmitter-containing vesicles with the presynaptic cell membrane, releasing neurotransmitters into the synaptic cleft.

On the postsynaptic side, neurotransmitters diffuse across the cleft and bind to neurotransmitter-specific sites on channels on the postsynaptic terminal. These receptors selectively allow ions to enter the postsynaptic terminal. The influx of positive or negative charge changes the voltage across the postsynaptic membrane, creating a small hyperpolarization or depolarization. Over time, the concentration of the neurotransmitters decreases to a level insufficient to activate the postsynaptic receptors.

Specifically, in this chapter we will focus on the neuromuscular junction, the site of innervation of skeletal muscle. We will also focus on the steps of neurotransmitter release and neurotransmitter diffusion.

*MATLAB® for Neuroscientists.*
DOI: http://dx.doi.org/10.1016/B978-0-12-383836-0.00026-6

# 26.3 EXERCISES

## 26.3.1 Modeling Neurotransmitter Release

In a classic experiment, Fatt and Katz demonstrated at the neuromuscular junction that spontaneous postsynaptic potentials occurred at voltages of 0.5 mV in the absence of presynaptic stimulation. Later work (del Castillo and Katz, 1954) further demonstrated that single acetylcholine channels produced much smaller responses than the 0.5 mV measured by Fatt and Katz. This result implied that the spontaneous results observed by Fatt and Katz not only recruited multiple acetylcholine channels, but also recruited roughly the same number of channels during each spontaneous postsynaptic potential del Castillo and Katz posited that synaptic transmission occurred in discrete units, termed quanta. Additional work (del Castillo and Katz) established that increasing $Ca^{+2}$ at the postsynaptic terminal produced postsynaptic responses in increments of the original spontaneous postsynaptic responses. The step-like response implicated neurotransmitter release as a quantized process. In other words, neurotransmitter release occurs in discrete steps rather than a continuous concentration in response to increasing calcium concentration. Synaptic vesicles contain a relatively constant number of neurotransmitter molecules. This fixed number of molecules per vesicle provides for the observed quantization of postsynaptic responses. At the neuromuscular junction, each vesicle contains approximately 5000 acetylcholine (ACh) molecules.

For purposes of this chapter, assume that the release of individual vesicles occurs independently with some probability $p$. In the presence of low calcium concentration, $p$ is low. After an action potential and subsequent calcium influx, $p$ increases. You can model the release of a single vesicle as a Bernoulli random variable.

## 26.3.2 Modeling Random Variables

A Bernoulli random variable $X$ takes on a value of 1 with probability $p$ or a value of 0 with probability $1 - p$. A coin toss is an excellent example of a Bernoulli process. Take a coin that lands on heads with probability 1/2. A Bernoulli random variable models a single coin flip, or a single trial.

Thus, in the case of the release of a single vesicle, the vesicle is released with probability $p$ or not with probability $1 - p$.

---

### EXERCISE 26.1

Using the **rand** function in the MATLAB® software, write a function titled **my_bernoulli_rnd()** to return the result of a Bernoulli trial, given $p$. You should be able to invoke the function like this:

```
>> p = 0.5;
>> my_bernoulli_rnd(p)
ans =
    1
```

---

You can model the process of multiple vesicles as the sum of multiple Bernoulli variables with probability $p$. The sum of $n$ Bernoulli trials with probability $p$ of success is termed a *binomial random variable* with parameters $n$ and $p$. Such a variable is called *binomial* because the probability of a certain number of successes in a trial can be calculated with the binomial coefficient:

$$f(k; n, p) = C(n, k)p^k(1-p)^k \tag{26.1}$$

where

$$C(n, k) = \binom{n}{k} = \frac{n!}{k!(n-k)!} \tag{26.2}$$

The function **f()**, called a *probability mass function*, yields the probability of a certain number of successes, $k$, in a single binomial trial with parameters $n$ and $p$. A single binomial trial with parameters $n$ and $p$ would model the number of successes for an experiment in which each trial has $n$ coin tosses with probability $p$ of success.

---

### EXERCISE 26.2

Generate a graph of the probability mass function for a binomial random variable with parameters $n = 10$ and $p = 0.5$.

---

Under conditions of calcium influx, approximately 150 quanta can be released at the neuromuscular junction in 1–2 milliseconds. In the absence of action potentials, only 1 quantum per second is released by the presynaptic terminal. For this chapter, model the release of multiple vesicles as a binomial random variable in which $p$ is either 0.001 or 150, depending on the state of calcium concentration. Then, the number of successes will be the number of vesicles released into the synaptic cleft.

The following code calculates $p$ for 1 second of time, choosing 10 random 1 ms intervals during which $p$ is high, representing the calcium influx following an action potential:

```
>> t = 0:999;
>> t_slices = length(t);
>> p = ones(t_slices,1)*0.001;
>> x = rand(10,1)*t_slices;
>> p(floor(x) + 1) = 150;
```

Note that $x$ holds the set of random intervals. When you evaluate **p(floor(x) + 1) = 150**, each element of $x$ corresponding to one of the random intervals is set to 150. The use of **floor()** forces the values of $x$ to integer values, the proper type for indices. (The **floor()** function truncates the decimal portion of a value.) You can see which indices contain large $p$ values with the following command:

```
>> find(p > 1)
```

---

**EXERCISE 26.3**

Evaluate the preceding code and graph p(*t*) against time. Write code to generate a binomial random variable at each time slice (use **binornd**). Graph the number of successes, or released vesicles, as a function of time.

---

### 26.3.3 Modeling the Motion of a Single Molecule

Upon release, the neurotransmitter molecules enter the synaptic cleft and diffuse across fairly quickly. On a microscopic level, diffusion is the aggregate effect of many particles moving randomly. As a first step, examine the motion of a single molecule:

```
xbounds = [0 10];
ybounds = [0 4];
xdata = [mean(xbounds)];
ydata = [0];
xgrid = 0.01;
ygrid = 0.01;
figure
handle = scatter(xdata, ydata, 'filled');
xlim(xbounds);

ylim(ybounds);
for t = 1:10000
    p = 0.5;
    dx = ((rand > p) - 0.5) * 2;
    dy = ((rand > p) - 0.5) * 2;
    xdata = xdata + dx*xgrid;
    % these two lines assure the molecule stays in x bounds
    xdata(find(xdata < xbounds(1))) = xbounds(1);
    xdata(find(xdata > xbounds(2))) = xbounds(2);
    ydata = ydata + dy*ygrid;
    % these two lines assure the molecule stays within y bounds
    ydata(find(ydata < ybounds(1))) = ybounds(1);
    ydata(find(ydata > ybounds(2))) = ybounds(2);
    set(handle, 'xdata', xdata, 'ydata', ydata);
    drawnow;
end
```

Some sample screenshots of the resulting animation are shown in Figure 26.1. In the preceding code, the handle of the scatterplot is stored in a variable. Much like matrices or scalar numbers, variables can store other types of information. In this case, you are storing a handle. Most of the graphics functions in MATLAB return a handle when invoked. You can use the handle to modify properties at a later time, as done previously with **set**.

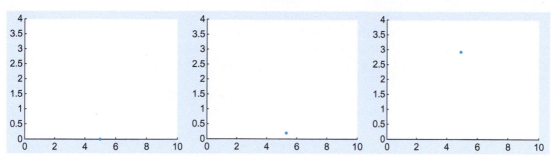

**FIGURE 26.1**    Screenshots of the animation for the diffusion of a single neurotransmitter molecule across the synaptic cleft for $t = 1$ (left), $t = 500$ (middle), and $t = 10,000$ (right).

The function **set** allows specifying properties of a graphics object. The first argument should be the handle. Following this should be a series of property, value pairs, consisting of the name of a property in single quotation marks and the desired value of the property. Here, the $x$ and $y$ data for the scatterplot are changed. Each time the variables *xdata* and *ydata* are modified, **set** is invoked to update the coordinates of the point in the scatterplot.

After **set** comes the function **drawnow**. Even though **set** updates the coordinates of the point in the scatterplot, MATLAB will not redraw the figure without notification. **drawnow** provides notification that the figure has changed and forces a redraw. Try running the preceding code with **drawnow** commented out.

---

**EXERCISE 26.4**

Modify the preceding code to model multiple particles (try 100).

---

As the number of particles increases, modeling the motion of individual particles becomes computationally limiting. Instead, you can model concentrations rather than particle counts.

### 26.3.4 Modeling Diffusion

To model the process as a concentration, you need to use the diffusion equation:

$$\frac{\partial \phi}{\partial t} = D \nabla^2 \phi \tag{26.3}$$

Or in two dimensions, you can use:

$$\frac{\partial \phi}{\partial t} = D \frac{\partial^2 \phi}{\partial x^2} + D \frac{\partial^2 \phi}{\partial y^2} \tag{26.4}$$

where $D$ is the constant of diffusion and $\phi$ is the concentration of acetylcholine, ACh.

In this chapter, you will use a *finite difference* approach to estimate the change in the probability distribution over time. The finite difference method estimates the infinitely small derivatives by finite differences. Because this approximates the continuous equation with discrete differences, we should mention that large time steps can produce unstable results.

To approximate $\phi$ over time and space, assume both are discrete yet partitioned into very small steps. Use the following notation to denote an element of phi in space and time:

$$\phi_{space}^{time} \tag{26.5}$$

Thus, a superscript denotes a *time* index, and a subscript denotes a *space* index.

To generate the finite difference equation, replace the derivatives with differences:

$$\frac{\phi_{x,y}^{t+1} - \phi_{x,y}^{t}}{\Delta t} = D \left[ \frac{((\phi_{x+1,y}^{t} - \phi_{x,y}^{t}) - (\phi_{x,y}^{t} - \phi_{x-1,y}^{t}))}{(\Delta x)^2} + \frac{((\phi_{x,y+1}^{t} - \phi_{x,y}^{t}) - (\phi_{x,y}^{t} - \phi_{x,y-1}^{t}))}{(\Delta y)^2} \right] \tag{26.6}$$

With some rearrangement, you get:

$$\phi_{x,y}^{t+1} = \phi_{x,y}^{t} + D\Delta t \left[ \frac{(\phi_{x+1,y}^{t} - 2\phi_{x,y}^{t} + \phi_{x-1,y}^{t})}{(\Delta x)^2} + \frac{(\phi_{x,y+1}^{t} - 2\phi_{x,y}^{t} + \phi_{x,y-1}^{t})}{(\Delta y)^2} \right] \tag{26.7}$$

which provides an expression for a concentration at a given spatial location and time in terms only of concentrations at previous time steps. If you use the same grid spacing for $x$ and $y$, you can simplify even further:

$$\phi_{x,y}^{t+1} = \phi_{x,y}^{t} + D\Delta t \left[ \frac{(\phi_{x+1,y}^{t} + \phi_{x,y+1}^{t} + \phi_{x-1,y}^{t} + \phi_{x,y-1}^{t} - 4\phi_{x,y}^{t})}{(\Delta x)^2} \right] \tag{26.8}$$

You already saw an efficient way to compute this spatial second derivative in MATLAB. As will be further discussed in Chapter 30, "Fitzhugh-Nagumo Model: Traveling Waves," this is found via a two-dimensional convolution of $\phi$ with the filter [0 1 0; 1 4 1; 0 1 0]/dx^2. To encode the dynamics of the ACh concentration diffusion in MATLAB, use a three-dimensional array: two spatial dimensions and one temporal dimension.

The following code implements iterations of the previous equation, using a three-dimensional array to track change in concentration. The time steps are in 10 ns increments, and the spatial steps are in 1 nm increments. The diffusion constant here is $4 \times 10^{-6}$ cm$^2$/sec, and free boundary conditions are used:

```
% phi : 100 x steps (100 nm), 40 y steps (40 nm)
% 100 t steps (1 t = 10 us)
clear all
close all
phi = zeros(100, 40, 100);
dt = 1e-10; % time in steps of 10 ns
dx = 1e-9; % space from 0 to 50
```

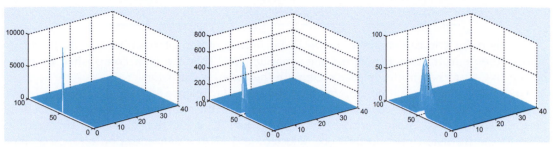

**FIGURE 26.2** Screenshots of the animation for the diffusion of the concentration of acetylcholine, ACh, in the synaptic cleft for $t = 10\,\mu s$ (left), $t = 250\,\mu s$ (middle), and $t = 1000\,\mu s$ (right).

```
D = 4e-6 * (1/100)^2;
phi0 = 5000/(dx^2);
phi(50,1,1) = phi0; % initial condition
F = [0  1  0; 1  -4  1; 0  1  0]/dx^2;
for t = 1:99
    phi(:,:,t + 1) = phi(:,:,t) +...
        D*dt*conv2(phi(:,:,t),F,'same');
    t
end
```

You can plot a time-slice of the concentration using **surf(phi(:, :, t))**. Sample screenshots of the concentration diffusing in the synaptic cleft using the **surf** function are shown in Figure 26.2. Similarly, you can visualize this using the command **imagesc(phi(:,:,t), [0 phi0])**, where the **[0 phi0]** input to the function ensures that the color scale is the same for all $t$.

---

**EXERCISE 26.5**

Generate an animated display to show the evolution of the ACh concentration diffusion using **surf()**. Capture the handle and use **set()** with the property **'zdata'**.

---

**EXERCISE 26.6**

The algorithm could be changed to save only the current and previous time steps, allowing an unlimited number of time steps to be calculated if so desired without exceeding the memory of MATLAB. Obviously, this approach is less than optimal if all the time step calculations are needed, but, for an animation, only the current time step is necessary. Modify the algorithm to store only the current and previous time step and animate the diffusion.

## 26.4 PROJECT

At this point you can combine the vesicular release of neurotransmitters with the diffusion of the neurotransmitters across the synaptic cleft. This can be accomplished by combining the neurotransmitter release code with the code for diffusion.

To do so, use a single time loop for both the vesicular release and the diffusion code. At any given time slice, the code will need to determine the number of vesicles released. If vesicles are released during a given time slice, then (1) a location along the presynaptic edge of the diffusion grid needs to be selected, and (2) the concentration of ACh in that square of the finite difference grid needs to be increased. Create an animated display of synaptic transmission by showing the evolution of the ACh concentration diffusion generated by multiple released vesicles over time.

## MATLAB FUNCTIONS, COMMANDS, AND OPERATORS COVERED IN THIS CHAPTER

**set**
**drawnow**
**surf**
**poissrnd**

# Modeling a Single Neuron

## 27.1 GOAL OF THIS CHAPTER

The goal of this chapter is to incorporate previous models of voltage-gated ion channels into a model of single neuron dynamics. This chapter will continue to follow work done by Hodgkin and Huxley (1952) resulting in a system of four ordinary differential equations that model action potential generation in neurons.

## 27.2 BACKGROUND

Neurons communicate with each other by transmitting and receiving electrochemical signals called *action potentials*. These action potentials are transient fluctuations in the cell's membrane potential, which propagate down a cell's axon without attenuation. In the central nervous system, action potentials have a duration on the order of milliseconds (1−2 msec usually) and can often be divided into three phases. The first phase of the action potential is a rapid depolarization of the membrane called the *rising phase* or *upstroke* of the action potential. This is followed by a repolarization of the membrane called the *falling phase* or *downstroke* of the action potential. The last phase follows a hyperpolarization of the membrane and is called the *undershoot*. A depiction of the action potential is shown in Figure 27.1.

Some of the earliest experiments to elucidate the mechanism underlying action potentials were performed by Hodgkin and Katz (1949), who showed that reducing the extracellular concentration of sodium led to a shorter upstroke phase of the action potential in giant squid axon. They inferred from this that the upstroke of the action potential depends on the cell increasing its permeability to sodium. They also suggested that the falling phase was due to an increase in potassium permeability. Therefore, they concluded that the action potential was generated by selective changes in membrane permeability to sodium and potassium. We now know that ion channels are responsible for this selective

*MATLAB® for Neuroscientists.*
DOI: http://dx.doi.org/10.1016/B978-0-12-383836-0.00027-8

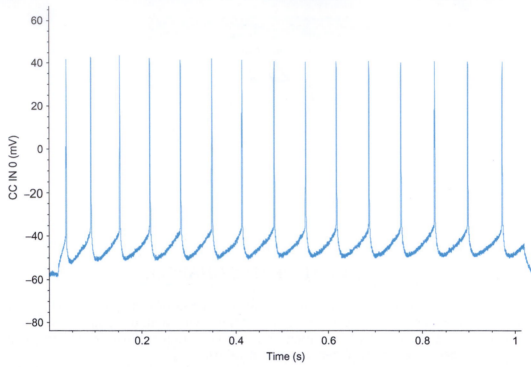

**FIGURE 27.1**   Intracellular action potential spike train from a deep pyramidal neuron recorded from the frontal cortex of a mouse. *(Courtesy of Amber Martell)*

permeability. These experiments were later followed up by Hodgkin and Huxley (1952), who performed voltage-clamp experiments to characterize the dynamics of these changes in permeability and then proposed the mathematical model of action potential generation outlined in the following section.

## 27.2.1  The Model

Neurons are incredibly complex. Like all eukaryotic cells, they are composed of many organelles, including a nucleus, mitochondria, an endoplasmic reticulum, etc. Each of these organelles has a role that enables the cell as a whole to perform its functions, including generating action potentials. Trying to capture all the complexity of a real neuron in a single model is impossible. Fortunately, it is also unnecessary, since, for purposes of this chapter, you are interested only in understanding action potential generation in neurons, and not any of the other complex processes that neurons undergo. Therefore, you should restrict your neuron model to include only those elements that contribute most directly to generating action potentials and ignore elements of a neuron that contribute less to action potential generation. In general, it is often not clear what elements of a complex biological system are most directly related to a behavior of interest, and the choices you make in

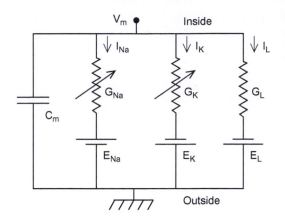

FIGURE 27.2  An electrical circuit diagram of a single axonal compartment of a neuron. *(Bower JM, Beeman D. The Book of Genesis: Exploring Realistic Neural Models with the GEneral NEural SImulation System, 2003)*

constructing a model are often not validated until the results of the model can be compared to experiments.

In this model, assume that action potential generation in neurons is mainly carried out by the electrical properties of the cell membrane. Several factors contribute to the electrical properties of the cell membrane. For instance, ion channels such as $Na_v$, $K_v$, and leak channels span the membrane and selectively pass ions across it. The voltage-gated channels are represented in Figure 27.2 as variable resistors (the resistors with an arrow going through them) because the amount of resistance to flow depends on the membrane potential, whereas the leak channel, which has a constant resistance to ion flow, is represented by an ordinary resistor. The phospholipids that comprise the membrane, which do not conduct electric charges, allow for most of the cell membrane to function as a dielectric, an insulative material that separates ions in the cytoplasm from those in the extracellular milieu. Although ions cannot flow through the phospholipid bilayer of the cell directly, charge can accumulate on one side of the cell membrane, inducing an opposed charge buildup on the opposite side of the membrane just as a capacitor does. This charge buildup involves charges moving toward the membrane and represents a capacitive current. Finally, the cell membrane contains many other transmembrane proteins such as the $Na^+/K^+$-ATPase that helps maintain ion concentration gradients across the cell membrane. The presence of a sodium concentration gradient, for example, ensures that when sodium ions have equilibrated across the $Na_v$ channels of the neuron, there will be a nonzero potential across the membrane. This potential is called the *sodium reversal potential*. Similarly, there will be a reversal potential for the $K_v$ channel. The electrical properties mentioned so far can be summarized by creating an electric circuit equivalent to the neuronal model. The circuit is shown in Figure 27.2. Notice that current flows from the inside of the cell (at the top of the electrical circuit) to the outside of the cell by either inducing a charge buildup at the membrane (represented by the capacitor) or by flowing through one of the three ion channels present in the membrane. From simple electrical circuit theory, you can represent the following circuit using a set of equations. In the next section, we will review the important concepts of electrical circuits needed to understand the circuit in Figure 27.2.

To express the circuit as a set of equations, you need to know four fundamental laws of electronics. The first is Ohm's law, which states that for some resistors (called *Ohmic resistors*) the voltage drop across the resistor, $V_R$, is related to the current flowing through the resistor, $I$, and the resistance of the resistor, $R$, by the equation:

$$V_R = IR \tag{27.1}$$

which can also be written as:

$$I = gV_R \tag{27.2}$$

where $g$ is the conductance of the resistor (note $g1/R$). The second law you will need states that the voltage drop across a capacitor, $V_C$, is related to the current induced by the capacitor, $I$, and the capacitance of the capacitor, $C$, by the equation

$$V_C = \frac{1}{C} \int I(t)dt \tag{27.3}$$

The last two laws that you will need are collectively known as *Kirchhoff's Loop Rules*. The first rule, Kirchhoff's Current Rule, states that the sum of current entering a circuit junction equals the sum of current exiting it, and a circuit junction is any intersection of wire where current has more than one path to flow down. The equation for this rule is given by:

$$\sum I_{\text{in}} = \sum I_{\text{out}} \tag{27.4}$$

The second rule, Kirchoff's Voltage Rule, states that the potential drop between any two points on a circuit is independent of what path was taken to arrive there. If you assume that the start and end point are the same, then this rule implies that the voltage drop across any closed loop is zero, and can be written as:

$$\sum_{\text{loop}} V = 0 \tag{27.5}$$

Now you can use these simple rules to calculate the membrane potential of the circuit in Figure 27.2. The membrane potential is defined as the potential difference between the inside and the outside of the cell. Therefore, in Figure 27.2 the membrane potential is the potential drop across any path from the inside of the cell to the outside. Beginning with the path that includes the capacitor, you see that the voltage drop across the capacitor is just the membrane potential, so Equation 27.3 becomes:

$$V_M = \frac{1}{C_M} \int I(t)dt \tag{27.6}$$

which can be rearranged to give:

$$I = C_M \frac{dV_M}{dt} \tag{27.7}$$

Now examine the potential drop across the second path (the sodium channel), which consists of two elements, a resistor and a battery. The total drop across both these elements is just the potential difference between the inside and outside of the cell, $V_M$, so:

$$V_M = V_R + E_{Na} \Rightarrow V_R = V_M - E_{Na} \tag{27.8}$$

Upon substitution into Equation 27.2, you have:

$$I_{Na} = g_{Na} * (V_M - E_{Na}) \tag{27.9}$$

Following the same process across the last two paths produces equations nearly identical to Equation 27.9 for $I_K$ and $I_L$:

$$I_K = g_K * (V_M - E_K) \tag{27.10}$$

$$I_L = g_L * (V_M - E_L) \tag{27.11}$$

Finally, use Kirchhoff's Current Rule to see that if you inject a current into the cell of $I_{inj}$, then:

$$I_{inj} = I + I_{Na} + I_K + I_L \tag{27.12}$$

Rearranging Equation 27.12 and substituting in Equations 27.7 and 27.9 through 27.11 gives:

$$C_M \frac{dV_M}{dt} = -g_{Na} * (V_M - E_{Na}) - g_K * (V_M - E_K) - g_L * (V_M - E_L) + I_{inj} \tag{27.13}$$

Recall that the sodium and potassium channels are voltage-gated, so their conductances are functions of voltage. In Chapter 25, "Voltage-Gated Ion Channels," you modeled the potassium conductance as:

$$g_K = \bar{g}_K * n \tag{27.14}$$

where

$$\frac{dn}{dt} = k_{1n} - (k_{1n} + k_{-1n})n \tag{27.15}$$

and the sodium conductance by:

$$g_{Na} = \bar{g}_{Na} * m * h \tag{27.16}$$

where

$$\frac{dm}{dt} = k_{1m} - (k_{1m} + k_{-1m})m$$

$$\frac{dh}{dt} = k_{1h} - (k_{1h} + k_{-1h})h \tag{27.17}$$

If you substitute Equation 27.14 and 27.16 into Equation 27.13 and collect Equations 27.15 and 27.17, you get the following system of equations:

$$C_m \frac{dV_M}{dt} = -\bar{g}_{Na}mh(V_M - E_{Na}) - \bar{g}_K n(V_M - E_K) - g_L(V_M - E_L) + I_{inj}$$

$$\frac{dn}{dt} = k_{1n} - (k_{1n} + k_{-1n})n$$

$$\frac{dm}{dt} = k_{1m} - (k_{1m} + k_{-1m})m \tag{27.18}$$

$$\frac{dh}{dt} = k_{1h} - (k_{1h} + k_{-1h})h$$

In the original Hodgkin-Huxley model, the final equations proposed were as follows:

$$C_m \frac{dV_M}{dt} = -\bar{g}_{Na} m^3 h (V_M - E_{Na}) - \bar{g}_K n^4 (V_M - E_K) - g_L(V_M - E_L) + I_{inj}$$

$$\frac{dn}{dt} = k_{1n} - (k_{1n} + k_{-1n})n$$

$$\frac{dm}{dt} = k_{1m} - (k_{1m} + k_{-1m})m \tag{27.19}$$

$$\frac{dh}{dt} = k_{1h} - (k_{1h} + k_{-1h})h$$

The changes to the first equation were made so that the model would better fit with the experimental data, although some explanation of the addition of these exponents has since been made from first principles.

Many of the parameter values needed to evaluate the system of Equation 27.19 are mentioned in Chapter 25, "Voltage-Gated Ion Channels." Table 27.1 identifies these parameter values along with some additional parameter values for the leak channel and capacitance of the membrane.

The functional forms for the transition rates between conformational states of the sodium and potassium channels are given in Equations 27.20–27.22. These rates were discussed in more detail in Chapter 25, "Voltage-Gated Ion Channels."

$$k_{1n} = \frac{0.01 * (10 - V_M)}{\exp\left(\dfrac{10 - V_M}{10}\right) - 1}$$

$$k_{-1n} = 0.125 * \exp\left(\frac{-V_M}{80}\right) \tag{27.20}$$

TABLE 27.1   Parameter Values for Hodgkin-Huxley Model

| Parameter | Value |
|---|---|
| $CM$ | $1\,\mu\text{F}/\text{cm}^2$ |
| $\bar{g}_K$ | $36\,\mu\text{S}/\text{cm}^2$ |
| $\bar{g}_{Na}$ | $120\,\mu\text{S}/\text{cm}^2$ |
| $g_L$ | $0.3\,\mu\text{S}/\text{cm}^2$ |
| $E_K$ | $-12\,\text{mV}$ |
| $E_{Na}$ | $115\,\text{mV}$ |
| $E_L$ | $10.613\,\text{mV}$ |

$$k_{1m} = \frac{0.1 * (25 - V_M)}{\exp\left(\dfrac{25 - V_M}{10}\right) - 1}$$

(27.21)

$$k_{-1m} = 4 * \exp\left(\frac{-V_M}{18}\right)$$

$$k_{1h} = 0.07 * \exp\left(\frac{-V_M}{20}\right)$$

$$k_{-1h} = \frac{1}{\exp\left(\dfrac{30 - V_M}{10}\right) + 1}$$

(27.22)

## 27.3 EXERCISES

Trying to write code to implement a set of equations such as Equation 27.19 while keeping track of all the rate functions and necessary parameters can seem daunting. The key to keeping larger coding projects manageable is to write many smaller functions first and then put them together to create larger functions until eventually the project is complete. For example, the following code is for a function **n_prime** that takes the current value of $n$ and the current membrane potential $V\_m$ and returns the derivative of $n$ according to the second equation of the Hodgkin-Huxley model:

```
function dn = n_prime(V_m, n)
% This function takes two arguments the membrane potential and the current value
% of the state variable n, and returns the value of the derivative of n for these values.
% First calculate the values of the forward and backward rate constants, k_1n and
% k_2n.
k_1n = 0.01*(10-V_m)/(exp((10-V_m)/10)-1);
k_2n = 0.125*exp(-V_m/80);
% Next calculate the value of the derivative.
dn = k_1n - (k_1n + k_2n)*n;
```

---

### EXERCISE 27.1

Write a function **m_prime(V_m, m)** similar to the one in the preceding example that calculates the derivative of $m$ given its current value and membrane potential.

---

## EXERCISE 27.2

Write a function **h_prime(V_m, h)** similar to the one in Exercise 27.1 that calculates the derivative of $h$ given its current value and membrane potential.

## EXERCISE 27.3

Write a function **V_prime(V_m, n, m, h, I_inj)** that calculates the derivative of $V_m$ given its current value, the values of the other state variables, and the injected current. *Hint:* Just repeat what you've done so far using the first equation of the Hodgkin-Huxley model.

## 27.4 PROJECT

In this project, you will model the voltage dynamics of a Hodgkin-Huxley neuron. You should perform the following:

1. Write a function **hodgkin_huxley(t, I_inj)** that takes a time series $t$ and a constant representing injected current and returns the value of $V$ at every point in $t$. Assume that the initial value for $V$ is $-10$ mV. Assume that all channels are initially closed. *Hint:* See Chapter 25, "Voltage-Gated Ion Channels," for a similar example.
2. Plot $V$ versus $t$ for injected currents of 5, 10, and 15 A/cm$^2$.
3. Determine what happens to the frequency of firing as the injected current increases.
4. Indicate how the action potential generated by this model compares to the result in Figure 27.1.

## MATLAB FUNCTIONS, COMMANDS, AND OPERATORS COVERED IN THIS CHAPTER

**length**
**for-loop**
**plot**
**hold on**

# 28

# Models of the Retina

## 28.1  GOAL OF THIS CHAPTER

The goal of this chapter is to understand the basic structure of the retina and to see how to create simple models of neuronal interactions. In this chapter you will build a simple model describing the interaction between cone cells and horizontal cells of the retina and solve it exactly by taking advantage of the capability of the MATLAB® software to easily manipulate matrices.

## 28.2  BACKGROUND

### 28.2.1  Neurobiological Background

The retina is the part of the eye that transforms light into an electrochemical message sent to the brain for processing. The mechanism is quite complicated, but we will give a brief overview. Light first contacts the cornea, a transparent tissue covering the pupil and the iris. The cornea helps converge light through the pupil. Next, light passes through the lens, where it is further focused onto the retina in the back of the eye. The retina has five distinct classes of neurons arranged into cell layers. Light first contacts the innermost layers of the retina, but it is the outermost layer that first processes the incoming light signal. The layer responsible for processing the incoming light is composed mainly of two different cell types: cones and rods. Rods are mainly responsible for sensing brightness, and cones are responsible for detecting color.

This chapter pertains to cones, which we will discuss exclusively from now on. Mammals such as humans have three types of cones. Each type is adept at "seeing" a certain color: red, green, or blue. Cones have a G-protein coupled receptor on their cell surface called *rhodopsin*. This receptor is closely associated with a chromophore called *11-cis-retinal*. When a photon of light hits the chromophore, it isomerizes to *11-trans-retinal*. This conformational change is detected by the rhodopsin molecule, and a G-protein is activated, which eventually closes ion channels on the cell membrane. The result is that ions (i.e., current) can no longer enter the cell, and the cell hyperpolarizes. Hyperpolarization decreases the cone's release of glutamate, a neurotransmitter that often has excitatory postsynaptic effects, which in turn decreases activity in postsynaptic cells in the next retinal layer. These postsynaptic cells are called *horizontal cells* (H-cells). H-cells normally maintain reciprocal synaptic connections with the cones that synapse onto them. H-cells release GABA, a neurotransmitter that has inhibitory postsynaptic effects. When the cones hyperpolarize in response to light and the H-cells decrease activity, the cones become disinhibited and begin to depolarize. This process is referred to as *negative feedback* because the initial light that induced hyperpolarization causes H-cells to feed back upon the cones in a way that counteracts the initial hyperpolarization. It is believed that this negative feedback is a regulatory mechanism to control color contrast. When the level of negative feedback of horizontal cells to cones is changed, the cones' response can be altered. Slight changes in feedback might be responsible for helping to determine changes in color. After all, although we have only three types of cones, humans can distinguish between millions of different colors! In addition to feeding back onto cones, horizontal cells also send signals to bipolar cells, which signal to amacrine cells, which finally signal to ganglion cells. The axons of these ganglion cells make up what anatomists call the *optic nerve*, the large nerve that connects the eye to the brain. The brain is then responsible for decoding the information sent from the retina to create what you "see" when you look at an object.

### 28.2.2 The Model

The model used in this chapter will be a system of two linear differential equations. The first will describe changes in the current leaving the cone of the retina, $C(t)$, and the second will describe the current leaving the horizontal cell, $H(t)$. We could build a larger system to account for the bipolar cells, amacrine cells, and ganglion cells, but we will keep it simple for now. The system is represented as follows:

$$\frac{dC}{dt} = \frac{1}{\tau_C}(-C - kH + L) \tag{28.1}$$

$$\frac{dH}{dt} = \frac{1}{\tau_H}(-H + C). \tag{28.2}$$

The first equation has three terms. The first indicates that the change in current is negatively proportional to the amount of current inside the cone, $C$. The second term represents the fact that the change in current is proportional to the current inside the horizontal cell, $H$, which negatively feeds back on the cell, and the third term indicates that the change in current into the cone is dependent on the light level, $L$. If the light level is high, then many photons will pass through the pupil, land on the retina, and activate the cones, resulting in a large change in current. The second equation states that the change in current in the horizontal cells depends negatively on the amount of current in the horizontal cells and the current of the cone cell that synapses onto the horizontal cell. Recall that the horizontal cells do not respond directly to light stimuli, so there is no term for the light intensity in the second equation. All other symbols in the preceding equations represent parameters (i.e., constants). Typical values for these parameters are $\tau_C = 0.025$ sec, $\tau_H = 0.08$ sec, and $k = 4$. Now also assume that the light intensity, $L$, is a constant, particularly $L = 10$. Finally, for the initial conditions, choose that $C(0) = H(0) = 0$. There is no current moving through either cell at $t = 0$.

The model equations as they are currently written can be simplified by a clever substitution. If you let:

$$\tilde{C} = C - \frac{L}{k+1} \quad \text{and} \quad \tilde{H} = H - \frac{L}{k+1}, \tag{28.3}$$

and substitute these equations into the previous equations, then you get:

$$\frac{d\tilde{C}}{dt} = \frac{1}{\tau_C}(-\tilde{C} - k\tilde{H}) \tag{28.4}$$

$$\frac{d\tilde{H}}{dt} = \frac{1}{\tau_H}(-\tilde{H} + \tilde{C}). \tag{28.5}$$

This is the model that you will study in its final form in this chapter. Note that the initial conditions now give:

$$\tilde{C}(0) = \tilde{H}(0) = \frac{L}{k+1} \tag{28.6}$$

### 28.2.3 Mathematical Background

Systems like the one in Equations 28.4 and 28.5 are especially suitable for study in MATLAB because they can be readily solved using simple matrix manipulations, as illustrated in the following simple example. Suppose you wanted to solve the system shown in Equations 28.7 and 28.8:

$$\frac{dx}{dt} = x + y \tag{28.7}$$

$$\frac{dy}{dt} = 4x + y \tag{28.8}$$

You begin by writing this system in matrix form to get:

$$\begin{bmatrix} \dfrac{dx}{dt} \\[2mm] \dfrac{dy}{dt} \end{bmatrix} = \begin{bmatrix} 1 & 1 \\ 4 & 1 \end{bmatrix} \begin{bmatrix} x \\ y \end{bmatrix}. \tag{28.9}$$

If you let the vector

$$\begin{bmatrix} x \\ y \end{bmatrix} = \vec{v} \quad \text{and} \quad A = \begin{bmatrix} 1 & 1 \\ 4 & 1 \end{bmatrix}, \tag{28.10}$$

then the system in Equations 28.7 and 28.8 becomes:

$$\frac{d\vec{v}}{dt} = A * \vec{v}. \tag{28.11}$$

Based on the Eigendecomposition Theorem (see Appendix B, "Linear Algebra Review"), you can substitute in for $A$ to get the following equation:

$$\frac{d\vec{v}}{dt} = VDV^{-1} * \vec{v}. \tag{28.12}$$

Next, you multiply across the left by $V^{-1}$ to get:

$$V^{-1}\frac{d\vec{v}}{dt} = V^{-1}VDV^{-1} * \vec{v} = DV^{-1} * \vec{v}. \tag{28.13}$$

If you let $V^{-1} * \vec{v} = \vec{u}$, then Equation 28.11 becomes:

$$\frac{d\vec{u}}{dt} = D * \vec{u} \tag{28.14}$$

This equation is similar to Equation 28.11, except for one very important exception: $D$ is diagonal. The eigendecomposition of:

$$A = \begin{bmatrix} 1 & 1 \\ 4 & 1 \end{bmatrix}$$

gives the eigenvalue matrix

$$D = \begin{bmatrix} 3 & 0 \\ 0 & -1 \end{bmatrix}$$

and the eigenvector matrix

$$V = \begin{bmatrix} 1 & 1 \\ 2 & -2 \end{bmatrix}.$$

If you substitute in for $D$ and convert Equation 28.14 into a system of equations, you get:

$$\frac{d\vec{u}}{dt} = \begin{bmatrix} \dfrac{du_1}{dt} \\ \dfrac{du_2}{dt} \end{bmatrix} = \begin{bmatrix} 3 & 0 \\ 0 & -1 \end{bmatrix} * \begin{bmatrix} u_1 \\ u_2 \end{bmatrix} = \begin{matrix} 3u_1 \\ -u_2 \end{matrix} \Rightarrow \tag{28.15}$$

$$\frac{du_1}{dt} = 3u_1$$

$$\frac{du_2}{dt} = -u_2.$$

This system is also a system of differential equations. However, each equation can be solved independently of one another to yield the solution:

$$\vec{u} = \begin{bmatrix} u_1 \\ u_2 \end{bmatrix} = \begin{bmatrix} C_1 e^{3t} \\ C_2 e^{-t} \end{bmatrix}. \tag{28.16}$$

Finally, recall that you let $V^{-1}\vec{v} = \vec{u}$, so that $V\vec{u} = \vec{v}$, and:

$$\vec{v} = \begin{bmatrix} x \\ y \end{bmatrix} = \begin{bmatrix} 1 & 1 \\ 2 & -2 \end{bmatrix} * \begin{bmatrix} C_1 e^{3t} \\ C_2 e^{-t} \end{bmatrix} = \begin{bmatrix} C_1 e^{3t} + C_2 e^{-t} \\ 2C_1 e^{3t} - 2C_2 e^{-t} \end{bmatrix} = C_1 \begin{bmatrix} 1 \\ 2 \end{bmatrix} e^{3t} + C_2 \begin{bmatrix} 1 \\ -2 \end{bmatrix} e^{-t} \Rightarrow$$

$$x(t) = C_1 e^{3t} + C_2 e^{-t}$$

$$y(t) = 2C_1 e^{3t} - 2C_2 e^{-t}. \tag{28.17}$$

Notice where the eigenvalues and eigenvectors appear in the preceding solution. The eigenvalues, $-1$ and $3$, appear in the exponents, and the eigenvectors appear as constant vectors multiplying the exponential with the corresponding eigenvalue. In general, the solution to any system of the form given in Equation 28.11 is:

$$\vec{v} = \begin{bmatrix} x \\ y \end{bmatrix} = C_1 * EV_1 * e^{\lambda_1 t} + C_2 * EV_2 * e^{\lambda_2 t} \tag{28.18}$$

where $\lambda_1$ and $\lambda_2$ are distinct (not equal) eigenvalues of the matrix $A$, and $EV_1$ and $EV_2$ are the corresponding eigenvectors. If $A$ has eigenvalues that are the same, then Equation 28.18 does not apply.

## 28.3 EXERCISES

You can see from Equation 28.18 that any set of equations that can be made into the form of Equation 28.11 can be solved by finding the eigenvalues and eigenvectors of the matrix $A$, as long as $A$ has distinct eigenvalues. You can do this easily in MATLAB with the following command:

>>[V, D] = eig(A);

For example, typing the command

>>[V, D] = eig([1 1;4 1])

produces the response:

```
V =
   0.4472   -0.4472
   0.8944    0.8944
D =
   3.0000   0
   0       -1.0000.
```

The matrix $D$ is a diagonal matrix with diagonal elements given by the eigenvalues of $A$. The columns of the matrix $V$ correspond to the eigenvectors of $A$. The matrix $D$ is the same as the one given earlier used to derive Equation 28.15, but the matrix $V$ produced by MATLAB is different from the one used to derive Equation 28.17. Nonetheless, you still have the relationship that $A = VDV^{-1}$. You can check this by typing the command

>> **V*D*inv(V)**
**ans =**
   1.0000   1.0000
   4.0000   1.0000.

The **inv()** function determines the inverse of a matrix.

Reexamining Equation 28.18 reveals that in order to find $x(t)$ and $y(t)$, you still need to solve for $C_1$ and $C_2$. It can be shown that if you have the initial conditions that $x(0) = x_o$ and $y(0) = y_o$, then:

$$\begin{bmatrix} C_1 \\ C_2 \end{bmatrix} = V^{-1} \begin{bmatrix} x_o \\ y_o \end{bmatrix} \tag{28.19}$$

Therefore, you can find these constants in MATLAB by typing the command

>> **[c_1; c_2] = inv(V)*[x_o; y_o];**

Finally, you can create $v$ according to Equation 28.18 by typing:

>> **v = c_1*V(:, 1)*exp(D(1,1)*t) + c_2*V(:, 2)*exp(D(2,2)*t);**

for some previously defined vector $t$. You can access the separate solutions $x$ and $y$ with $v$ (1, :) and $v(2, :)$, respectively.

---

### EXERCISE 28.1

Use MATLAB to check that if you let $A = [1\ 1;\ 4\ 1]$, $V = [1\ 1;\ 2\ -2]$, and $D = [3\ 0;\ 0\ -1]$, then $A = VDV^{-1}$ also holds.

---

### EXERCISE 28.2

Put the system of equations in Equations 28.4 and 28.5 into matrix form. What is the matrix $A$?

---

## 28.4 PROJECT

In this project, you will solve the retinal feedback system described previously. Specifically, you should do the following:

1. Write a function **solution(A,init)** that takes a matrix with distinct eigenvalues and a set of initial conditions, and plots the solutions $x(t)$ and $y(t)$ on the same graph. The function should return an output message if the input matrix does not have distinct eigenvalues. *Hint:* See the **error()** function help for producing output messages for functions that you want to throw an error message under certain conditions.
2. Use the function **solution(A,init)** to plot the solution of the retinal feedback system. In low light levels ($L = 3$), it takes longer for cells to respond. The measured parameters under low light level conditions are $\tau_C = 0.1$ sec, $\tau_H = 0.5$ sec, and $k = 0.5$. Plot the solution to the system with these parameters and compare with the previous plots. Under which conditions do the cones respond more strongly? Does this make sense?

## MATLAB FUNCTIONS, COMMANDS, AND OPERATORS COVERED IN THIS CHAPTER

**eig**
**inv**
**error**

# Simplified Model of Spiking Neurons

## 29.1 GOAL OF THIS CHAPTER

The goal of this chapter is to study a computationally efficient spiking cortical neuron model first introduced by Izhikevich (2003), and to generalize this model to a network of neurons. Ultimately, you will obtain and examine a raster plot of modeled network activity.

## 29.2 BACKGROUND

The task of understanding how different areas of the brain interact with each other to perform higher level functions such as motor coordination and speech is a major interest of modern neuroscience, but also an extremely difficult one. Many factors contribute to the global dynamics of neural networks. First, neurons isolated from a network exhibit a variety of patterns and behaviors. Some examples include regular spiking neurons, fast spiking neurons, intrinsic bursting neurons, and subthreshold membrane oscillations, as shown in Figure 29.1. Some of these behaviors are more common than others. For example, under normal conditions there are more regular spiking neurons in the cortex than intrinsic bursting ones. How these different dynamics are manifested at the network level remains an open area of current research. Second, the synaptic coupling between neurons can have a large impact on the network's dynamics leading to synchronization among the neurons and network oscillations.

Network oscillations in the brain are often categorized by their frequency. Oscillations with a frequency less than 4 Hz are called *delta rhythms*. Oscillations between 4 and 8 Hz are called *theta rhythms*. Rhythms from 8 to 12 Hz are called *alpha rhythms*, and rhythms from 12 to 30 Hz are called *beta rhythms*. Rhythms above 30 Hz are called *gamma rhythms*.

*MATLAB® for Neuroscientists.*
DOI: http://dx.doi.org/10.1016/B978-0-12-383836-0.00029-1

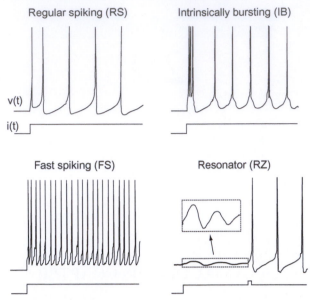

**FIGURE 29.1**   Known types of neurons. (An electronic version of the figure and reproduction permissions are freely available at www.izhikevich.org/publications/spikes.htm.)

## 29.2.1 The Model

The model for this chapter is a two-dimensional system of ordinary differential equations with a reset condition, as shown in Equations 29.1–29.3:

$$\frac{dv}{dt} = 0.04v^2 + 5v + 140 - u + I \tag{29.1}$$

$$\frac{du}{dt} = a(bv - u) \tag{29.2}$$

The reset condition is:

$$\text{if: } v \geq 30 \text{ then } \begin{cases} v \leftarrow c \\ u \leftarrow u + d. \end{cases} \tag{29.3}$$

The variable $v$ represents the membrane potential of the neuron while $u$ represents a generic recovery variable that feeds back negatively onto $v$. There are five additional parameters in the model: $I$, $a$, $b$, $c$, and $d$. The parameter $I$ represents external input to the neuron. This input could be thought of as external input to the neuron from outside the network or even synaptic input from a neuron within the network. The parameter $a$ controls the rate of recovery of $u$, and $b$ controls the sensitivity of recovery to subthreshold fluctuations of the membrane potential. The parameters $c$ and $d$ control the after-spike reset values for $v$ and $u$, respectively. If you choose certain parameter combinations, this simple model can exhibit all the firing patterns and behaviors shown in Figure 29.1.

To model a whole network of neurons, you will have to couple many neurons together where each neuron behaves according to Equations 29.1–29.3. This means that you will have to select a value for each parameter for every neuron you model. Additionally, since the network is coupled (i.e., the neurons are connected to each other), the input $I$ to a particular neuron will depend on other neurons in the network that synapse onto it. Therefore, you will need to model the connectivity of the network. You could choose to make every neuron in the network be connected to every other neuron in the network, or possibly make neurons connect only to neurons that are close. Additionally, you need to choose whether a connection between neurons will be excitatory or inhibitory and how strong the connection will be.

## 29.3 EXERCISES

The code for implementing Equations 29.1–29.3 is not very complicated. You can solve the equations using Euler's method, as introduced in Chapter 25, "Voltage-Gated Ion Channels." The script for implementing a single neuron is as follows (adapted from Izhikevich, 2003):

```
%These are some default parameter values
I = 10;
a = 0.02;
b = 0.2;
c = -65;
d = 8;
%The initial values for v and u
v = -65;
u = b*v;
%Initialize the vector that will contain the membrane potential time series.
v_tot = zeros(1000, 1);

for t = 1:1000
    %set v_tot at this time point to the current value of v
    v_tot(t) = v;
    %Reset v and u if v has crossed threshold. See Eq. 29.3 earlier.
    if (v >= 30)
        v = c;
        u = u + d;
end;
    %Use Euler's method to integrate Eqs. 29.1 and 29.2 from earlier. Here v is
    %calculated in 2 steps in order to keep the time step small (0.5 ms step in the
    %line below).
    v = v + 0.5*(0.04*v&circ;2 + 5*v + 140-u + I);
    v = v + 0.5*(0.04*v&circ;2 + 5*v + 140-u + I);
    u = u + a*(b*v-u);
end;
```

%This line uses the function find to locate the indices of v_tot that hold elements
%with values greater than or equal to 30 and then sets these elements to 30.
%This normalizes to heights of the action potential peaks to 30.
v_tot(find(v_tot >= 30)) = 30;
%Plot the neuron's membrane potential.
plot(v_tot);

---

### EXERCISE 29.1

Before going on to generalize this model to a network of neurons, you should explore this model thoroughly. See if you can discover what parameter sets lead to regular spiking, fast spiking, or intrinsically bursting behavior. What kinds of behaviors do the following parameter sets produce?

**a.** $[a, b, c, d] = [0.02, 0.2, -65, 8]$
**b.** $[a, b, c, d] = [0.02, 0.2, -55, 4]$
**c.** $[a, b, c, d] = [0.1, 0.2, -65, 2]$
**d.** $[a, b, c, d] = [0.1, 0.25, -65, 2]$

---

Now consider how to modify this script to model a network of neurons where each neuron is described by the dynamics of Equations 29.1–29.3. First, convert the parameters from numbers to vectors. The vectors will hold the value of each parameter for each neuron in the network. Since you want some neurons in the network to be regular spiking and others to be intrinsically bursting, you will need to have different values for the elements of the vectors. The modified code should begin as follows:

```
% The number of excitatory neurons in the network. The mammalian cortex has
% about 4 times as many excitatory neurons as inhibitory ones.
Ne = 800;
%The number of inhibitory neurons in the network.
Ni = 200;
%Random numbers
re = rand(Ne, 1);
ri = rand(Ni, 1);
%This will set the value of a for all excitatory neurons to 0.02 and the value of a
%for inhibitory neurons to a random number between 0.02 and 0.1
a = [0.02*ones(Ne, 1); 0.02 + 0.08*ri];
%This will allow b to range from 0.2–0.25
b = [0.2*ones(Ne, 1); 0.25-0.05*ri];
%This will allow the spike reset membrane potential to range between -65 and -50
c = [-65 + 15*re.^2; -65*ones(Ni,1)];
%This will allow the recovery reset value to range between 2 and 8
d = [8-6*re.^2; 2*ones(Ni, 1)];
    ⋮
```

Before you continue with the code, it is worthwhile to consider how these definitions of the parameters impact the composition of the network.

---

### EXERCISE 29.2

What parameter sets are most neurons likely to possess? Is there any correlation between excitatory neurons as represented in this model and regular spiking neurons, for example?

---

The next line of code should create a weight matrix that holds the strength of the connectivity between every pair of neurons in the network. Since the network has $Ne + Ni$ neurons, then the weight matrix will be a square matrix with these dimensions. The code to implement this is:

**S = [0.5\*rand(Ne + Ni, Ne), -rand(Ne + Ni, Ni)];**

Notice that this definition allows the strength of connections of excitatory neurons onto other neurons to range from 0 to 0.5, whereas inhibitory neurons have a synaptic strength between 0 and −1. According to this definition, a single inhibitory neuron can have, in general, a stronger effect on the neurons it contacts than a single excitatory neuron, which is supported by current experimental research. Also notice that very few elements of $S$ will be exactly 0, so in this model almost every neuron has synaptic contacts with all other neurons in the network. The rest of the code for the network model is:

```
%The initial values for v and u
v = -65*ones(Ne + Ni,1);
u = b.*v;
%Firings will be a two-column matrix. The first column will indicate the time that a
%neuron's membrane potential crossed 30, and the second column will be a number
%between 1 and Ne + Ni that identifies which neuron fired at that time.
%firings = [];
for t = 1:1000
    %Create some random input external to the network
    I = [5*randn(Ne, 1); 2*randn(Ni,1)];
    %Determine which neurons crossed threshold at the current time step t.
    fired = find(v >= 30);
    %Add the times of firing and the neuron number to firings.
    firings = [firings; t*ones(1, length(fired)), fired];
    %Reset the neurons that fired to the spike reset membrane potential and
    %recovery variable.
    v(fired) = c(fired);
    u(fired) = u(fired) + d(fired);
    %Add to the input, I, for each neuron a value equal to the sum of the synaptic
```

%strengths of all other neurons that fired in the last time step connected to that
%neuron.

```
I = I + ;sum(S(:,fired), 2);
%Move the simulation forward using Euler's method.
v = v + 0.5*(0.04*v^2 + 5*v + 140-u + I);
v = v + 0.5*(0.04*v^2 + 5*v + 140-u + I);
u = u + a*(b*v-u);
```
end;
%Plot the raster plot of the network activity.
```
plot(firings(:,1), firings(:,2),'.');
```

## 29.4 PROJECT

In this project, you will examine the behavior of a cortical network of spiking neurons. Specifically, you are asked to do the following:

According to the definition of the parameter values for $c$ in the network model, determine whether an inhibitory neuron can be an intrinsically bursting neuron. Will there be more regular spiking neurons in the network or intrinsically bursting neurons? Examine the raster plot produced by the preceding code. Are there any oscillations present in the network? If so, are they delta rhythms, theta rhythms, alpha rhythms, etc.?

Modify the code by redefining $c$ and $d$ to allow for more bursting neurons to be present in the network. What effect, if any, does this have on the presence of network oscillations?

Alter the weight matrix so that there are fewer connections between neurons of the network. What effect does this have on the network dynamics?

## MATLAB FUNCTIONS, COMMANDS, AND OPERATORS COVERED IN THIS CHAPTER

rand
randn
plot
find

# Fitzhugh-Nagumo Model: Traveling Waves

## 30.1 GOALS OF THIS CHAPTER

The purpose of this chapter is to learn how to model traveling waves in an excitable media. This entails the solution of a partial differential equation involving a first derivative in time coordinates and a second derivative in spatial coordinates. You will learn how to compute a second derivative in the MATLAB® software, and use a modification of the Fitzhugh-Nagumo model introduced in Chapter 15, "Exploring the Fitzhugh-Nagumo Model," to generate traveling waves in both one and two dimensions.

## 30.2 BACKGROUND

The Fitzhugh-Nagumo model is often used as a generic model for excitable media because it is analytically tractable. You will use it as a simple model to generate traveling waves by the addition of a diffusion term: a second derivative in spatial coordinates. In this chapter, you will modify the Fitzhugh-Nagumo model introduced in Chapter 15 in this way, and study its behavior in one and two dimensions. Thus you can simulate action potential wave propagation along the axon of a single neuron or the spreading of electrical potential waves in a network of cortical neurons.

There are many forms of the equations for the voltage, $v$, and recovery, $r$, variables in the Fitzhugh-Nagumo model. In general they are given by:

$$\frac{\partial v}{\partial t} = f(v) - r + I + \frac{\partial^2 v}{\partial x^2} \tag{30.1}$$

*MATLAB® for Neuroscientists.*
DOI: http://dx.doi.org/10.1016/B978-0-12-383836-0.00030-8

$$\frac{\partial r}{\partial t} = av - br \tag{30.2}$$

The function $f(v)$ is a third order polynomial that provides positive feedback, whereas the slower recovery variable $r$ provides negative feedback. By making the voltage and recovery variables functions of spatial coordinates as well as time, you can model dynamics in a spatially extended regime. The final term in the first equation introduces diffusion into the system, and thus the first equation is known as a *reaction-diffusion equation*.

*A note of caution:* As with ordinary differential equations, whenever you attempt to solve partial differential equations computationally, you must be careful that the various errors that can be introduced, such as truncation errors and roundoff errors, are not significant and that the necessary conditions for stability are met. See Strauss (1992) for a more in-depth discussion of such matters. If you are not careful, then the solutions produced by your code may stray quite significantly from the true solutions you seek.

## 30.3 EXERCISES

### 30.3.1 Second Derivative Operator

How do you model a second derivative computationally in MATLAB? There are a few approaches to this, but the one you will use here is the simplest computational approximation known as the *centered second difference*:

$$\frac{d^2v(x)}{dx^2} \sim \frac{v(x + \Delta x) - 2v(x) + v(x - \Delta x)}{(\Delta x)^2} \tag{30.3}$$

This approach can be justified by combining the Taylor expansions for $v(x + \Delta x)$ and $v(x\Delta x)$ (Strauss, 1992). If the mesh size of the spatial variable is represented by $\Delta x$, then the $j^{th}$ element of the array $v$, $v_j$, is the value of $v$ for $x = j\Delta x$, so you have:

$$\frac{d^2v_j}{dx^2} \sim \frac{v_{j+1} - 2v_j + v_{j-1}}{(\Delta x)^2} \tag{30.4}$$

The second derivative can thus be computed by convolving the array $v$ with the second derivative operator filter $F = [1 \ -2 \ 1]/(\Delta x)^2$. This can be extended to two dimensions as well. Assuming equal mesh spacing along both directions ($\Delta x = \Delta y$), then the two-dimensional second derivative operator filter is given by $F = [0 \ 1 \ 0; \ 1 \ -4 \ 1; \ 0 \ 1 \ 0]/(\Delta x)^2$.

Now create a function in MATLAB for the second derivative operator in one dimension, and name it **secDer.m**. Its input will be the one-dimensional array $v(x)$ and the spatial mesh size, $dx$, and the output will be the second derivative, $v''(x)$. This function will use the convolution function **conv**, which, by default, introduces undesirable edge effects. Also include an option to improve the edge effects by making the boundary conditions periodic. You'll do this by adding a third input to the function, $BC$, which, if set to 1, will return the default **conv** output found by padding the input matrix with zeros, also known as *free boundary conditions*, and if set to 2, then you will have *periodic boundary conditions*.

Periodic boundary conditions means that the boundaries of the input array (i.e., the first and last elements) are considered neighboring points. You could use the **if, elseif** control structure to carry out the options for the boundary conditions, but instead we will introduce you to another useful control structure that MATLAB offers: **switch**.

```
function V = secDer(v,dx,BC)
%
%F is the discrete 2nd derivative filter in 1D
F = [1 −2 1 ]/dx^2;
%
%BC determines your boundary conditions
switch BC
  case 1 %free bc's
    V = conv(v,F);
    V = V(2:end-1); %return an array the same size as the input array
  case 2 %periodic bc's
    %since the convolution filter is of length 3 then we only have to
    %pad the input array v by 1 element on either side
    pv = zeros(1,length(v) + 2); %extend the input array by 2
    pv(2:end-1) = v;
    %now we fill in these two padded points with the values that the extended
    %input array would have if the first element of v and the last element of v
    %were neighbors
    pv(1) = v(end);
    pv(end) = v(1);
    V = conv(pv,F);
    V = V(3:end-2); %return the valid portion of the convolution
end
```

Give this a try and see how it works by testing it on a function whose second derivative is well known: cosine. If $f(x) = \cos(x)$, then $f''(x) = -\cos(x)$. You can compare the output of the second derivative function, **secDer**, with the analytic solution by running the following script, whose output is shown in Figure 30.1.

```
x = linspace(-pi,pi,100); %forces even spacing in array of 100 pts from −pi to pi
dx = x(2)-x(1); %determines this spacing, the spatial mesh size
x = x(2:end); % for periodicity knock of 1st term in x array so that we don't have repeat
          % value of cos(x) at the endpoints of x (since cos(-pi) = cos(pi))
f = cos(x); %input array
d2f = secDer(f,dx,2); %computational solution to the second derivative
                % of f with periodic BC
d2fA = -cos(x); %analytic solution to the second derivative of f
plot(x,f,'k','LineWidth',3) %plot input f
hold on
plot(x,d2f,'b','LineWidth',5) %plot computational result of f"
plot(x,d2fA,'k:','LineWidth',3) %plot analytic result of f"
axis([-pi pi -1 1]); set(gca,'fontsize',20)
```

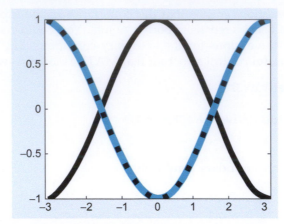

**FIGURE 30.1**   Testing the second derivative function **secDer.m**. The solid black line is the input, **cos(x)**, and the blue line is the output of the **secDer** function with periodic boundary conditions. This matches exactly with the analytic solution to the second derivative, **-cos(x)**, shown as the dotted black line.

---

**EXERCISE 30.1**

Compare this with the result you get if you use free boundary conditions. You can do this by setting *BC* to 1 when calling the **secDer** function. Remove or comment out the last line of the script above, which sets the limits for the axes in the plot, so that you can see the effect of changing the boundary conditions.

---

**EXERCISE 30.2**

Rewrite the **secDer.m** function using the **if, elseif** control structure rather than the **switch** control structure. Test your function by using $f = \sin(x)$ and compare your result with the known solution $f'' = -\sin(x)$.

---

With this second derivative operator, you can now model a traveling wave in 1D. For the one-dimensional problem, you will use the following form of the Fitzhugh-Nagumo equations (Wilson, 1999):

$$\frac{\partial v}{\partial t} = 10\left[v - \frac{1}{3}v^3 - r + D\frac{\partial^2 v}{\partial x^2}\right] + I \qquad (30.5)$$

$$\frac{\partial r}{\partial t} = p[a + 1.25v - br] \qquad (30.6)$$

The variables $v(x,t)$ and $r(x,t)$ are the voltage and the recovery variables at position $x$ at time $t$. If modeling a pulse traveling along a nerve fiber, you can think of $x$ as the position along the nerve fiber. Similarly, if you want to model a traveling wave of activity across a one-dimensional network of neurons, then $x$ indicates the neuron location in the one-dimensional population. Now consider the latter case. You will use the following parameter values: $D = 1$, $a = 1.5$, $b = 1$, and $p = 0.8$. For this set of parameters, the steady state values are $v0 = -1.5$ and $r0 = -3/8$, which you will use as your initial conditions for these variables. The driving stimulus is given by $I$. To solve this system, you are going to need an ODE solver.

## 30.3.2 Built-in ODE Solvers

In previous chapters, you solved differential equations through manually written ODE solvers using the Euler method and the Runge-Kutta method, which had the advantage of complete transparency in the mechanisms behind the operations. In this chapter, we will introduce you to the most practical and commonly used of the built-in ODE solvers in MATLAB: the function **ode45**. This solver is based on an explicit Runge-Kutta formula and has been optimized to adaptively find the most efficient time steps to produce a solution within a certain allowed relative error tolerance ($10^{-3}$ by default) and absolute error tolerance ($10^{-6}$ by default). Look at the help section for **ode45** for more information on how to adjust these options as well as to learn about the other built-in ODE solvers offered by MATLAB and the conditions under which to use them.

To familiarize yourself with the proper syntax for using the **ode45** ODE solver, first consider the simpler case of solving this system of equations for one point in space (i.e., for one neuron). First, you must code the system of first-order ODEs as a function that the solver can use. You will represent the Fitzhugh-Nagumo system in a function called **F_N1**. The **F_N1** function assumes that $v$ and $r$ become elements $V(1)$ and $V(2)$ of the two-element input vector $V$. Although $t$ and $V$ must be the function's first two arguments, the function does not need to use them. The output $vdot$, the derivative of $V$, must be a column vector, as shown in the following code:

```
function vdot = F_N1(t,V)
%
%set parameters of the model
a = 1.5; b = 1; p = .08; I = 1.5;
%dv/dt:
vdot(1) = 10*(V(1) - (V(1).^3)/3 - V(2) + I);
%dr/dt
vdot(2) = p*(1.25*V(1) + a - b*V(2));
vdot = vdot'; %make correct dimensions for ODE solver: must be a column vector
```

Note that the diffusion term is left out, since there is only one point in space and a spatial derivative makes no sense in this case. For this one neuron system, you set the input to 1.5 so that the model will initiate a series of action potentials. You can then generate and plot the solution as follows:

v0 = [ − 1.5; − 3/8]; %initial conditions for V variable
tspan = [0 100]; %beginning and end values of time
[t,v] = ode45('F_N1', tspan, v0);
plot(t,v(:,1),'k*','LineWidth',5);

Now look at what was produced by running the preceding script:

>> whos

| Name | Size | Bytes | Class |
| --- | --- | --- | --- |
| t | 2597 × 1 | 20776 | double |
| tspan | 1 × 2 | 16 | double |
| v | 2597 × 2 | 41552 | double |
| v0 | 2 × 1 | 16 | double |

Note that the time, which goes from 0 to 100, is a column vector of 2597 points. These time points are not evenly distributed between 0 and 100; rather, the mesh size varies and has been selected by the solver to most efficiently compute the differential equation within the tolerated error. For example, consider how the spacing between time points varies in just the first 10 time points:

>> t(1:10)
ans =
   0
   0.0015
   0.0031
   0.0046
   0.0061
   0.0138
   0.0214
   0.0291
   0.0367
   0.0451
>> diff(t(1:10))
ans =
   0.0015
   0.0015
   0.0015
   0.0015
   0.0077
   0.0077
   0.0077

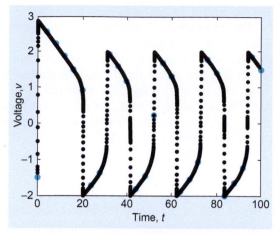

**FIGURE 30.2** Voltage, $v$, versus time, $t$, output of Fitzhugh-Nagumo system of equations for one point in space found using the **ode45** solver with **tspan** = [1 100] (black dots) and with **tspan** = [1:4:100] (blue dots).

> **0.0077**
> **0.0084**

The first column of the output vector, $v(:,1)$, represents the voltage values at the corresponding times starting with $v0(1)$ at the first time point. Similarly, the second column of the output vector, $v(:,2)$, represents the recovery variable values for the corresponding times, starting with $v0(2)$.

When you specify specific time points in **tspan,** the ODE solver will still use its most efficient time mesh to solve the differential equation; however, it will now return the values of the outputs at the specific times indicated in **tspan.** Compare the previous result with that found by specifying the time to be from 0 to 100 at intervals of 4:

```
hold on
tspan = [0:4:100];
[t,v] = ode45('F_N1',tspan,v0);
plot(t,v(:,1),'b*','LineWidth',5)
```

This result is shown in Figure 30.2.

### 30.3.3 Fitzhugh-Nagumo Traveling Wave

You are now ready to tackle the full problem of simulating a propagating wave along a line of neurons. Let the number of neurons be given by $N$. A naïve approach might be to allow the initial conditions to be a $2 \times N$ matrix with the first row representing the $N$ initial voltages and the second row the $N$ values of the initial recovery variables. However, recall that the **ode45** solver will accept only a single column array for the initial conditions. What you will do to satisfy this requirement is let the initial value column vector be of length $2N$ and let the first $N$ elements represent the initial voltages of the $N$ neurons and the last $N$ elements represent the initial recovery values. The ODE solver will produce as its output a $t \times 1$ time vector, and a $t \times 2N$ matrix, whose first $N$ columns represent the evolution of the voltage of the population of neurons as time progresses and whose second $N$ columns represent the evolution of the recovery variables.

The stimulus to initiate the wave that you will use is $I = 6$ for the first 0.5 s; then the stimulus will be off, $I = 0$, for the rest of the time. You can choose where along the line of neurons to initiate the wave. In the following example, you stimulate the center cells. The following script, **FNmain.m**, will produce a traveling wave of activity along a one-dimensional population of $N$ neurons whose dynamics are governed by the Fitzhugh-Nagumo equations, as shown in Figure 30.3.

```
%FNmain.m
clear all; close all
%
global N I BC %by making these variables global they can exist within
%the workspace of functions without explicitly being input to the functions
N = 128; %number of neurons
v0(1:N) = −1.5; %initial conditions for V variable
v0(N + 1:2*N) = −3/8; %initial conditions for R variable
I = 6; %the input stimulus value
BC = 2; %set to 1 (free) or 2 (periodic boundary conditions)
%
tspan = [0:.1:.5]; %time with stimulus
[t1,v1] = ode45('F_N',tspan,v0);
tspan = [.5:.1:25]; % time without stimulus
I = 0;%turn off stimulus
[t2,v2] = ode45('F_N',tspan,v1(end,:)'); %note: initial cond are final v1 values
%piece together (concatenate) time (t1 and t2) and solution (v1 and v2)
%variables without double counting the seam values
t = [t1; t2(2:end)]; v = [v1; v2(2:end,:)];
%spacetime plot of v variable w/ neurons along y axis, time along x-axis
figure(1); imagesc(v(:,1:N)') ; colorbar
%spacetime plot of r variable w/ neurons along y axis, time along x-axis
figure(2); imagesc(v(:,N + 1:end)'); colorbar
```

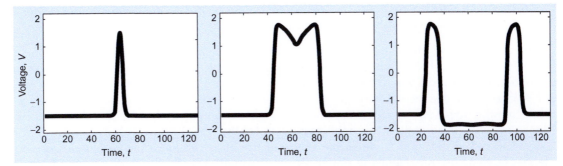

**FIGURE 30.3**    Traveling Fitzhugh-Nagumo wave in one dimension for $t = 1$ (left), $t = 5$ (center), and $t = 10$ (right).

```
%create a movie of the traveling wave
figure(3)
for ii = 1:length(t)
    plot(v(ii,1:N))
    axis([0 N −2.1 2.2])
    pause(.05)
end
```

When you run the preceding **FNmain.m** script, make sure that this M-file is in the same directory as the **secDer** function previously created as well as the following function **F_N**, since the main script calls these functions. The function **F_N** describes the coupled system of differential equations used to model the dynamics.

```
function vdot = F_N(t,v)
global N I BC
%
%set parameters of the model
D = 10; a = 1.5; b = 1; p = .08;
%dv/dt:
vdot(1:N) = 10*(v(1:N) - (v(1:N).^3)/3 - v(N + 1:end)) + D*secDer(v(1:N),1,BC)';
%add input I to the center five cells
vdot(round(N/2)-2:round(N/2) + 2) = vdot(round(N/2)-2:round(N/2) + 2) + I;
%dr/dt:
vdot(N + 1:2*N) = p*(1.25*v(1:N) + a − b*v(N + 1:end));
%
vdot = vdot'; %make correct dimensions for ODE solver
```

---

### EXERCISE 30.3

Use periodic boundary conditions, as in the preceding example, but change the location of the wave initiation to some point off center. For example, rather than stimulate the center cells, stimulate cells a fourth of the way from the edge. Watch the two waves initiated annihilate each other.

---

### EXERCISE 30.4

Change the boundary conditions to make them free boundary conditions and start the wave at the left end of the array to create a traveling wave that goes from left to right. Play with the parameters to see how they affect the dynamics.

## 30.4 PROJECT

In this project, you will simulate a traveling wave of activity in a two-dimensional $N \times N$ array of cortical neurons. This time you will use these versions of the Fitzhugh-Nagumo equations (Murray, 2002):

$$\frac{\partial v}{\partial t} = -v(a - v)(1 - v) - r + \frac{D \partial^2 v}{\partial x^2} \tag{30.7}$$

$$\frac{\partial r}{\partial t} = bv - gr \tag{30.8}$$

with the parameters taking on the values $a = 0.25$, $b = 0.001$, $g = 0.003$, and $D = 0.05$. You will compute the solution using a similar method as that used for the one-dimensional problem. This time the initial value column vector will be of length $2N^2$, where the first $N^2$ elements will represent the initial voltages of the $N \times N$ neuron array and the last $N^2$ elements will represent the initial recovery values. The ODE solver will produce as its output a $t \times 1$ time vector, and a $t \times 2N^2$ matrix, whose first $N^2$ columns represent the evolution of the voltage of the population of neurons as time progresses and whose second $N^2$ columns represent the evolution of the recovery variables. For a given row of this output matrix, i.e., a particular time point, you can reconstruct the $N \times N$ array of voltage variables from the $1 \times N^2$ array using the *reshape* function, whose input is the matrix to be reshaped as well as the number of rows and columns desired in the output.

```
>> A = [1  2  3  4  5  6  7  8  9]
A = 1  2  3  4  5  6  7  8  9

>> a = reshape(A,3,3)
a = 1  4  7
    2  5  8
    3  6  9
```

You will need to do this when you call the two-dimensional second derivative function as well as when you wish to visualize the results. There are more elegant and efficient ways to handle the issue of programming the two-dimensional partial differential equation, but here we will present a way to solve the problem in MATLAB that is conceptually simple, allowing you to use the tools previously used while not introducing any more complicated commands.

In the following script, **FN2main.m**, you will create an .avi file of the simulation and name it **TravelingWave.avi**:

```
%FN2main.m
clear all; close all
% next 4 lines are to create the movie file of the wave
fig = figure;
set(fig,'DoubleBuffer','on');
set(gca,'NextPlot','replace','Visible','off')
```

```
mov = avifile('TravelingWave.avi')
%
tspan = [0:5:800]; %simulation time
global N BC
BC = 1; %boundary conditions: 1 free, 2 periodic
N = 32; %number of neurons
v0(1:N^2) = 0; %initial conditions for V variable
v0(N^2 + 1:2*N^2) = 0; %initial conditions for R variable
v0(1:N)  = .6; %initially stimulate all cells along left edge of population
[t,v] = ode45('F_N2',tspan,v0);
%obtain min and max values of output
clims = [min(min(v(:,1:N^2))) max(max(v(:,1:N^2)))];
%generate the movie of the voltage
for ii = 1:size(v,1)
   figure(1)
   im = imagesc(reshape(v(ii,1:N^2),N,N), clims);
   axis square
   set(im,'EraseMode','none');
   Frame = getframe(gca);
   mov = addframe(mov,Frame);
end
mov = close(mov);
```

Specifying the same *clims* in the **imagesc** option for each time frame of the imaged voltage array ensures that the same colormap is used throughout the movie, which is analogous to using consistent z-axis limits. The preceding main script calls on a number of functions that need to be created and stored in the same directory as the main script. These functions, **F_N2**, **SecDer2**, and **conv2periodic**, follow. Note that you will need to complete the **F_N2** function. The **SecDer2** function is just an extension of the **SecDer** function to two dimensions and employs the reshape function to carry out the 2D convolutions. The **conv2_periodic** function has been written in a generic form so that it can handle input matrices of various sizes for future applications.

```
function vdot = F_N2(t,v)
global N BC
%
D = .05; a = 0.25; b = .001; g = .003; %set the parameters
%
%dV/dt
vdot(1:N^2) = -(v(1:N^2)).*(a-(v(1:N^2))).*(1-(v(1:N^2)))-...
              v(N^2 + 1:end) + D.*secDer2(v(1:N^2),1,BC);
%dR/dt
vdot(N^2 + 1:2*N^2) = ???
%
vdot = vdot';
```

```
function V = secDer2(v,dx,BC)
global N
%
F = [0 1 0; 1 −4 1; 0 1 0]/dx^2;
%determines your boundary conditions
switch BC
    case 1 %free bc's
        V = conv2(reshape(v,N,N)',F,'same');
        V = reshape(V',N*N,1);
    case 2 %periodic bc's
        V = conv2_periodic(reshape(v,N,N)',F);
        V = reshape(V',N*N,1);

end

function sp = conv2_periodic(s,c)
% 2D convolution for periodic boundary conditions.
% Output of convolution is same size as leading input matrix
[NN,M] = size(s);
[n,m] = size(c); %% both n & m should be odd
%enlarge matrix s in preparation convolution with matrix c via periodic edges
padn = round(n/2) - 1;
padm = round(m/2) - 1;
sp = [zeros(padn,M + (2*padm)); ...
        zeros(NN,padm) s zeros(NN,padm); zeros(padn,M + (2*padm))];
%fill in zero padding with the periodic values
sp(1:padn,padm + 1:padm + M) = s(NN + 1-padn:NN,:);
sp(padn + 1 + NN:2*padn + NN, padm + 1:padm + M) =  s(1:padn,:);
sp(padn + 1:padn + NN,1:padm) =  s(:,M + 1-padm:M);
sp(padn + 1:padn + NN,padm + M + 1:2*padm + M) =  s(:,1:padm);
sp(1:padn,1:padm) =  s(NN + 1-padn:NN,M + 1-padm:M);
sp(padn + NN + 1:2*padn + NN,1:padm) =  s(1:padn,M + 1-padm:M);
sp(1:padn,padm + M + 1:2*padm + M) = s(NN + 1-padn:NN,1:padm);
sp(padn + NN + 1:2*padn + NN,padm + M + 1:2*padm + M) = s(1:padn,1:padm);
%
%perform 2D convolution
sp = conv2(sp,c,'same');
% reduce matrix back to its original size
sp = sp(padn + 1:padn + NN,padm + 1:padm + M);
```

In this project, you should do the following:

1. Create the main script and the functions given here in the same working directory.
   Complete the **F_N2** function by replacing the symbols **???** with the proper quantities so
   that **F_N2** implements the Fitzhugh-Nagumo model equations as given in this section.
   Run the main script to generate a plane wave from the left, as shown in Figure 30.4.

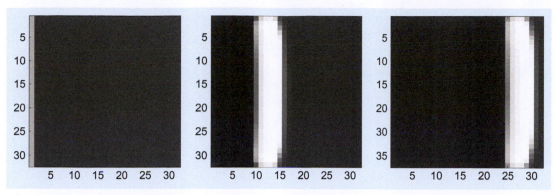

**FIGURE 30.4**  Two-dimensional traveling wave produced by the Fitzhugh-Nagumo equations for $t = 0$ (left), $t = 300$ (center), and $t = 600$ (right).

2. Alter the code so that the wave is initiated from one of the corners or in the center of the array and forms a propagating ring. Also try running the simulation with periodic boundary conditions. Play with the various parameters to see how they affect the wave dynamics.

3. Now use the model to generate a spiral by resetting the upper half of the voltage and recovery variables to 0 when the traveling plane wave is approximately halfway across the neural network. Start with the original code, as in the first part of the project so that a plane wave is initiated along the entire left edge of the array. In the main script, change the name of the .avi file that will be created to **SpiralWave.avi**. Let $N = 60$ and *tspan = [0:5:1800]*. This will cause the simulation to take more time to run but will allow you to view the spiral well.

    In the main script, create a variable *Reset* that can take on the value 0 or 1. Use the **switch** control structure so that when *Reset* is equal to 0, it runs the original script, and when it is equal to 1, the line **[t,v] = ode45('F_N2',tspan,v0);** will not be executed, and in its place the following lines will be executed:

```
[t,v] = ode45('F_N2',tspan,v0);
[t1,v1] = ode45('F_N2',tspan(1:round(length(tspan)/3)),v0);
vR = Vreset(v1(end,:));
[t2,v2] = ode45('F_N2',tspan(round(length(tspan)/3):end),vR);
v = [v1; v2(2:end,:)];
t = [t1; t2(2:end,:)];
```

Create the following **Vreset** function in the same directory as the main file:

```
function vR = Vreset(v)
global N
%
%reset half of voltage variables to zero
```

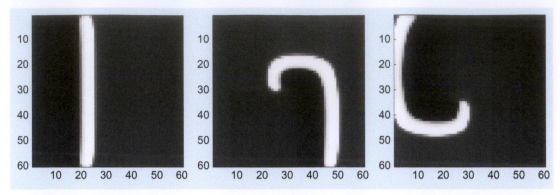

**FIGURE 30.5**  Spiral wave produced by the Fitzhugh-Nagumo equations for $t = 500$ (left), $t = 1000$ (center), and $t = 1500$ (right).

**VR = reshape(v(1:N^2),N,N)';**
**VR(:,1:round(N/2)) = 0;**
**%reset half of recovery variables to zero**
**RR = reshape(v(N^2 + 1:end),N,N)';**
**RR(:,1:round(N/2)) = 0;**
**%**
**vR = [reshape(VR',N*N,1);reshape(RR',N*N,1)];**
Explain how the **Vreset** function works. Submit the main script and set of functions that give the option to generate spiral waves. Plot screenshots of the spiral wave generated at various timesteps $t(i)$ by using the command:

**imagesc(reshape(v(i,1:N^2),N,N), clims); axis square**

for various values of $i$, as shown in Figure 30.5. Again, you can play with the various parameters to see how they affect the wave dynamics.

# MATLAB FUNCTIONS, COMMANDS, AND OPERATORS COVERED IN THIS CHAPTER

**switch**
**ode45**
**global**
**imagesc**
**reshape**

# 31

# Decision Theory

## 31.1 GOALS OF THIS CHAPTER

In this chapter, you will learn how to implement progressively more comprehensive mathematical models of decision making using MATLAB®. The exploration of decision models will introduce solving partial differential equations as finite differences, focusing on the diffusion equation. A simple model accounting for perceptual decisions and corresponding activity in cortical areas LIP and MT will be discussed.

## 31.2 BACKGROUND

One of the fundamental behavior characteristics of choice and reaction time (RT) is a trade-off between accuracy (selecting the correct choice) and speed (e.g., Swensson, 1972). Research participants asked to make decisions limited by time make more errors as the time allotment grows shorter. Any serious model for reaction time should account for this phenomenon.

However, accuracy does not improve infinitely. Participants still err, even when given large time windows. Not only must any model handle the inverse relationship between accuracy and speed, but it must also still maintain a small probability of error.

There are many possible ways of addressing this aspect of decision-making behavior. For decades, one of the most successful models has been the diffusion drift model (DDM) (Ratcliff, 1978). In the diffusion drift model, decision processes are described in terms of evidence accumulation over time, with larger durations allowing greater amounts of evidence to accumulate.

Neurobiological investigations of choice in sensory systems have proposed putative mechanisms for the behavior described by the DDM and similar evidence accumulation models of decision. Later in this chapter, we will discuss a biological model for decision behavior in the perception of motion.

# 31.3 SIMPLE ACCUMULATION OF EVIDENCE

To begin, we can write a naïve model of evidence accumulation for a simple task (Go/No Go) in discrete time, in which each time step allows for the accumulation of a small amount of evidence. We will use a basic, discrete version of the diffusion drift model (DDM), a well-established model of decision processes (Ratcliff, 1978).

In this formulation of the accumulation, the evidence will take on continuous values. We can represent this as a differential:

$$dX = Bdt + \sigma dW$$

Here, $dX$ represents the change in evidence during the time step $dt$. $B$ is a constant bias that directs the evidence total over time. The magnitude and sign of $B$ influences the long term behavior of the evidence accumulation process. Larger values cause faster accumulation and smaller magnitude values cause slower accumulation.

The last term, $dW$, represents a discretized Brownian motion term. Simply put, a Brownian motion is a random walk in which steps follow a Gaussian distribution. (The dW label originates from the Wiener process, another name for Brownian motion.) Formally, the Brownian motion term is characterized by three properties:

1. $W(0) = 0$
2. $W(t)$ is continuous in $t$
3. For any two values of $t$, $t_1$, and $t_2$, the difference between $W(t_1)$ and $W(t_2)$ follows an independent normal distribution with mean 0 and variance equal to the difference $t_2 - t_1$.

Consequently, one important aspect of the Brownian motion term is its scaling with respect to time. Since $dW$ follows an $N(0, dt)$ distribution, increasing or decreasing the time step size increases or decreases the variance of the random portion of the walk. The scalar scaling constant $\sigma$ allows for adjustments to uncertainty.

To complete our simple model, we need a start point $x_0$. The start point will represent the evidence accumulated prior to the experiment. Often, we can treat this as completely undecided. An evidence total above the start point implies a positive choice, and an evidence total below the start point will imply a negative choice. In most cases, we will choose a value of 0 for $x_0$, so that a positive accumulation will indicate a positive choice.

We now have enough to write a simple simulation for a Go/No Go task with a fixed time interval. This would correspond to presenting a stimulus and requiring the research participant to select a choice at the end of the time interval.

```
function choice = simple_model(bias, sigma, dt, time_interval)
    x = [0];
    time = 0;
    while time < time_interval
        time = time + dt;
        dW = randn * (dt^0.5); % randn is always N(0,1)
        dX = bias * dt + sigma * dW;
        % add dx to the most previous value of x
        x = [x ; x(length(x)) + dX];
    end
```

```
% time is up
    choice = x(length(x)) > x(1);
end
```

---

### EXERCISE 31.1

Simulate 20 choice experiments, each 1 second long, with $B = 1$ per second, $\sigma = 1$, $dt = 0.1$ second. What is the distribution of results? Modify **simple_model** to return the evidence ($x$) as well as the resulting choice. Examine the time course of the evidence for $B = 1$, $B = 10$, $B = 0$, $B = 0.1$, and $B = -1$ over a 10-second trial.

---

Under this testing paradigm, in which the participant is interrogated at the end of a fixed time interval, we cannot directly evaluate reaction time. However, we can examine error rate (ER) as a function of the time interval.

---

### EXERCISE 31.2

Explore how the time interval influences the error rate. Generate 10 trials each for tasks ranging in duration from 0.5 to 10 seconds. Use parameters $B = 0.1$ per second, $\sigma = 1$, $dt = 0.1$ second. Assume that a positive response is correct. How does ER change with time?

---

As the time interval increases, the influence of the bias parameter affects the evidence accumulation more and more strongly. To explore the probability distribution of the evidence total, $X$, at some time $t$, we can integrate the previous difference equation. It is important to note that, being stochastic, the Brownian motion term cannot be integrated using standard integration techniques. Here, the integral is the sum of $t/dt$ independent normally distributed variables with equal mean (0) and variance ($dt$). Thus, the total should be distributed normally with mean 0 and variance $(t/dt)(dt) = t$. This corresponds to the distribution of $W(t)$ as defined earlier.

$$\int dX = \int B\,dt + \int \sigma\,dW$$
$$X(t) = Bt + \sigma W(t)$$

With this expression of $X(t)$, we can calculate an expectation of the first moment (i.e., the mean).

$$E[X(t)] = E[Bt + \sigma W(t)]$$
$$E[X(t)] = E[Bt] + E[\sigma W(t)]$$
$$E[X(t)] = Bt + 0$$
$$E[X(t)] = Bt$$

---

### EXERCISE 31.3

Show that the variance (the expectation of the centered second moment or $E[(X(t) - E[X(t)])^2]$ is $\sigma^2 t$. Generate evidence trajectories for 10 trials over 10 seconds with parameters $B = 0.1$ per second, $\sigma = 1$, $dt = 0.1$ second. Calculate mean and variance over time. Plot the trajectories, mean trajectory, and one standard deviation above and below the mean. Compare the trajectories to the expected value.

---

Knowing the distribution of evidence at time $t$ under the fixed time paradigm, we can determine the probability of being above threshold and the expected error rate. $X(t)$ follows a normal distribution with mean $Bt$ and variance $\sigma^2 t$. The proportion of this distribution above the threshold is the probability of exceeding the threshold at time $t$. The standard function for expressing this is $\Phi(z)$, the cumulative distribution function for the standard normal distribution. The value of $\Phi(z)$ is defined as the integral of the standard normal probability density from negative infinity to $z$. This is also the probability that a random variable with a standard normal distribution will have a value less than or equal to $z$.

Because the value of interest here is the probability of exceeding the threshold, the probability of exceeding the threshold at time $t$ is equivalent to

$$1 - \Phi\left(\frac{x_0 - Bt}{\sigma\sqrt{t}}\right)$$

The Statistics Toolbox supplies a function to calculate the value of the cumulative standard normal distribution, but we can easily define it using a function available in the MATLAB core functions, specifically the error function **erf**. Defined in terms of the error function,

$$\Phi(z) = \frac{1}{2}\left[1 + erf\left(\frac{z}{\sqrt{2}}\right)\right]$$

With **erf**, finding a value of $\Phi(z)$ is as simple as

```
>> phi = 0.5 * ( 1 + erf(1/2^0.5) )
phi =
    0.8413
```

---

### EXERCISE 31.4

Write a MATLAB function for the cumulative standard normal distribution, $\Phi(z)$. Write a function **simple_model2** that accepts bias, time limit, and start point, and that returns the choice. **simple_model2** should use phi and the equations above to calculate the probability of being above or below the start point without iterating through a trajectory. Once a probability of the evidence total being above or below the start point is calculated, a draw from a uniform random distribution can be used to choose an option.

---

## 31.4 FREE RESPONSE TASKS

By adding positive and/or negative thresholds, we can extend our model to simulate a testing paradigm that allows a free response. Under such a paradigm, if the evidence exceeds a threshold, then the trial ends and a choice is made.

A model may have both a positive and a negative threshold or only a single threshold. Which a model has will depend on the task to be simulated. A Go/No Go task in which a participant must respond before an interval when a signal is perceived would be a good match to a simulation paradigm with a single positive threshold. Evidence accumulation would be towards the Go response. Such a simulation would also have a time limit. Trials failing to accumulate enough evidence by the time limit would produce a No Go result.

Two thresholds, positive and negative, could be appropriate for simulating a two alternative forced choice (2 AFC) task. In the simplest case, the two thresholds are equidistant from the evidence start point, $x_0$, but the distances from the start can be asymmetric. Even with two thresholds, there is a single accumulator with only one bias parameter.

---

### EXERCISE 31.5

Write a function **two_choice_trial** that accepts five parameters: positive and negative thresholds $\theta_+$ and $\theta_-$, a variance for accumulation error $\sigma$, a start value $x_0$, and a bias $B$. The function should use the discrete time representation of the decision process by calculating values for the total evidence at successive time intervals until evidence exceeds a threshold. The time at the threshold crossing is the reaction time. The function should return both the reaction time and 1 or $-1$ for the response.

---

Now that our model allows an immediate return once sufficient evidence is accumulated, we can investigate the relationship between error rate (ER) and reaction time (RT) more stringently.

---

### EXERCISE 31.6

Generate a large number of trials using a free response decision paradigm. Choose a relatively small bias. Plot a red point at $(RT, 1)$ for each correct response (assuming that a response in the direction of the bias is correct), where $RT$ is the response time. Plot a blue point at $(RT, -1)$ for each incorrect response. Does the distribution of points say anything about the relationship between $RT$ and $ER$?

---

## 31.5 MULTIPLE ITERATORS: THE RACE MODEL

The discrete version of the diffusion drift model can accommodate a two threshold paradigm, but even with two thresholds, the single iterator allows for only a single set of bias and variance values. Under some two choice scenarios, the race model may be a better match.

Under the race model, each of the two choices has an independent iterator and threshold. The first process whose accumulated evidence exceeds its threshold is the choice of the system. This choice "wins the race."

The function below simulates a multiple choice task with free response. Parameters are vectors with length equal to the numbers of choices in the task. To specify a set of choices, set the parameters to vectors whose length is equal to the number of choices desired for the task.

It should be noted that while the race model has been demonstrated to account accurately for experimental results in tasks with two choices, the appropriateness of multiple choice models is a subject of active debate.

```
function [choice, rt] = race_trial(dt, biases, sigmas, thetas, initial_values)
    X = initial_values;
    t = 0;
    while X < thetas
        t = t + dt;
        % draw from Weiner process
        dW = randn(size(biases))*dt;
        dX = biases * dt + sigmas.*dW;
        X = X + dX; end
    choice = find(X > thetas);
    rt = t;
end
```

---

### EXERCISE 31.7

Extend **race_trial** to accept a fixed time interval, and to return the best choice if the total time reaches the maximum interval without a clear winner.

---

## 31.6 CORTICAL MODELS

Any neurobiological model should account for the perceived successes of the diffusion model in describing the characteristics of reaction time in decision processes under psychometric testing. One such model has been proposed (Shadlen and Newsome, 2001; Mazurek et al., 2003) to account for interactions between visual areas MT and LIP. Area MT is a cortical area sensitive to visual motion, and area LIP is a cortical area implicated in decision processes. Neurons in area MT have been found to respond strongly to motion, with directional specificity. Many neurons appear to have a characteristic direction, and respond preferentially to stimuli moving in that direction. Mean rates of neurons varying with direction relative to preferred orientation can be found in Figure 31.1.

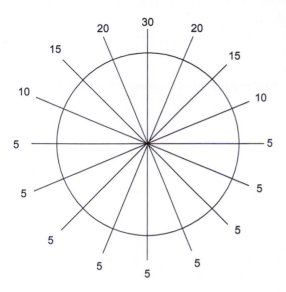

FIGURE 31.1 Mean firing rates for an idealized MT neuron, tuned to respond to upward motion.

Under the model proposed by Shadlen and Newsome, the activity of individual neurons in LIP reflects the activity of associated neurons in MT over a long time period. Moreover, area MT neurons directionally tuned to opposite directions inhibit the LIP neuron of the oppositely directed cell. Thus, under this model, LIP neurons integrate the difference in activity between sensory cells with opposing preferred directions.

If we equate evidence with neural activity, we can take spike counts as evidence. This is a subtle difference in from the discrete models we have used thus far, in that evidence will be discrete counts of spikes rather than continuous. Another difference is the absence of a mechanism for retrograde accumulation in the evidence count. Once a spike is counted, it remains counted.

We can represent the spike counts from an MT neuron as $N$. At any time interval $dt$, we can simulate activity by drawing from a uniform random distribution and comparing to the mean rate per time interval $dt$. The mean rate will vary with both time and the current stimulus, if any.

Likewise, we can simulate activity in LIP by adjusting the probability of firing by the current evidence count. The model described above also has a feed-forward inhibitory term to account for strong evidence in the opposite direction. To account for both connections, we will adjust probability of firing by the difference between the two evidence counts. Neural activity above a certain threshold implies sufficient evidence for a choice.

The function **lip_activity** ahead models the activity of a single LIP neuron with fixed probabilities for its corresponding excitatory and inhibitory MT neurons.

```
function rt = lip_activity(MT_p_values, ...
                           LIP_weights, ...
                           LIP_threshold)
% Parameters:
% MT_p_values - a vector with 2 elements, firing probabilities for the
```

```
%        excitatory and inhibitory neurons, resp.
% LIP_weights - a length 2 vector of weighting factors for the evidence
%        from the excitatory (positive) and
%        inhibitory (negative) neurons
% LIP_threshold - the LIP firing rate that represents the choice threshold
%        criterion
% use fixed time scale of 1 ms
dt = 0.001;
N = [0 0]; % plus is first, minus is second
rate = 0.0;
LIP_event_times = [];
while rate < LIP_threshold
    t = t + 0.001;
    dN = rand(2) < MT_p_values;
    N = N + dN;
    p_LIP = sum(N.*LIP_weights);
    LIP_event = rand < p_LIP;
    LIP_event_times = [LIP_event_times t];
    % check LIP mean rate for last M spikes
    rate = M/(t - LIP_event_times(N_LIP - M));
end
rt = t;
end
```

The function expects MT neuron probabilities, weights for the LIP neurons, and a firing threshold for the LIP neuron in impulses per second (ips).

---

### EXERCISE 31.8

Modify **lip_activity** to return the event times for the simulated LIP neuron. Modify the function to capture and return the event times for the two MT neurons. Generate a sample set of coordinated MT and LIP simulated event times and examine them as a raster. How do the patterns of activity differ?

---

## 31.7 PROJECT

In this project, you are asked to write a simulation of MT-LIP neurons using the Shadlen-Newsome model discussed in the previous section. Specifically, you are asked to do the following:

- Write code to simulate the effect of presenting a directionally oriented stimulus during intervals of time. This will involve generating a time series where the value at each step is an orientation or a value that indicates the absence of a stimulus.

- You need to allow the probability of firing for the two MT neurons in the model to change with time, based on the orientation of any presented stimulus.
- The model should include two MT neurons and two LIP neurons. Each MT neuron should have feed forward connections to both LIP neurons, one excitatory and one inhibitory.
- Generate activity patterns for both MT neurons and both LIP neurons for various stimulus presentations.

# MATLAB FUNCTIONS, COMMANDS, AND OPERATORS COVERED IN THIS CHAPTER

**erf()**

## 32.1 GOAL OF THIS CHAPTER

In this chapter, you will learn about modeling sequential phenomena using Markov processes. Simple Markov models will be introduced to characterize sequences in behavior. Hidden Markov models will be introduced with the HMM functions within the Statistics Toolbox in MATLAB®. Finally, hidden Markov models will be used to extract timing data from electrophysiological data by taking advantage of the sequential pattern in the waveform shape.

## 32.2 INTRODUCTION

From the sequential depolarization then hyperpolarization of a neural spike to chains of motor gestures, sequential data is quite common in neuroscience. How can we model data within the context of preserving the sequential relationship?

A *Markov model* describes a system as a set of discrete states and transition probabilities of moving between states. Additionally, Markov models are characterized by adherence to the *Markov property*, which states that the transition probability from any state in the network depends only on some finite set of prior states. Thus, only a limited, recent subset of the state transition history of the model is necessary to determine transition probabilities for the next state. Formally, given random variables $X_1, X_2, X_3, \ldots X_t$ taking on values $x_1, x_2, x_3, \ldots x_t$, a Markov model asserts that

$$P(X_t = x_t | X_1 = x_1, X_2 = x_2, X_3 = x_3, \ldots X_{t-1} = x_{t-1})$$
$$= P(X_t = x_t | X_{t-1-n} = x_{t-1-n}, X_{t-n} = x_{t-n}, \ldots X_{t-1} = x_{t-1})$$

The preceding equation indicates that, given a history of random variable values for a system $x_1, x_2, x_3, \ldots x_t$ where $x_n$ denotes the state at time step $n$, the probability of the state at time $t$ given the entire history of the system is equivalent to the probability given only the $n$

*MATLAB® for Neuroscientists.*
DOI: http://dx.doi.org/10.1016/B978-0-12-383836-0.00032-1

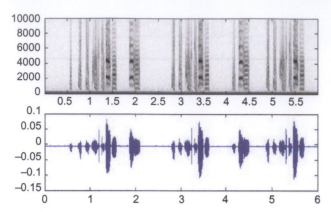

**FIGURE 32.1** Sonogram (top; frequency components over time) and sound amplitude (bottom) of three motifs from a zebra finch adult.

previous states. The number of prior states $n$ necessary to characterize the model is the *order* of the Markov model. In this chapter, we will focus exclusively on first-order Markov models, in which the knowledge of the current state only is sufficient to determine the transition probabilities for the possible successor state. For the first-order Markov model,

$$P(X_t = x_t | X_1 = x_1, X_2 = x_2, X_3 = x_3, \ldots X_{t-1} = x_{t-1}) = P(X_t = x_t | X_{t-1} = x_{t-1})$$

We will characterize a model $M$ as $\{S, T, s, O, E\}$, where

$S$ is a set of all states in the model of cardinality (size) $N$
$T$ is an $N \times N$ matrix of probabilities for transition between pairs of states
$s$ is the initial state
$O$ is a set of output values of cardinality $M$
$E$ is a set of pairs $(s, o)$, mapping a state $s$ with an emitted output value $o$

As a first example, we will examine syllable order in birdsong as model behavior. Songbird behavior is widely studied in the context of neuroscience as a model for sensori-motor learning and auditory perception. For our purposes, the sequential, repetitive, and hierarchical structure of song lends itself quite well to a Markov model representation. A sample song from a zebra finch can be found in Figure 32.1. The lower graph shows amplitude variation over time. The upper graph shows a spectrogram of the data, which shows the frequency content of the amplitude signal over time.

Within the structure of the song, there is a clear substructure of elements separated by relatively quiet intervals. These larger groupings are termed *motifs*. Within the motifs are smaller discrete elements, termed *syllables*. The division of the song into these parts might be clearer in the spectrogram. Note that the syllable order within a motif is fairly regular from motif to motif. During the analysis of the song, noting the sequence of syllables is often of interest. A plausible annotation is shown in Figure 32.2. Note that syllable 2 repeats, and syllables 1 and 8 are optional.

From this annotation, we can attempt to generate a Markov model for a single motif. We can design a fairly straightforward model by incorporating a separate state for each syllable in sequence. Figure 32.3 illustrates the state transitions for such a model.

In Figure 32.3, each circle represents a separate state of the model. The numbers in the states represent the syllable's output according to the syllable annotation in Figure 32.2.

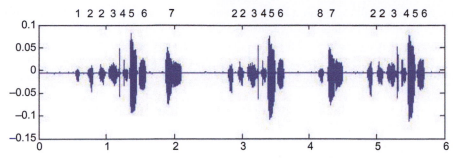

FIGURE 32.2  Annotated song.

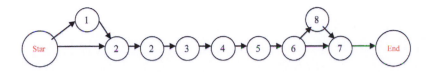

FIGURE 32.3  State transitions for the annotated song.

Note that we have two states that output syllable 2, we will denote these as 2a and 2b. This accounts for the repetition in the natural sequence within the song. For those two states, it is important to distinguish between the two *states*, which transition to different succeeding states (2b or 3), and the *output*, which is the same.

Also, the transition diagram currently lacks transition probabilities. An accurate estimation of transition probabilities would require examining transitions throughout a large sample of annotated song. As an initial example, however, we will limit the sample to the three motifs in the annotated Figure 32.2. The state transition diagram in Figure 32.3 has 11 separate states (don't forget the start and end states!). With the set of states and the transition probabilities estimated from the song sample, we can generate a transition matrix $T$. Here, the column indicates the current state and the row indicates the putative next state, and the matrix has been filled with transition probabilities calculated from our admittedly small data set above.

$$T = \begin{bmatrix} 0 & 0 & 0 & 0 & 0 & 0 & 0 & 0 & 0 & 0 & 0 \\ 1/3 & 0 & 0 & 0 & 0 & 0 & 0 & 0 & 0 & 0 & 0 \\ 2/3 & 1 & 0 & 0 & 0 & 0 & 0 & 0 & 0 & 0 & 0 \\ 0 & 0 & 1 & 0 & 0 & 0 & 0 & 0 & 0 & 0 & 0 \\ 0 & 0 & 0 & 1 & 0 & 0 & 0 & 0 & 0 & 0 & 0 \\ 0 & 0 & 0 & 0 & 1 & 0 & 0 & 0 & 0 & 0 & 0 \\ 0 & 0 & 0 & 0 & 0 & 1 & 0 & 0 & 0 & 0 & 0 \\ 0 & 0 & 0 & 0 & 0 & 0 & 1 & 0 & 0 & 0 & 0 \\ 0 & 0 & 0 & 0 & 0 & 0 & 0 & 2/3 & 0 & 1 & 0 \\ 0 & 0 & 0 & 0 & 0 & 0 & 0 & 1/3 & 0 & 0 & 0 \\ 0 & 0 & 0 & 0 & 0 & 0 & 0 & 0 & 1 & 0 & 0 \end{bmatrix}$$

For example, looking down the first column (i.e., in the start state), there is a probability of 1/3 of transitioning to the 1 state, and a probability of 2/3 of transitioning to the 2 state.

Another example worth examining is the two columns corresponding to states 7 and 8. In the column corresponding to state 7, all values are zero except the bottom row, indicating that the succeeding state is always the end state. In the column corresponding to state 8, only the row corresponding to state 7 is non-zero, denoting a transition to state 7 from state 8. The sum of probabilities within a column must equal 1, but the sum of probabilities within a row often do not.

In MATLAB, we can build this matrix easily by starting with a matrix of zeros and inserting all the non-zero values.

```
>> T = zeros(11, 11);
>> T(2,1) = 1/3;
>> T(3,1) = 2/3;
>> T(2:8, 1:7) = eye(6); % this adds the diagonal ones
>> T(9,8) = 2/3;
>> T(10,8) = 1/3;
>> T(11,9) = 1;
>> T(9,10) = 1;
```

To complete the specification of the Markov model, we need to supply $S$ (the set of states), $s$ (the starting state), $O$ (the set of outputs), and $E$ (the pairs mapping a state to its output). The set of states is simply the set of values from 1 to 11, and the starting state is state 1.

Now, given a state $s$ (valued 1–11), we can obtain the transition probabilities by defining a state vector $\mathbf{s}$, where the elements of $s$ are defined as:

$$s_i = \begin{cases} 1 & i = n \\ 0 & i \neq n \end{cases}$$

and then the vector of transition probabilities is simply the product of the transition matrix $T$ and the state vector $\mathbf{s}$.

$$p = Ts$$

In MATLAB syntax, if we let $n$ be the current state, we can choose an appropriate next state using **rand**.

```
% n is the current state
% build the state vector
s = zeros(11,1);
s(n) = 1;
p = T*s;
% p now contains probabilities for transition
p
% build a cumulative probability function
cdf = cumsum(p);
new_n = 0;
choice = rand();
for ii = 1:11
        if choice < cdf(ii) then
                new_n = ii;
```

```
            break;
        end
end
% new_n now contains the new state

%
% NB: the above for loop could be replaced by
% new_n = min(find(cdf > choice));
% The comparison returns a vector of 1 or 0 values, find locates the indices of
% all non-zero values (i.e. indices where the comparison is true), and min returns
% the first one.
% Both bits of code are equally correct. The solution with find is more succinct
% is will likely execute more quickly for longer arrays. The solution with the for
loop
% shows the algorithm more explicitly.
%
```

---

### EXERCISE 32.1

Write a function markov_sequence that accepts as parameters a transition matrix $T$, a start state start, and end start end and generates a plausible sequence of states. Remember that this is a sequence of **states**, not syllables, so the start state will be 1, the two states corresponding to productions of syllable 2 will have different numbers, etc.

---

### EXERCISE 32.2

Modify the code written in Exercise 32.1 to accept a vector of syllables to emit for each state. Choose appropriate values for the start and end states (e.g., $-1$ or 0).

---

### EXERCISE 32.3

Load a larger recording of the same bird, and create a Markov model for multiple motifs. A longer recording can be found in file zf_y89.wav on the web site repository. Load the file using wavread().

# 32.3 FINDING THE MOST PROBABLE PATH: THE VITERBI ALGORITHM

For a set of outputs, what is the most probable path through a corresponding model?

Given the probabilistic nature of a Markov model, it's entirely feasible that a sequence of outputs might correspond to multiple sets of states. For small models, all possible paths can be explored, but what do we do for larger, more complex models? With greater interconnectivity, traversing all possible paths can be rather costly computationally, especially for models with recurrent loops.

There is an established algorithm, the Viterbi algorithm, which can answer this question fairly efficiently. Initially, we start with the first observation and state. We will denote states with the variables $q_i$, time steps with $t$, and observations with the variable $x_i$. The probability of the most probable path (the Viterbi path) at time $t$ for observation $x$ will be denoted by $V_{t,x}$.

So, for the first observation, we have:

$$V_{0,0} = P(x_0|s_0)$$

This states that the most probable path to generate the first observation is the probability of producing the first observation from the initial state. At the initial time step, no transitions have occurred, so the state at time $t$ must be the starting state.

The algorithm specifies steps later in time as a function of the steps immediately prior.

$$V_{t+1,i} = P(x_{t+1}|S_{t+1} = s_i) \max_j (V_{t,j} P(S_{t+1} = s_i | S_t = s_j))$$

The first term is the probability of the observation $x_{t+1}$ at time step $t+1$ from state $s_i$. The second term calculates the probability of reaching $s_i$. The Viterbi algorithm assumes that the optimal path to a state will include the optimal path to one of its predecessors. This allows breaking the problem into simpler, yet similar subproblems.

This inherent structure in the problem space allows us to break the problem up in a coherent way and build the overall solution from the components of the smaller subproblems. In our case, we will use a matrix to store values of the Viterbi probability as they are calculated so that the expression inside the max operator becomes a simple element-wise multiplication. This type of decomposition of a problem space is called dynamic programming, and is a common approach to reducing the complexity of amenable problems.

The most probable path is the set of $j$ values that maximize the Viterbi probability. Thus, it is necessary to save the value in a matrix for later. Once the full probability is calculated, the path can be output by stepping through the values of $j$.

```
function [path, P] = simple_viterbi(seq, start, T, E)
    % Calculates the path for the most probable route through a Markov
    % model that produces a set of observations.
    % Parameters:
    %   seq : sequence of observations, numbered 1..M
    %   T   : transition matrix, SxS, where
```

```
%        T(n,m) is the probability of a transition
%        from state m to n
%   E  : emission matrix, MxS, where
%        E(m,s) is the probability of emitting
%        output m when in state s
%   start : the start state as a scalar value
%
% Returns:
%   path : the most likely sequence of states
%          to produce input parameter seq
%   P    : the probability of the returned path

% create matrix for storing calculations and path
state_count = size(T);
state_count = state_count(1);
time_count = length(seq);
V = zeros(time_count, state_count);
max_state = zeros(time_count, state_count);

% at t = 1, only V(1, start) has value
% there is no probability for other states
V(1, start) = E(seq(1), start);

for t = 2:time_count
    for s = 1:state_count
        P_arrival = T(s, :).*V(t-1,:);
        max_P_arrival = max(P_arrival);
        argmax_P_arrival = find(P_arrival == max_P_arrival);
        emission = E(seq(t), s);
        V(t, s) = max_P_arrival*emission;
        max_state(t, s) = argmax_P_arrival;
    end
end

% now, grab the best probability
P = max(V(time_count, :));
path = [find(V(time_count, :) == P)];
% build path
for t = time_count:-1:2
    prev_best = max_state(t, path(1));
    path = [prev_best; path];
end
end
```

---

**EXERCISE 32.4**

With a large model and multiple transitions, the probability can "underflow," becoming indistinguishable from zero as represented with the floating point numbers used by MATLAB. One way to circumvent this is to convert take the logarithm of the probability and work in a logarithmic scale, converting the value back before returning. Update the above function to calculate the probabilities as logarithms. You will need to address zero-valued probabilities. This may be as simple as using negative infinity (in MATLAB, $\log(0) = -\text{Inf}$).

---

## 32.4 HIDDEN MARKOV MODELS

Up to now, we have only considered models with simple correspondences between the output and the state. Many systems require models with a more computationally expressive output. We can formally extend the simple Markov model to allow each state to output one of multiple possible outputs, each with a distinct emission probability. Our more comprehensive model $M$ is $\{S, T, s, O, E\}$, where

$S$ is a set of all states in the model of cardinality (size) $N$
$T$ is an $N \times N$ matrix of probabilities for transition between pairs of states
$s$ is the initial state
is a set of output values of cardinality $M$
$E$ is an $N \times M$ matrix of probabilities for emitting output $m$ from state $n$

We can view our prior Markov models as special cases of the more comprehensive model in which for every row $i$ of matrix $E$, there is a value $E_{i,j} = 1$. (In other words, every row has a single output whose emission probability is unity.)

Because in practice these more comprehensive models are applied to problem domains where only partial information about the state and output sequences is available, this more comprehensive model is usually called a *hidden Markov model* (HMM). Typically, it is the sequence of output values that is available, and the corresponding sequence of states is hidden. This is the scenario that we will address here.

## 32.5 TRAINING AN HMM: THE BAUM-WELCH ALGORITHM

Given a putative HMM and a sample data set, the Baum-Welch algorithm provides a means of estimating transition and emission probabilities for the model. While an extensive discussion of the algorithm is beyond the scope of this chapter, we will briefly explain the algorithm.

The Baum-Welch algorithm is an iterative algorithm, in that it is run repeatedly until a desired convergence is met. The iterative nature of the algorithm is guaranteed to converge under most circumstances, but that convergence is not necessarily the best solution, merely a local optimum. To calculate new probability estimates, we need to calculate *forward* and *backward* probabilities. These are probabilities at each time step of the model's current state given an output sequence. Forward probabilities are calculated working from the start of the sequence, and backward probabilities are calculated from the end.

$$\alpha_{t+1,i} = E_{i,out(t)} \left( \sum_j \alpha_{t,j} T_{i,j} \right)$$

$$\beta_{t,i} = \sum_j \beta_{t+1,j} T_{j,i} E_{j,out(t)}$$

Forward and backward probabilities thus defined, the probability of being in state $i$ at time $t$ can be calculated as the product of the forward and backward probabilities for state $i$ and time $t$, normalized by the total probabilities for all such states at time $t$.

$$\gamma_{t,i} = \frac{\alpha_{t,i} \beta_{t,i}}{\sum_i \alpha_{t,i} \beta_{t,i}}$$

To estimate probabilities, we will also need to estimate the probability of transition between states $i$ and $j$ at time $t$ given the sequence of outputs. This value can be expressed as

$$\xi_{t,i,j} = \frac{\alpha_{t,i} T_{i,j} \beta_{t+1,j} E_{j,out(t)}}{\sum_i \alpha_{t,i} \beta_{t,i}}$$

This probability is the product of (1) the probability of a valid output sequence from the start state to time step $t$, (2) the probability of a valid output sequence from the state $j$ to the end of the model, (3) the transition from state $i$ to state $j$, and (4) the probability of emitting the observed value from state $j$, all normalized by the total probability of all states at time $t$.

Our new transition probabilities can be estimated by dividing the probability of transitioning between $i$ and $j$ by the probability of being at state $i$ at time $t$:

$$T'_{i,j} = \frac{\sum_t \xi_{t,i,j}}{\sum_t \gamma_{t,i}}$$

Similarly, new emission probabilities can be estimated from the proportion of the probability of reaching state $i$ when output $x$ was emitted:

$$E'_{i,x} = \frac{\sum_{\{t | out(t) = x\}} \gamma_{t,i}}{\sum_t \gamma_{t,i}}$$

The Statistics Toolbox provides an implementation of the Baum-Welch algorithm as the function **hmmtrain()**. We will use this function in the next section.

## 32.6 A SIMPLE EXAMPLE

As a simple initial illustration, we will demonstrate how to apply HMMs to the automated annotation of syllables in a large corpus of song. In the domain of syllable classification, we will assume that our HMM model states denote the "true" syllable in the underlying sequence, and the auditory recording is the output sequence. The sequence of states will then be the annotated sequence of syllables.

To simplify the example here, we will rate syllables using only one metric: duration. A more rigorous effort would make use of multiple metrics, likely including spectral characteristics of the sound.

We will use the file **training-segments.txt**, which contains an $N \times 3$ matrix stored as a text file. Columns 1 and 2 are the start and stop of sound segments that roughly correspond to syllable start and stop times in milliseconds from a common $t = 0$ point. Column 3 is a number that uniquely groups segments that are in the same bout of song. All rows with the same value in column 3 we will treat as in the same sequence.

In addition to working only with duration, we will discretize the continuous duration value into bins of 5 ms each. The Statistics Toolbox function **hmmtrain** expects either a single sequence or a cell array of sequences. For our data, we will supply a cell array of sequences. The function **make_sequences** ahead will transform the loaded matrix from **training_segments.m** to a cell array. Since the minimum segment size is 30 ms, the values of 1 and 2 will be used to denote the beginning and end of a sequence respectively.

```
function seqs = make_sequences(seg_array, max_duration)
% From an Nx3 segment array, this function produces
% a cell array of sequences
    seqs = {};
    current_seg = [];
    bouts = unique(seg_array(:,3));
    sample = 1;
    granularity = 5;
    max_duration = max_duration/granularity;
    for bout = 1:length(bouts)
        in_bout = find(seg_array(:,3) == bouts(bout));
        dur = seg_array(in_bout,2) - seg_array(in_bout,1);
        dur = floor(dur/granularity);
        if max(dur) < max_duration
            seqs{sample} = [1 dur' 2];
            sample = sample + 1;
        end
    end
end
```

Load the segment timing data and generate sequences. Examine the first sequence.

```
>> segments = load('training-segments.txt');
>> seqs = make_sequences(segments, 400);
```

```
>> seqs{1}
```

**ans =**

Columns 1 through 30:

```
1  12  10  10  11   6  10  12  23  21  14  25  21  15  24  21  16  48  15  24
22  16  24  23  15  25  22  15   8  11
```

Columns 31 through 41:

```
25  22  15  25  23  16  24  23  16   6   2
```

The last step before estimating probabilities with **hmmtrain** is the construction of initial transition and emission matrices. We could build a feasible model, but in many cases the Baum-Welch algorithm will find a reasonable set of transition probabilities with random starting data.

Here, we start with entirely random transition probabilities using a uniform distribution, and normalize them. **hmmtrain** expects that the matrix element at $T(a,b)$ is the transition probability between state $a$ and state $b$. Likewise, the matrix element at $E(a,b)$ is the probability of emitting output $b$ at state $a$.

```
>> T = rand(12);
>> normed_T = T./repmat(sum(T,2), 1, 12);
```

The initial emissions matrix will also be nearly random. Because the sequences have explicit start and stop values, the initial emissions probabilities will reflect this. The emissions probabilities for rows 1 and 12, corresponding to states 1 and 12, will be set to all zeros except in columns 1 and 2, which will have values of 1. This sets the probability for any state but 1 to output a 1 to zero, and the probability for any state but 12 to output a 2 to zero.

```
>> E = rand(12,400/5);
>> E(:,1:2) = 0;
>> E(1,:) = 0;
>> E(12,:) = 0;
>> E(1,1) = 1;
>> E(12,2) = 1;
>> E(:,1:6)
```

**ans =**

| 1.0000 | 0 | 0 | 0 | 0 | 0 |
|--------|---|--------|--------|--------|--------|
| 0 | 0 | 0.6744 | 0.3785 | 0.9552 | 0.6392 |
| 0 | 0 | 0.2424 | 0.1119 | 0.6278 | 0.2518 |
| 0 | 0 | 0.1684 | 0.8595 | 0.8191 | 0.4199 |
| 0 | 0 | 0.8611 | 0.7132 | 0.1890 | 0.1718 |
| 0 | 0 | 0.3266 | 0.5750 | 0.7515 | 0.8845 |
| 0 | 0 | 0.3252 | 0.7411 | 0.2216 | 0.3410 |
| 0 | 0 | 0.4002 | 0.7537 | 0.4164 | 0.2051 |

|   |        |        |        |        |        |
|---|--------|--------|--------|--------|--------|
| 0 | 0      | 0.7635 | 0.5551 | 0.7909 | 0.7954 |
| 0 | 0      | 0.1641 | 0.8009 | 0.4000 | 0.0245 |
| 0 | 0      | 0.6883 | 0.6703 | 0.0451 | 0.0398 |
| 0 | 1.0000 | 0      | 0      | 0      | 0      |

```
>> normed_E = E./repmat(sum(E,2), 1, 400/5);
```

Now, we can invoke **hmmtrain** and estimate probabilities.

```
>> [estT, estE] = hmmtrain(seqs, normed_T, normed_E);
>> estT

estT =
```

| 0.5000 | 0 | 0.5000 | 0 | 0.0000 | 0.0000 | 0 | 0 | 0 | 0 | 0.0000 | 0 |
|--------|---|--------|---|--------|--------|---|---|---|---|--------|---|
| 0 | 0 | 0 | 0 | 0 | 0 | 0 | 1.0000 | 0 | 0 | 0 | 0 |
| 0 | 0.0000 | 0.6085 | 0.1456 | 0.1500 | 0.0000 | 0 | 0 | 0.0477 | 0.0000 | 0.0481 | 0 |
| 0 | 0 | 0.0000 | 0.0744 | 0.2843 | 0 | 0 | 0 | 0.5689 | 0 | 0.0000 | 0.0724 |
| 0 | 0.1008 | 0.0000 | 0.0000 | 0.0000 | 0.0000 | 0.0274 | 0 | 0.0000 | 0.0000 | 0.8718 | 0 |
| 0 | 0.4870 | 0.0000 | 0.2570 | 0.0104 | 0 | 0 | 0 | 0.0435 | 0.0881 | 0 | 0.1140 |
| 0 | 0.0000 | 0.0000 | 0.0000 | 0.0000 | 0 | 0 | 0.0000 | 0.0000 | 1.0000 | 0 | 0.0000 |
| 0 | 0 | 0 | 0 | 0 | 0.9671 | 0.0266 | 0 | 0 | 0 | 0 | 0.0063 |
| 0 | 0.0000 | 0.0000 | 0.0000 | 0.0570 | 0.0000 | 0 | 0 | 0.2180 | 0 | 0.7250 | 0.0000 |
| 0 | 0 | 0 | 0 | 0 | 0.7265 | 0.2549 | 0 | 0 | 0 | 0 | 0.0185 |
| 0 | 0.7401 | 0.0000 | 0.0427 | 0.0000 | 0 | 0 | 0 | 0.0000 | 0.2172 | 0.0000 | 0 |
| 0 | 0 | 0 | 0 | 0 | 0 | 0 | 0 | 0 | 0 | 0 | 1.0000 |

The actual values that **hmmtrain** produces will depend on the set of starting values. In this case, the probabilities indicate a fairly straightforward network.

With a viable set of transition and emission probabilities, we can use the Viterbi algorithm to calculate the most probable set of states. For example, the most probable state sequence for segment sequence 5 indicates that the state sequence 2-8-6 is likely a motif.

```
>> hmmviterbi(seqs{5},estT, estE)

ans =
```

Columns 1 through 30:

```
1  3  4  9  9  11  2  8  6  2  8  6  2  8  6  4  9  9  11  2  8  6  4  9  11
2  8  6  2  8
```

Columns 31 through 40:

```
6  2  8  6  9  11  2  8  6  12
```

**EXERCISE 32.5**

Look at the distribution of duration probabilities for each your states (each row of *estE*). Do they make sense? Is it reasonable to assert that the states correspond to distinct segments?

## 32.7 PROJECT

The sequential nature of HMMs has been used to locate spike waveforms as an alternative to threshold-based waveform identification (Herbst et al., 2008). As a project, you will use an approach similar to that in (Hahnloser et al.), albeit simpler, to extract event times from an extracellular electrophysiological recording of spontaneous neural activity.

Appropriate neural data can be found in ch32_project_data.mat. The file contains data sampled at 20 kHz. Each state will approximate a single sample, or .05 ms. The initial model will have two recurrent loops. The first involves a single recurrent state to handle noise. The second is the bulk of the model, a 60 state loop to model the shape of the spike waveform over 1 ms. Thus, state 1 will have two outgoing transitions: one to state 2 and a recurrent one back to state 1. The remaining states will each have one outgoing transition to the next state until state 60, which will have an outgoing transition to state 1.

```
>> RA_T = zeros(60);
>> RA_T(1,1) = 1.0;
>> RA_T(1:59,2:60) = eye(59);
>> RA_T(60,1) = 1.0;
>> normed_RA_T = RA_T./repmat(sum(RA_T,1), 60, 1);
```

To use the HMM functions included in the Statistics Toolbox, the output values must be discrete. Much like the duration data, the sampled data can be binned. A rescaling of the signal by a factor of 50 and shifting will allow a reasonable number of bins (320).

When setting initial values for the emission matrix, the values in row 1 (i.e., for state 1) should be highest around the baseline. To catch the waveform, there should be higher probabilities for the larger indices (i.e., larger output values, greater signal) for the first few indices and higher probabilities for smaller indices (lower than baseline) for the next few indices. The following are feasible choices for the initial emission matrix (remember to normalize the two matrices!).

```
>> RA_E = rand(60,320)/10;
>> RA_E(1,140:180) = 1.0;
>> RA_E(2:7,180:320) = 1.0;
>> RA_E(8:18,1:140) = 1.0;
```

If **hmmtrain** runs extremely slowly, it may be best to cut out the 100 samples immediately around 3-4 different spikes to acquire the waveform in the HMM. If this is done, it may be necessary to adjust the emission probability distribution for state 1. Setting this to a normal distribution centered on the baseline after the training process might be necessary for the Viterbi algorithm to work properly.

# MATLAB FUNCTIONS, COMMANDS, AND OPERATORS COVERED IN THIS CHAPTER

**hmmtrain**
**hmmviterbi**

# 33

# Modeling Spike Trains
# as a Poisson Process

## 33.1 GOALS OF THIS CHAPTER

This chapter focuses on point process models for characterizing and simulating trains of actions potentials generated by neurons. Initially, a simple homogeneous Poisson process model will be introduced to capture fundamental characteristics. Models with greater sophistication will be introduced to incorporate more complex activity, such as refractory periods and bursting.

## 33.2 BACKGROUND

Often when attempting to characterize extracellular activity, plausible values for simulating the underlying intracellular phenomena driving the activity are not available. In such cases, a stochastic description of the observed activity can be a useful model for the event train produced by the cell.

Consider the spike train in Figure 33.1. How can the sequence of neural events be characterized? A very simple characterization is the mean event rate per unit time. In this case, that value is 11.3 ips. This value has the advantage of being a scalar: easily computed and easily manipulated. However, reducing all the activity in a spike train to a single number discards any temporal structure in the events.

Still, even a simple characterization can lead to useful insights. If we choose a temporal interval substantially smaller than the mean rate, we can treat the mean rate as an event probability during that event. Such a model, with discrete intervals of time, is described by a Bernoulli process.

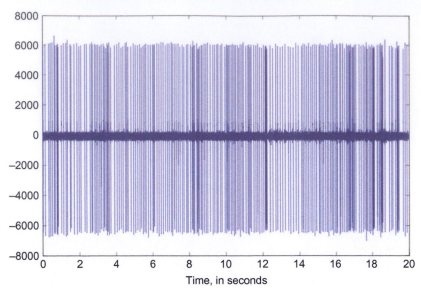

FIGURE 33.1    An extracellularly recorded spike train.

## 33.3  THE BERNOULLI PROCESS: EVENTS IN DISCRETE TIME

A Bernoulli process generates a sequence of event times corresponding to positive events from a series of Bernoulli trials. However, in a Bernoulli process, time is measured discretely, so that the event "time" is the index of the corresponding discrete time interval. Each discrete interval corresponds to a Bernoulli trial with probability $p$ of success or $1 - p$ of failure. For example, taking heads as a success, a sequence of coin tosses in which the time intervals of the successful (i.e., heads) tosses are noted constitutes a Bernoulli process.

Taking each time interval as a trial, we can treat the simple model just described as a Bernoulli process.

### EXERCISE 33.1

Use a Bernoulli process to simulate a spike train from the mean activity rate for the spike train in Figure 33.1. Take the mean rate of 11.3 ips, and use a time slice of 1 ms. Generate a 20 s train (i.e., 20,000 time slices). Treat each time slice as a Bernoulli random variable with a small probability of a neural event. The $p$ value (probability of an event) should be chosen so that the mean successes over 1000 trials equals the mean rate, or $Np = 11.3$ events, where $N$ is the number of time slices in one second. A success will correspond to a neural event occurring during the time slice. To pick a

| | |
|---|---|
| uniform random number between 0 and 1, rand can be used: | `>> rand(2,2)` |
| | `ans =` |
| `>> rand` | 0.9134   0.0975 |
| `ans =` | 0.6324   0.2785 |
| 0.1270 | |

Unfortunately, the Bernoulli process falls short in some important ways. Most importantly, as mentioned earlier, the Bernoulli process is discrete: a consequence of this is that the event times are no more precise than the time interval. Looking at a distribution of the interspike intervals with a bin size smaller than the time interval used for the Bernoulli process illustrates this quite clearly. Fortunately, there is an analogous process for continuous time: the Poisson process. Exploring the Poisson process will be the focus of the next section.

## 33.4 THE POISSON PROCESS: EVENTS IN CONTINUOUS TIME

Like the Bernoulli process, a Poisson process is a counting process (i.e., the event count increases with increasing time), but instead of counting discrete events, a Poisson process describes events in continuous time.

A Poisson process adheres to the following criteria:

1. Event counts in disjoint intervals are statistically independent, and depend only on the size of the interval.
2. No two events occur simultaneously.

From these simple properties, the form of the Poisson distribution can be derived. Below is the probability mass function for the Poisson distribution.

$$P(x = k) = \frac{\lambda^k}{k!} e^{-\lambda}$$

A Poisson distribution is parameterized by a single value, lambda. Lambda is often interpreted as the rate of success. This interpretation falls out of the expectation of the distribution.

$$E[X] = \sum_{x=0}^{\infty} x \frac{\lambda^x}{x!} e^{-\lambda}$$

$$E[X] = e^{-\lambda} \sum_{x=0}^{\infty} x \frac{\lambda^x}{x!}$$

$$E[X] = e^{-\lambda} \left[ 0 + \sum_{x=1}^{\infty} \frac{\lambda^x}{(x-1)!} \right]$$

$$E[X] = e^{-\lambda} \sum_{x=0}^{\infty} \frac{\lambda^{x+1}}{x!}$$

$$E[X] = e^{-\lambda} \sum_{x=0}^{\infty} \lambda \frac{\lambda^{x}}{x!}$$

$$E[X] = \lambda e^{-\lambda} \sum_{k=0}^{\infty} \frac{\lambda^{x}}{x!}$$

The term on the far right hand side is the Taylor expansion at 0 for $e^{\lambda}$, which when multiplied by the adjacent $e^{-\lambda}$ cancels, leaving

$$E[X] = \lambda(e^{-\lambda})(e^{\lambda}) = \lambda$$

as the expectation. In other words, the mean event count of a Poisson distribution with value lambda is lambda. The parameter lambda can also be interpreted as the mean event count over an interval $T$, for which $\lambda$ can be written as $\lambda = rT$. When expressed as the mean event count over an interval of length $T$, $r$ represents the mean event count per unit interval. Thus, for an interval $T' = 2T$, the corresponding lambda $\lambda'$ is $2\lambda$.

From the criteria above, the distribution of time intervals between events can be shown to follow an exponential distribution. Unlike the Poisson distribution of event counts, the exponential distribution is a continuous distribution, with probability density function (PDF) instead of a probability mass function. Here is the PDF for an exponential distribution with parameter $\lambda$:

$$f(x; \lambda) = \lambda e^{-\lambda x}, \ x \geq 0$$

---

### EXERCISE 33.2

Demonstrate that the variance of the Poisson distribution is the same as the mean, $\lambda$. Demonstrate that the mean of the exponential distribution is $1/\lambda$.

---

It should be noted that while the distribution of time intervals between consecutive events follows an exponential distribution, the distribution of event *times* follows a uniform distribution. This implies that any one event is no more probable at one time than any other.

## 33.4.1 Simulating an Event Train Using a Poisson Model

Here, we will introduce two approaches for simulating an event train with a Poisson model. A simple approach is to draw intervals between consecutive events from an exponential distribution. This approach works quite well when needing to generate consecutive events without an explicit time bound.

Alternatively, one can use a different approach if working with a fixed time bound. Building on the principle that event times are uniformly distributed, one can draw a sample event count for the interval from an appropriate Poisson distribution, and then draw times for each event by picking from an appropriate random distribution for each event.

1. Take an interval $T$ and a rate $r$
2. Draw a sample event count $N$ for interval $T$ from a Poisson distribution with parameter $\lambda = Tr$
3. Draw $N$ times, $t_1, t_2, t_3, \ldots t_N$, from a uniform distribution ranging from 0 to $T$
4. Return $t_1, t_2, t_3, \ldots t_N$ as the event times.

### 33.4.2 Picking Poisson and Exponentially Distributed Values

In most cases, the Statistics Toolbox will be available. The Statistics Toolbox offers a wide selection of functions to calculate probability density functions, cumulative distribution functions, inverse cumulative distribution functions, and generate random variables. Naming conventions for various distributions typically add a suffix denoting the type of calculation, such as cumulative distribution function or random variable generation, to an abbreviated name of the distribution. For example, here are functions for the Poisson distribution:

**poissrnd**—Poisson distributed random variables
**poisspdf**—Poisson probability density function
**poisscdf**—Poisson cumulative distribution function
**poissinv**—inverse Poisson cumulative distribution function

---

**EXERCISE 33.3**

Demonstrate that both methods of generating event trains produce similarly distributed values. For each method, generate at least 5 minutes of event data for a cell with a mean rate of 10 ips. Compare empirical distributions of ISI and event counts per second. The Statistics Toolbox provides the functions **poissrnd**, **exprnd**, and **unifrnd** to draw from Poisson, exponential, and uniform distributions respectively.

---

(If the Statistics Toolbox is not available, see the next section for alternatives.)

## 33.5 PICKING RANDOM VARIABLES WITHOUT THE STATISTICS TOOLBOX

If the Statistics Toolbox is not available, there are a number of methods for drawing exponential and Poisson distributed random variables using only a uniformly distributed random variable between 0 and 1. The MATLAB® function **rand**, introduced earlier, is not part of the Statistics Toolbox and should be available under even a basic installation.

### 33.5.1 Exponential Distributions

A simple method for generating values with an exponential distribution, the inversion method, relies on the uncomplicated form of the inverse of the cumulative probability function. For a given value of a random variable, the cumulative probability function returns the probability that a random variable will be less than the given value. The cumulative probability function is calculated from the probability density function $f(x)$ through the integral

$$F(x) = \int_{-\infty}^{x} f(t)dt$$

The cumulative probability function maps a value of the distribution to a probability. The inverse of the cumulative probability function maps a probability to a corresponding random variable in the distribution. With an inverse cumulative probability function and a uniform distribution from 0 to 1, we can generate exponentially distributed values (this is actually valid for any inverse cumulative probability function).

The inverse cumulative function takes the form

$$X = -\frac{\ln U}{\lambda}$$

where $U$ is uniformly distributed over $0\ldots1$.

---

**EXERCISE 33.4**

Derive the inverse cumulative probability function from the probability density function. *Hint*: the quantity $1-U$, where $U$ is a uniformly distributed random variable from $0\ldots1$, is itself also a uniformly distributed random variable from $0\ldots1$.

---

To generate the next event time for a Poisson process with parameter $\lambda$:

1. Select a random value, $U$, uniformly distributed from $0\ldots1$
2. Calculate $\Delta t = -(\ln U)/\lambda$ (This is the interval between events, not the event time itself!)
3. Return event time $t_n = t_{n-1} + \Delta t$

The following MATLAB function implements the above algorithm, generating a valid time for the next event given a lambda value and the time of the previous event.

```
function t = next_event(lambda, event)
    U = rand;
    dt = -log(U)/lambda;
    t = event + dt;
end
```

## 33.5.2 Poisson Distributions

If a viable generator for exponentially distributed variates is available, successive exponential values can be generated until the combined time exceeds the interval over which a Poisson trial is needed. Then, the Poisson success count is the number of exponential variables whose summed time is less than the interval.

1. Start with parameter lambda, $\lambda$, the interval $T$
2. Let $T'$, the cumulative time, and $N$, the event count, be set to 0
3. Select $t$, an exponentially distributed interval with parameter $\lambda$
4. If $T' + t > T$, return $N$ as the number of events in the interval
5. $T' = T' + t$
6. $N = N + 1$

The inversion method (i.e., using the inverse cumulative distribution) can also be used with the Poisson distribution. Since the Poisson distribution is discrete, we do not necessarily need to derive the inverse cumulative distribution in closed form to attain accurate results.

1. Select a value $U$ from a uniform random distribution over $0\dots1$
2. Let $k$, the number of Poisson successes, and $P$, the cumulative probability, both be set to 0
3. Calculate $p_i = \dfrac{\lambda^k}{k!} e^{-\lambda}$
4. $P = P + p_i$
5. If $P$ exceeds $U$, return $k$ as the number of successes in the trial
6. If $P$ does not exceed $U$, increment $k$ and return to step 3.

---

### EXERCISE 33.5

Write a MATLAB function that accepts two parameters, a time interval $T$ and a lambda value $\lambda$, and returns a scalar event count and a vector of event times for events randomly generated from appropriate Poisson/exponential distributions. Use only **rand** to generate random values.

---

## 33.6 NON-HOMOGENEOUS POISSON PROCESSES: TIME-VARYING RATES OF ACTIVITY

In the formulation of the Poisson process, we have thus far limited our discussion to those processes having a constant mean rate of events, represented as a constant parameter $\lambda$. Such Poisson processes are termed *homogeneous*. To capture fluctuations in rate, we can generalize the $\lambda$ of the Poisson process to a time-varying function. Processes in which lambda varies with time are called *non-homogeneous*.

The non-homogenous Poisson process provides a convenient extension of our existing model to handle time-varying rates. Fortunately, generating variates for the non-homogeneous Poisson process does not require substantial additional complexity beyond our existing methods.

A fairly straightforward approach to event-time generation is an application of the acceptance-rejection approach, in which a number of possible event times are generated and then pruned (von Neumann, 1951). In the case of a non-homogeneous Poisson function, we can generate event times using an appropriate homogeneous Poisson process with constant $\lambda$, then eliminate a proportion of the events, depending on the relationship between $\lambda$ and $\lambda(t)$ at the given time $t$.

1. Select a constant $\lambda_c$ such that $\lambda_c \geq \lambda(t)$ for all $t$. This is the constant rate that will be used in a homogeneous Poisson process to generate events. Because we will generate excess events and then prune them, the constant rate must exceed the time-varying rate at all time points.
2. Generate the next event time $t$ using the constant rate $\lambda_c$.
3. Determine the time-varying rate at time $t$ from $\lambda(t)$.
4. Accept the next event time $t$ with probability $\lambda(t)/\lambda_c$ (i.e., pick a value $v$ from a uniform distribution 0...1—if v exceeds $\lambda(t)/\lambda_c$, then the event time must be discarded).
5. If not accepted, return to step 2 and generate a new potential event time. Otherwise, return the event time $t$.

---

**EXERCISE 33.6**

In the web site repository, you can find the file MT.mat, which represents the responses of a neuron in visual area MT. Cells were exposed to a stimulus paradigm which incorporated 4 seconds of idle activity, after which a visually relevant stimulus was presented for 500 ms. Formulate a plausible lambda function $\lambda(t)$ and simulate additional spike trains. A simple approach to estimating $\lambda(t)$ is to construct a piecewise function consisting of mean rates in each of the three time regimes (pre-presentation, presentation, post-presentation). Compare to the spike train in MT.mat.

---

Up until now, we have not addressed refraction or burstiness. We can further generalize our time-varying lambda to a conditional lambda, $\lambda(t|\theta)$, where $\theta$ represents a parameterization of the lambda value. Creating an intensity function conditional on having proximity to the most recent event allows incorporation of the refractory period into our model. For example, we could define $\lambda(t|\theta)$ to return 0 for values of $t$ less than the refractory period immediately following an event.

Alternatively, we could estimate $\lambda(t|\theta)$ from the cell itself. To do this, we will calculate the ISI distribution. Once the ISI distribution is obtained, this can be used to estimate the intensity for an offset $t$ from the most recent spike.

## 33.7 PROJECT

In the web site repository, you will find the file RA_test.mat, which contains a full signal of spontaneous neural activity from area RA, a motor nucleus in the birdsong system. Identify the locations of neural events. (*Hints*: Use a relational operator to obtain a 1 or 0 signal, with a 1 reserved for periods where a threshold is exceeded. Use **diff** to identify only the transitions across the threshold. Use **find** to locate the indices corresponding to the transitions.)

From the event times, create an ISI histogram and estimate the ISI distribution. From the ISI estimate, obtain an appropriate $\lambda(t|\theta)$ that accounts for the refractory period. Generate sample event trains and compare to the original data file.

## MATLAB FUNCTIONS, COMMANDS, AND OPERATORS COVERED IN THIS CHAPTER

**rand**
**poissrnd**
**exprnd**
**unifnd**

# Exploring the Wilson-Cowan Equations

## 34.1 GOAL OF THIS CHAPTER

In this chapter, we will continue to apply phase plane analysis to a model of two interacting neuronal populations, an excitatory and an inhibitory population known as the Wilson-Cowan (W-C) equations. Rather than relying on the pplane7 program used in Chapter 15, "Exploring the Fitzhugh-Nagumo Model," we will apply the knowledge gained thus far to write our own rudimentary phase plane code for nonlinear systems. It is highly recommended that Chapter 14, "Introduction to Phase Plane Analysis," be completed prior to beginning this chapter.

## 34.2 BACKGROUND

Understanding the interplay between populations of neurons can be incredibly difficult because of the vast number of dynamic variables inherent in the problem. For example, suppose that one wants to model a population of 100 excitatory neurons and 100 inhibitory neurons all interconnected to each other. One possible method to proceed is to model each individual neuron; for example, by using the Hodgkin-Huxley formalism (see Chapter 27, "Modeling a Single Neuron"), and then connecting the neurons together in some fashion to form interacting populations. This method is the essence behind biophysical models often seen in current literature. One drawback of this method is the necessary computational power required to simultaneously solve the hundreds of coupled differential equations that result from treating each neuron individually as a complicated nonlinear dynamical system. From this perspective, it would be advantageous to have a model that describes the evolution of the average activity of an entire population of excitatory neurons and couples this to the dynamics of the average activity of an entire inhibitory population, thereby reducing the system down from hundreds of dynamical variables to just two. Additionally, such a simple system would be amenable to phase plane analysis, making it much easier to predict the behavior of such a complicated network of interconnected neurons.

## 34.3 THE MODEL

The most famous population-based dynamical model is the Wilson-Cowan (W-C) model (Wilson and Cowan, 1972). The basic dynamical variables of this model are $E$, the fraction of neurons in the excitatory population which are currently active, and $I$, the fraction of active inhibitory neurons within the population. The equations governing the interaction between these variables are

$$\frac{dE}{dt} = -\alpha E + (1-E) * f(s_E)$$

$$\frac{dI}{dt} = -\alpha I + (1-I) * f(s_I),$$

where $\alpha$ is the time constant for the decrease in activity in a population with no input current to the population to sustain activity, $s_E$ and $s_I$ are the input current to the excitatory and inhibitory populations respectively, and $f$ is the gain function of the system. A detailed derivation of these equations requiring the temporal coarse-graining of a set of integro-differential equations can be found in the original paper by Wilson and Cowan (1972); however, a heuristic understanding of the final results can be appreciated as follows.

In the absence of any input current to a population the total amount of activity in that population will decay to 0 with rate alpha. This idea is captured by the first term of the W-C equations. If there is an external source of current driving a population, then the population activity level will increase at a rate that depends upon the fraction of neurons within the population that are sensitive to incoming current, and therefore not already active (represented by the $1-E$ and $1-I$ factors), that are receiving above threshold excitation by the total input current to that population (represented by the $f(s_E)$ and $f(s_I)$ factors).

If we consider the input current to the excitatory population in more detail, then we realize that there are three main sources of current input into the excitatory population: action potentials from excitatory neurons to other excitatory neurons within the same population, action potentials from neurons within the inhibitory population synapsing onto neurons of the excitatory population, and current from neurons or external sources outside either neuronal population. A similar set of arguments applies to the total current entering the inhibitory population, and so we can express the total input currents as a sum from these three sources as

$$s_E = W_{EE} * E - W_{EI} * I + h_E$$

$$s_I = W_{IE} * E - W_{II} * I + h_I.$$

The various $W_{xy}$ constants represent the strength of synaptic contacts between neurons in the two different populations; so for instance, $W_{EI}$ represents how strongly the inhibitory neurons synapse onto the excitatory ones.

Note that the form of these current equations makes sense because the amount of current provided to the excitatory population from other neurons within the same population, $s_E$, should be proportional to the amount of activity within the population. In other words, if the excitatory population is completely inactive and $E$ is 0, then there should be no contribution to $s_E$ from the excitatory population.

The only part of the equations left to specify is the gain function, $f$. Since the fraction of active neurons within a population cannot exceed 100%, the gain function should asymptotically approach 1 for large input currents. Similarly, there can never be less than zero active neurons within a population, so $f$ should asymptotically approach 0 for negative total input currents. For intermediate amounts of input current, the gain function should monotonically increase from 0 to 1 as the input current is increased. A set of mathematical functions that obey these rules are called *sigmoidal functions*. The most common sigmoidal functions are the hyperbolic tangent function (**tanh**), the error function (**erf**), and the Boltzmann sigmoid function.

## 34.4 EXERCISES

Let's plot an example sigmoidal function, in particular, the **tanh** function. In order to ensure that this function saturates at 0 and 1 as discussed above, it must be shifted and scaled. Type the following at the command prompt to produce the appropriate function.

```
>> x = -5:0.1:5;
>> y = 0.5*(1 + tanh(x));
>> plot(x,y);
```

Similarly, we can plot a scaled, shifted error function (**erf**) in MATLAB® as follows:

```
>> x = -5:0.1:5;
>> y = 0.5*(1 + erf(x));
>> plot(x,y);
```

If you plot the two functions together, you can see that they are not equivalent (as shown in Figure 34.1).

The **tanh** function rises more steeply towards 1 than the **erf** function, indicating that the response of a population with the **tanh** gain function has a higher level of activation for the same total input current than another population whose gain function is described by the **erf** function. In general, one can modify the W-C equations to allow for different sigmoidal gain functions for the two populations.

In the projects that follow, we will use yet another gain function closely related to the **tanh** function. Rather than rescaling and shifting the **tanh** function, we will multiply it by the Heaviside step function. Although there is a Heaviside step function command in

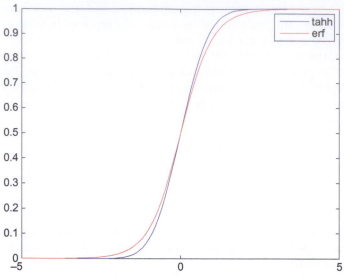

**FIGURE 34.1**    **tanh** and **erf** functions plotted in MATLAB.

MATLAB, it does not return a number if the argument is 0; so the easiest way to generate this gain function is with the following command:

>> y = **tanh(x)\*(x > 0)**;

Plot this function alongside the other gain functions and compare them.

## 34.5  PROJECTS

In order to write our own version of **pplane8**, a function we'll call **nonlin_phase_plane**, we are going to rely upon several already completed projects, and a new MATLAB data structure called a cell array. The general layout of our function will parallel the function **phase_plane** written in Chapter 14, "Introduction to Phase Plane Analysis." A sketch of this function is shown ahead.

function stable = phase_plane(A, init)

%Determine nullclines of the linear system described by the matrix A
x = -5:0.1:5;
x_null = -1\*A(1,2)/A(1,1)\*x;
y_null = -1\*A(2,1)/A(2,2)\*x;

%Determine stability of fixed point of system which depends upon the eigenvalues of A.
%Note that for a linear system there is only one fixed at the origin.
eigvals = eig(A);
stable = classify_fixed_pts(eigvals);

%Determine the vector fields for the phase plane
x = -5:0.1:5;
y = x;
[X, Y] = meshgrid(x,y);
F = A(1,1)*X + A(1,2)*Y;
G = A(2,1)*X + A(2,2)*Y;

%Determine a trajectory from an initial condition using rk4 see Chapter 19,
%"Voltage-Gated Ion Channels
trajectory = rk4(*necessary input arguments*);

%Plot everything including nullclines, vector field (using the quiver command), and the trajectory.

Updating this code to be able to handle nonlinear systems such as the W-C system requires several modifications, which we will now consider. The first part of **phase_plane** determines the nullclines based on the matrix A. In the nonlinear case, the nullclines are described by the curves where the derivatives are zero, so for the general nonlinear system

$$\frac{dx}{dt} = f(x, y)$$

$$\frac{dy}{dt} = g(x, y)$$

The x- and y-nullclines are the curves defined by $f = 0$ and $g = 0$, respectively. In order to make our **nonlin_phase_plane** generic enough to handle different nonlinear systems, we will employ the **feval** command and a function handle to call upon another function to determine the nullclines. Since the nullclines are no longer linear, they may intersect never, once, or more than once, producing a variable number of fixed points. Therefore, unlike in the linear case, **nonlin_phase_plane** will also have to determine where the nullclines intersect, and potentially analyze the stability of multiple fixed points.

Another crucial difference in analyzing the stability of fixed points for nonlinear systems is that the Jacobian of the system must be determined and evaluated at each fixed point in order to classify its stability (for a refresher, see Chapter 15, "Exploring the Fitzhugh-Nagumo Model"). Since different nonlinear systems will have different Jacobians, another function handle will be needed here. Depending upon how you write your code, there may be more function handles needed to keep **nonlin_phase_-plane** generic enough to handle multiple nonlinear systems.

Obviously, these various function handles must be passed into **nonlin_phase_plane** as input arguments, but in order to keep the total number of input arguments to **non-lin_phase_plane** at a minimum, we can create a cell array input where each element of the array is a function handle. Cell arrays are MATLAB's way of grouping dissimilar data structures such as character strings of different lengths or certain data types such as function handles. Working with cell arrays is similar to working with arrays that hold other data types, except that the syntax for cell arrays uses the curly bracket symbols instead of the square bracket ones. For instance, the following line of code creates a cell array with four elements, each a function handle:

```
>> system_file = {@WCnullclines, @calculate_Jacobian, @WC_determ, @WC_plot};
```

You can also construct a cell array by using the **cell** command. The syntax would be as follows:

```
>> system_file = cell(1,4);
>> system_file{1} = @WCnullclines; %etc.
```

Accessing the first element of the cell array for use with the **feval** command would look something like this:

```
>> feval(system_file{1}, input_args_to_WCnullclines);
```

A general sketch of the code for **nonlin_phase_plane** is shown below. Feel free to use it as a guide when writing your own phase plane analysis code.

```
function [x0, y0, net_act] = nonlin_phase_plane(system_file, params, init, t, x, y)
%function [x0, y0, net_act] = nonlin_phase_plane(system_file, params, init, t, x, y)
%The input arguments are
%1) system_file is a cell array where each element of the array is a function
%handle. The first function handle calls a function for plotting the
%nullclines of the system. The second calls a function to determine the
%Jacobian of the system. The third calls a function containing the differential
%equations that define the system and are used to determine the vector
%field of the phase plane. The last function handle calls a function to
%control plotting the system. Example,
%system_file = {@WCnullclines, @calculate_Jacobian, @WC_determ, @WC_plot};
%2) params contains all parameters needed by the system. Example for WC system,
%params =  [W_EE, W_EI, W_IE, W_II, h_E, h_I, alpha];
%3) init contains the initial condition of the system for calculating
%trajectories. Example,
%init = [0.1 0];
%4) Contains the time interval that the trajectory should be simulated for.
%t = 0:0.01:500;
%5&6) The last two arguments specify the spacing of the vector fields on the
%phase plane in the x and y-directions. Example,
%x = 0:0.1:1;
%y = x;

%Determine nullclines for the system
[Enull, Inull, vec] = feval(system_file{1}, params);

%Find fixed points where nullclines intersect.
[x0,y0] = intersections(vec,Enull,Inull,vec);
```

```
%Determine the stability of the fixed points by calculating
%the Jacobian and evaluating it at each fixed point.
    stable = cell(1, length(x0));
    for k = 1:length(x0)
    %Returns the Jacobian, J, evaluated at the k-th fixed point
    J = feval(system_file{2}, x0(k), y0(k), params);
    %Find eigenvalues and eigenvectors of Jacobian
    eigvals = eig(J);
%Classify fixed point based on eigenvalues of J
    stable{k} = classify_fxd_point(eigvals);
end;

%Determine the vector field for the phase plane.
lx = length(x);
ly = length(y);
u = zeros(lx, ly);
v = u;
for ii = 1:lx
    for jj = 1:ly
        point = [x(ii), y(jj)];
        %Determine x and y component of vector field
        sol = feval(system_file{3}, point, params);
        u(ii,jj) = sol(1);
        v(ii,jj) = sol(2);
    end;
end;
[X, Y] = meshgrid(x, y);

%Find trajectories and plot them.
%Use 4th order Runge-Kutta to simulate system.
net_act = rk4(system_file{3}, init, params, t);

%Handle details of plotting the phase plane for the particular system in
%question.
feval(system_file{4}, vec, Enull, Inull, x0, y0, stable, X, Y, u, v, net_act, t);
```

1. Use the sketch above while writing the necessary supporting functions to create your own version of **pplane8**.
2. Create functions to handle the W-C system nullclines, Jacobians, and vector fields. Study the phase plane under the following conditions:
   For all three parameter sets we will fix $\alpha = 0.1$, and the initial condition will be $(E_o, I_o) = (0.1, 0)$.
   a. W_EE = W_IE = 0.5, W_EI = W_II = 0.3, h_E = h_I = 0.001.
   b. W_EE = W_IE = W_II = 1, W_EI = 1.6, h_E = 0.325, h_I = 0.175.
   c. W_EE = W_IE = W_EI = 1.2, W_II = 0.8, h_E = 0.1, h_I = 0.001.

What type of fixed points do each of these parameter sets exhibit? Based on these findings is it possible that synaptic plasticity could drive a population of neurons to produce periodic activity in the population average?

3. Create the system files necessary for the F-N model studied in Chapter 15, "Exploring the Fitzhugh-Nagumo Model," and compare results to those obtained with **pplane8**.

# MATLAB FUNCTIONS, COMMANDS, AND OPERATORS COVERED IN THIS CHAPTER

**tanh()**
**erf()**
**cell()**

# Neural Networks as Forest Fires: Stochastic Neurodynamics

## 35.1 GOALS OF THIS CHAPTER

The purpose of this chapter is to familiarize you with simulating a stochastic process. We will be modeling a large-scale network of firing neurons through the analogy of forest fires. In the process, you will become more adept at using convolutions and creating movies of neural population dynamics.

## 35.2 BACKGROUND

The brain is made up of approximately $3 \times 10^{10}$ neurons, each supporting up to $10^4$ synaptic connections. Even after allowing for the specializations of neural circuitry that must exist to achieve coordinated activity, it is remarkable that such activity is not completely random. It seems reasonable, however, that large-scale neocortical activity has some random component, and thus its appropriate representation must be probabilistic. In creating a large-scale model of neuronal population dynamics that includes this probabilistic nature, we can use the analogy of a well-studied stochastic model of forest fires.

There exists a close correspondence between the dynamics and properties of large networks of spiking neurons and that of forest fires. In 1990, Bak, Chen, and Tang introduced the forest fire model, a probabilistic cellular automaton defined on a lattice. Each lattice site is occupied by either a *green* tree, a *burning* tree, or a *burnt* tree. The state of the system is updated by the following rules: (1) a *burning* tree becomes a *burnt* tree with probability 1; (2) a *burnt* tree grows into a *green* tree with probability $p$; (3) a *green* tree becomes a *burning* tree if at least one of its neighbors is burning, or if it is hit by lightning (with probability $f$.)

Just as a forest contains *green*, *burning*, or *burnt* trees, neural networks contain *quiescent*, *active*, or *refractory* neurons. When the membrane voltage $V(t)$ of a neuron is above

*MATLAB® for Neuroscientists.*
DOI: http://dx.doi.org/10.1016/B978-0-12-383836-0.00035-7

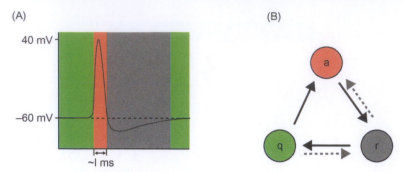

**FIGURE 35.1** A. The membrane potential of a typical spiking neuron. When the membrane potential of the neuron is at or above its resting potential (dotted line) and below the threshold potential, the neuron is quiescent (green). After the neuron fires the action potential (red), the neuron enters a refractory state (black), during which the membrane potential of the neuron is less than the membrane potential. B. A neuron cycles through these states with time, given its membrane properties and input from connecting neurons (dotted lines indicate transitions that can occur with inhibitory input to the neuron).

its resting potential (typically around $-60\,\mathrm{mV}$) and below the threshold potential (around $-40\,\mathrm{mV}$), the neuron is said to be in the *quiescent* state. It can be excited, just as a green tree can be ignited. When the membrane potential of the neuron exceeds the threshold, the neuron is *active*, and the membrane potential spikes to a large positive value for about a millisecond. Immediately afterwards, the neuron enters a *refractory* state, during which the membrane potential of the neuron is hyperpolarized. With a characteristic time constant determined by the membrane properties, the membrane potential returns to the resting potential and thus the neuron can return to the *quiescent* state (see Figure 35.1).

Models of neuronal activity, both as standard forest fires as well as networks of integrate-and-fire neurons, are capable of displaying a variety of interesting behavior. These include long-range (power law) temporal correlations in the inter-spike interval histograms, stochastic resonance in the noise-driven appearance of spirals of neural activity, and traveling waves.

## 35.2.1 Neural Analysis

The key conjecture (introduced by Bak, Chen, and Tang (1990) and elaborated by Drossel and Schwabl (1992)) is that in the limit $p \to 0$, $f/p \to 0$ the dynamics of the forest fire become *critical*. In words, the first condition says that the state is critical as long as trees grow slowly, and the second condition guarantees that a lightning strike will destroy a large number of trees. A critical state of a driven dynamical system is one in which interactions between all its elements occur so that correlations develop at all length scales in the system. Such a state is marginally stable and can support large fluctuations. In the forest fire, this means that power law distributions of such quantities as the size of clusters of burning trees would develop as the forest dynamics tend to the critical limit.

In a neural network, $1/p$ is interpreted as the mean time it takes for a refractory neuron to recover sensitivity, and thus return to the quiescent state, while $1/f$ is the mean time between spontaneous activations of a neuron (produced, for example, by noise). An analysis in the context of stochastic neurodynamics was carried out by Buice and Cowan (2007). At first, we will consider a network of these simplified stochastic neurons that are connected in a very simple way, having either nearest neighbor or next-to-nearest neighbor excitatory connections.

## 35.3 EXERCISES

Our goal is to create a movie of the activity of an $N \times N$ array of neurons for $T$ time frames, i.e., the spatiotemporal dynamics of the neurons will be modeled in an $N \times N \times T$ matrix. Each pixel of the movie will either be green, red, or black to represent a neuron at that site and time in the quiescent, active, or refractory state, respectively. We begin by writing a function called **ForestFireModel** that will generate an avi file of the forest fire simulation that can be viewed outside of MATLAB® and that will take the following inputs: $T$, the number of iterations (time); $p$, the growth rate; $f$, the probability of spontaneous ignition; $N$, the lattice size; $th$, the threshold value. Furthermore, we will have the input $BC$ to be able to set the boundary conditions for our simulation to be either free or periodic, and finally the input *conn* to be either 1 for nearest neighbor or 2 for next-to-nearest neighbor connectivity. This is a simple approach, without going so far as to design a GUI interface, to create a modeling environment in which you have numerous parameters that you may want to alter.

The downside to so many parameters, though, is that it could be inconvenient to populate them all of the time. Therefore it is useful to include a series of **nargin** statements at the top of our **ForestFireModel** function, providing a default set of parameters. Inside a function, **nargin** indicates how many input arguments have been given in the function call. We will use this to encode a good, working set of default parameters, and furthermore allow the function to be called without argument for convenience. Note the order in which these parameters are given.

```
function Forest = ForestFireModel(T,p,f,N,th,BC,conn)
% This function simulates a neural network as a forest fires

if nargin < 7, conn = 2; end %connectivity (1 nn, 2 nnn)
if nargin < 6, BC = 'periodic'; end %Boundary Cond. ('free' or 'periodic')
if nargin < 5, th = .1; end %threshold
if nargin < 4, N = 64; end %lattice size
if nargin < 3, f = 2e-4; end %prob of spontaneous ignition
if nargin < 2, p = .1; end %growth rate
if nargin < 1, T = 100; end %number of iterations (time)

mov = avifile('Forest.avi'); %initialize the movie file that will be created
```

We are going to assign numerical values to the three possible states of the neuron:

```
%a: 1 <-- active state
%q: 0 <-- quiescent state
%r: 100 <-- refractory state
```

Next, we create our initial "forest," F, of neurons, an $N \times N$ matrix. The command **rand(N)** creates an $N \times N$ matrix of numbers uniformly distributed between 0 and 1, so **rand(N) < 0.4** creates a $N \times N$ matrix of 0s and 1s with a site being populated by 1 with a probability of 0.4. We can multiply this by 100 to make this a matrix of 0s and 100s—a forest with mainly quiescent trees (0s) and about 40% in the refractory state (100 s):

```
%% initial conditions
F = 100*(rand(N) < 0.4); %initialize forest with trees (q and r) here and there
                         %N*N matrix of 0's and 100's randomly placed
```

The connectivity matrix gives the weight of the connections of a given neuron to its neighbors. The center of the matrix is the location of the source neuron, and it will thus have a value of zero since we do not want to model self-stimulation. The weight of the connection to a cell is proportional to the distance from the source neuron. Here we use another MATLAB control structure, the **switch/case** statements:

```
%% -------------------------------
%% connectivity array
%% -------------------------------
switch conn
%% Nearest neighbor interactions
    case 1
        W = [0 1 0; 1 0 1; 0 1 0];
%% Next-nearest neighbor interactions
    case 2
        W = [ 0 0 1 0 0; 0 sqrt(2) 2 sqrt(2) 0; ...
            1 2 0 2 1; 0 sqrt(2) 2 sqrt(2) 0; 0 0 1 0 0];
        W = W/sum(sum(W)); %normalize for a net
end
%% ---end connectivity array -----
```

We are now ready for the main loop of the simulation. This will use the built-in **conv2** function for free boundary conditions, and the **conv2_periodic.m** function introduced in Chapter 16, "Convolution," for the periodic boundary conditions.

```
%% -------------------------------
%% main loop
%% -------------------------------
for ii = 1:T
    Ac = (F == 1); %id's active fires in previous timestep
    Qu = (F == 0); %id's quiescent trees in previous timestep
    Rs = (F == 100); %id's refractive trees in previous timestep
```

```
if length(BC) == 4 %free
   CoA = conv2(double(Ac), W, 'same'); %convolve fires w/ connectivity array
elseif length(BC) == 8 %periodic
   CoA = conv2_periodic(Ac,W);
end
```

At this point we have defined three identifying matrices, **Ac**, **Qu**, and **Rs**, that are $N \times N$ matrices of 0s taking on the value of 1 only if that neuron is in the active, quiescent, or refractory state respectively. We have also created a matrix **CoA** that is the net input to all of the neurons. It was found by convolving the active neurons (those elements of the forest **F** with a value of 1 at time $i$) with the connectivity matrix.

Now we create a temporary $N \times N$ matrix, **Fr**, which is all 0s and only takes on the value of 1 for those neurons who were in the quiescent state and whose input exceeded the threshold value. We also create another $N \times N$ temporary matrix, **Frn**, which is all 0s and only takes on the value of 1 with probability $f$ if the neuron was in the quiescent state, but its input did not exceed threshold. The use of element-wise matrix multiplication of the identifying matrix **Qu** in the definition of these matrices ensures that only those sites that were in the quiescent state (having a value of 0 in the forest matrix **F**) take on a non-zero value. These two matrices identify those neurons in the forest that will transition from the quiescent state to the active state.

```
Fr = (CoA > th).*Qu; % id's q trees whose input exceeded threshold
Frn = (Fr == 0).*(rand(N) < f).*Qu; %with prob f selects trees from set q
                    %whose input does not exceed threshold
```

The forest matrix **F** is now updated by the transition rules:

```
F = 100*Ac + 100*(rand(N) > p).*Rs + Fr + Frn; %updates forest by rules
%first term: a->r w/ prob 1; second term: r->q w/ prob p
% final terms: q->a if input exceeds th or w/ prob f; ;
```

We can now redefine the identifying matrices:

```
Ac = (F == 1); % redefine matrices of q,a & r tree locations
Qu = (F == 0);
Rs = (F == 100);
```

The next few lines generate the movie of the simulation, and save it to an AVI file. The use of **colormap(colorcube(12))** will cause a matrix element with a value of 7 to appear as a green pixel, an element with a value of 5 as red, and one with a value of 10 as black (see Figure 35.2):

```
figure(1)
colormap(colorcube(12));
im = image(Qu*7 + Ac*5 + Rs*10);
axis square
Frame = getframe(gca);
mov = addframe(mov,Frame);
Forest(:,:,ii) = F;
```

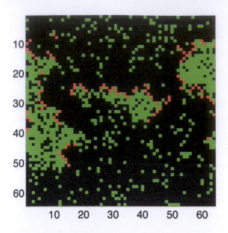

**FIGURE 35.2** A snapshot of the neural network simulated as a forest fire. Each pixel represents a single neuron, and it is colored green, red, or black to represent the neuron being in the quiescent, active, or refractory state, respectively.

**end %ends the main loop**
**save Forest Forest**
**mov = close(mov);**

The entire spatio-temporal dynamics of this network is saved in the $N \times N \times T$ matrix, **Forest**, which can be loaded once the program has run using the command **load Forest**.

---

### EXERCISE 35.1

Enter the **ForestFireModel** function given above on your own computer. Make sure that it runs, and that you understand each part of the program. Run the model with the following command: **ForestFireModel (100,0.1,0,64,.1,'periodic',2)**. Why does the forest turn all green in the end?

---

### EXERCISE 35.2

Now let $p = .5$ and $f = .5$, and describe the dynamics. Is this a realistic set of parameters for a neural network?

---

### EXERCISE 35.3

Now let $p = .01$ and $f = .0075$. How do the dynamics differ? Continue to play with the various input parameters to see how they impact the dynamics of the network. How many different behaviors can you get out of the "forest" of neurons?

## 35.4 PROJECTS

1. Of course, true neural networks are more complicated than forests. Neural dynamics are more complex than the simple ignition of a tree. In addition, there are two types of neurons: excitatory and inhibitory. About 25% of the neurons in the cortex of the brain are inhibitory; they act to prevent other neurons from activating. To some extent the first complexity doesn't matter too much, but the existence of inhibition can radically alter neural network dynamics. There is also some evidence that the spatial extent of inhibition differs from that of excitation. This has profound consequences for the dynamics of neural activation.

   Forest fire models have been developed by Drossel, Clar, and Schwabl (1994) in which trees have a certain immunity to fire, which acts like inhibition. Modify the forest fire model to incorporate inhibition. Create simulations for a network of randomly connected lattice sites in which proximal neighbor connections tend to be excitatory and more distal ones tend to be inhibitory (i.e., a Mexican hat distribution of excitatory and inhibitory interactions). You can take a peek at Chapter 16, "Convolution," to help setup the connectivity matrix. What new patterns of activity are you able to produce? Can you find a parameter set that will create blobs or stripes?

2. The stochastic nature of the neuronal population can be introduced to the model in a number of places. Incorporate additional noise to the system by making the threshold follow Brownian motion within reflecting boundaries. *Hint:* you will need to introduce an $N \times N$ matrix that contains the threshold value for each of the neurons and is updated at each timestep. You will use the **rand** function to generate the amount by which the threshold value is increased or decreased at each step, and impose a minimum and maximum value that the threshold can be. If the random increment exceeds the boundary by an amount $x$, then the new value of the threshold is a distance $x$ within the boundary.

3. For a more advanced project, modify the above simulations using noisy integrate-and-fire neurons to populate the forest. See what interesting population dynamics you can produce.

## MATLAB FUNCTIONS, COMMANDS, AND OPERATORS COVERED IN THIS CHAPTER

**nargin**
**avifile**
**switch**
**colormap**

# Neural Networks Part I: Unsupervised Learning

## 36.1 GOALS OF THIS CHAPTER

This chapter has two goals that are of equal importance. The first goal is to become familiar with the general concept of unsupervised neural networks and how they may relate to certain forms of synaptic plasticity in the nervous system. The second goal is to learn how to build two common forms of unsupervised neural networks to solve a classification problem.

## 36.2 BACKGROUND

Neural networks have assumed a central role in a variety of fields. The nature of this role is fundamentally dualistic. On the one hand, neural networks can provide powerful models of elementary processes in the brain, including processes of plasticity and learning. On the other hand, they provide solutions to a broad range of specific problems in applied engineering, such as speech recognition, financial forecasting, or object classification.

### 36.2.1 But What is a Neural Network?

Despite its "biological" sounding name, neural networks are actually quite abstract computing structures. In fact, they are sometimes referred to as *artificial neural networks*. Essentially, they consist of rather simple computational elements that are connected to each other in various ways to serve a certain function. The architecture of these networks was inspired by the mid- to late-20th-century notion of brain function, hence the term "neural."

*MATLAB® for Neuroscientists.*
DOI: http://dx.doi.org/10.1016/B978-0-12-383836-0.00036-9

At its conceptual core, a unit in a neural network consists of three things (Figure 36.1):

1. A set of inputs that can vary in magnitude and sign coming from the outside world or from other neurons in the network.
2. A set of weights operating on these inputs that can vary in magnitude and sign (implementing synaptic efficiency and type of synapse on a neuron). There is also a bias weight $b$ that operates on an input that is fixed to a value of 1.
3. A transfer function that converts the sum of the weighted inputs or net input, $n$, to an output, $o$.

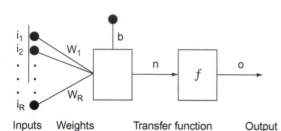

**FIGURE 36.1**    The concept of a neural network.

The output of a unit can then become the input to another unit. Individual units are typically not very functional or powerful. Neural networks derive their power from connecting up large numbers of neurons in certain configurations (typically called layers) and from learning (i.e., setting the weights of these connections).

Neural networks are extremely good at learning a particular function (such as classifying objects). There are several different ways to train a neural network, and we will become acquainted with the most common ones in this and the next chapter.

Such trained multi-layer networks are extremely powerful. It has been shown that a suitable three-layer (i.e., three layers of units) network can approximate any computable function arbitrarily well. In other words, neural networks that are properly set up can do anything that can be done computationally. This is what makes them so appealing for applied engineering problems, because the problem solver might not always be able to explicitly formulate a solution to a problem, but he might be able to create and train a neural network that can solve the problem, even if he doesn't understand how it works. For example, a neural network could be useful to control the output of a sugar factory given known inputs.

## 36.2.2 Unsupervised Learning and the Hebbian Learning Rule

Despite the fact that neural networks are very far from real biological neural networks, the learning rules that have been developed to modify the connections between computing elements in neural networks nevertheless resemble the properties of synaptic plasticity in the nervous system. In this chapter, we will focus on *unsupervised learning rules* (in contrast to supervised or error-correcting learning rules), which turn out to be very similar to Hebbian plasticity rules that have been discovered in the nervous system. Unsupervised learning tries to capture the statistical structure of patterned inputs to the network without an explicit teaching signal. As will be clear in a moment, these learning rules are sensitive to correlations between components of patterned inputs; they strengthen connections between components that are correlated, and weaken connections that are uncorrelated. These learning rules serve at least three computational functions: 1) to form associations between two sets of patterns; 2) to group patterned inputs that are similar into particular categories; and 3) to form content-addressable memories such that partial patterns that are fed to the network can be completed.

Donald Hebb was one of the first to propose that the substrate for learning in the nervous system was synaptic plasticity. In his book *The Organization of Behavior*, Hebb stated, "When an axon of cell A is near enough to excite a cell B and repeatedly or persistently takes part in firing it, some growth process or metabolic change takes place in one or both cells such that A's efficiency, as one of the cells firing B, is increased" (1949, p. 62). In fact, William James, the father of American psychology, formulated the same idea almost sixty years earlier in his book *The Principles of Psychology* when he stated, "When two elementary brain-processes have been active together or in immediate succession, one of the them tends to propagate its excitement into the other" (1890, p. 566). Nevertheless, the concept of synaptic plasticity between two neurons that are co-active is usually attributed to Donald Hebb. Mathematically, Hebbian plasticity can be described as:

$$\Delta w_{ij} = \varepsilon \cdot pre_i \cdot post_j \tag{36.1}$$

where $\Delta w_{ij}$ denotes the change in synaptic weight between a presynaptic neuron $i$ and a postsynaptic neuron $j$, $pre_i$ and $post_j$ are the activities of presynaptic neuron $i$ and postsynaptic neuron $j$, respectively, and $\varepsilon$ is a learning constant that determines the rate of plasticity. The Hebbian learning rule states that a synapse will be strengthened when the presynaptic and postsynaptic neuron are active. Neurophysiologically, this means that the synapse will be potentiated when the presynaptic neuron is firing and the postsynaptic neuron is depolarized. In 1973, Bliss and Lomo first showed that the synapses between the perforant pathway and the granule cells in the dentate gyrus of the hippocampus could be artificially potentiated using a stimulation protocol that followed the Hebbian learning rule (Bliss and Lomo, 1973). The effects of the stimulation protocol they used has been termed *Long-Term Potentiation* because the synapses appear to be potentiated indefinitely. Since that time, may other experiments have shown that Long-Term Potentiation could be implemented in many parts of the brain, including the neocortex. Long-Term Potentiation (with initial capital letters), which refers to an artificial stimulation protocol, should be distinguished from *long-term potentiation* (all lowercase), which refers to the concept that synapses may be potentiated naturally when some form of associative learning takes place.

From a computational perspective, the simple Hebbian rule can be used to form associations between two sets of activation patterns (Anderson et al., 1977). Imagine a feedforward network consisting of an input (presynaptic) and output (postsynaptic) set of neurons, $f$ and $g$ respectively, that are fully connected as shown in Figure 36.2. The transfer function of the $f$ neurons is assumed to be linear and, thus, this network is referred to as a *linear associator* (Anderson et al., 1977). The activation of each set of neurons can be viewed

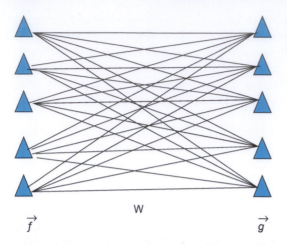

FIGURE 36.2 A simple linear associator network composed of an input and output set of neurons that are fully connected.

as column vectors, $\vec{f}$ and $\vec{g}$. Assume we want to associate a green traffic light with "go," a red traffic light with "stop," and a yellow traffic light with "slow." Furthermore, assume that the green, red, and yellow traffic lights are coded as mutually orthogonal and normal (i.e., unit length) $\vec{f}$ vectors.

The Hebbian learning rule can be used to form these associations. Mathematically, the learning rule generates a weight matrix as an outer-product between the $f$ and $g$ vectors:

$$W = \vec{g}\,\vec{f}^{\,T}$$

$$W = \vec{g}_{go}\vec{f}^{\,T}_{green} + \vec{g}_{stop}\vec{f}^{\,T}_{red} + \vec{g}_{slow}\vec{f}^{\,T}_{yellow}$$

Now, if we probe the network with the green presynaptic pattern, we get the following:

$$W\vec{f}_{green} = \vec{g}_{go}\vec{f}^{\,T}_{green}\vec{f}_{green} + \vec{g}_{stop}\vec{f}^{\,T}_{red}\vec{f}_{green} + \vec{g}_{slow}\vec{f}^{\,T}_{yellow}\vec{f}_{green}$$

Given that the input patterns are orthogonal to each other and normal, the output of the network is "go":

$$W\vec{f}_{green} = \vec{g}_{go}\cdot 1 + 0 + 0$$

## 36.2.3 Competitive Learning and Long-Term Depression

Despite its elegant simplicity, the Hebbian learning rule as formulated in Equation 36.1 is problematic because it only allows for potentiation, which means that the synapse will only grow stronger and eventually saturate. Neurophysiologically, it is known that synapses can also depress using a slightly different stimulation protocol. Fortunately, there is a neural network learning rule that can either potentiate or depress. It was proposed by Rumelhart and Zipser (1985) and is referred to as the competitive learning rule:

$$\Delta w_{ij} = \varepsilon \cdot pre_i \cdot post_j - \varepsilon \cdot post_j \cdot w_{ij} \tag{36.2}$$

The first term on the right-hand side of Equation 36.2 is exactly the Hebbian learning rule. The second term, however, will depress the synapse when the postsynaptic neuron is active regardless of the state of the presynaptic neuron. Therefore, if the presynaptic neuron is not active, the first term goes to zero and the synapse will depress. Also, notice that depression is proportional to the magnitude of the synaptic weight. This means that if the weight is very large (and positive), depression will be stronger. Conversely, if the weight is small, depression will be weaker. The competitive learning rule can be described equivalently as follows:

$$\Delta w_{ij} = \varepsilon \cdot post_j \cdot (pre_i - w_{ij}) \tag{36.3}$$

The formulation in Equation 36.3 makes clear what the learning rule is trying to do. Learning will equilibrate (i.e., terminate) when the synaptic weight matches the activity of the presynaptic neuron. On a global scale, what this means is that the learning rule is trying to develop a matched filter to the input, and can be used to group or categorize inputs. To make this clearer, consider two kinds of patterned inputs corresponding to apples and oranges. Each example of an apple or orange is described by a vector of three numbers that describe features of the object such as its color, shape, and size. Consider the problem of developing a neural network to categorize the apples and oranges (Figure 36.3).

Imagine the apple and orange vectors clustered in a three-dimensional space. The weights feeding into either the apple unit or orange unit can also be viewed as vectors with the same dimensionality as the input vectors. Before learning, the apple and orange weight vectors are pointing in random directions (Figure 36.4). However, after learning, the weight vectors will be pointing toward the center of the apple and orange input vector clusters because the competitive learning rule will try to move the weight vectors to match the inputs (Figure 36.5). Finally, notice how the two category units are mutually inhibiting each other (the black circles indicate fixed inhibitory synapses which do not undergo plasticity). This mutual inhibition allows only one unit to be active at a time, so that only one weight vector is adjusted for a given input vector.

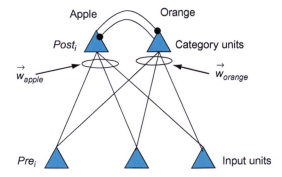

**FIGURE 36.3** A two-layer neural network that takes input vectors corresponding to apples and oranges, and categorizes them by activating one of the two category units.

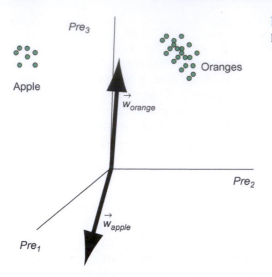

**FIGURE 36.4**   The weight vectors in an untrained competitive learning neural network.

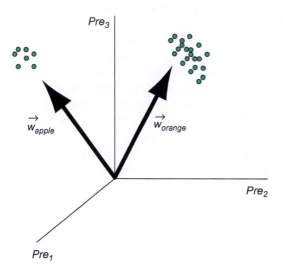

**FIGURE 36.5**   The weight vectors in a competitive learning neural network that has been trained.

## 36.2.4 Neural Network Architectures: Feedforward vs. Recurrent

As with real neural circuits in the brain, artificial neural network architectures are often described as being *feedforward* or *recurrent*. Feedforward neural networks process signals in a one-way direction and have no inherent temporal dynamics. Thus, they are often described as being static. In contrast, recurrent networks have loops and can be viewed as a dynamic system whose state traverses a state space and possesses stable and unstable equilibria. The linear associator described above is an example of a feedforward network. The competitive learning network is a sort of hybrid network because it has a feedforward component leading from the inputs to the outputs. However, the output neurons are mutually connected

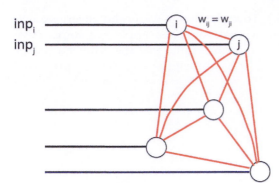

FIGURE 36.6 The architecture of a Hopfield neural network. The weights (i.e., connections) between neurons $i$ and $j$ are symmetric.

and, thus, are recurrently connected. An example of a purely recurrent neural network is the *Hopfield network* (Figure 36.6). A Hopfield network uses a Hebbian-like learning rule to generate stable equilibria corresponding to patterns that need to be stored. By providing partial patterns of stored patterns to a Hopfield network, the network's state will tend to progress toward its stable equilibrium corresponding to the stored pattern, and, thus, complete the pattern. This is an example of a content-addressable memory. The Hopfield network is a fully interconnected network with the constraint that connections between pairs of neurons are symmetric. Also, there are no self-connections. These constraints result in a symmetric weight matrix with zeros along the main diagonal.

## 36.3 EXERCISES

### 36.3.1 Competitive Learning Network

Suppose you want to categorize the following 6 two-dimensional vectors into two classes:

```
>> inp = [0.1  0.8  0.1  0.9  0.2  0.7;0.7  0.9  0.8  0.8  0.75  0.9];
```

Each column of this matrix represents one two-dimensional input vector.

There are three vectors near the (0,1) and three vectors near (1,1). What now?

First, create a two-layer network with two input "feature" neurons and two output "category" neurons. To do this, create a random $2 \times 2$ matrix of weights:

```
>>   W = rand(NCAT,NFEATURES);
```

where NCAT = 2 and NFEATURES = 2.

Competitive learning works optimally if the weight vector associated with each output neuron is normalized to 1:

```
>>   W = W./repmat(sqrt(sum(W.^2,2))1,NFEATURES)
```

This network needs to be trained to classify properly. In this network, the output neurons compete to respond to the input in a winner-take-all fashion such that only the weight vector feeding into the winning output neuron is trained. We could implement this by including inhibitory connections between the two output neurons, and let the dynamics of the network find the winner as in Figure 36.3. To make things easier, however, we will use the **max** function in MATLAB® to find the winner, and then apply the competitive learning rule to the winner's weight vector. We will also assume that the winner's activation equals 1. Create a function that implements the competitive learning training rule with a learning rate parameter **lr**:

```
>> function [Wout] = train_cl(W,inp,lr)
>> %competitive learning rule
>> out = W*inp;
>> [mx ind] = max(out);
>> W(ind,:) = W(ind,:) + lr*(inp'-W(ind,:));
>> Wout = W./repmat(sqrt(sum(W.^2,2))1,size(W,2));
end
```

Notice that the weight vectors associated with the winning output neuron are renormalized after training.

Using this training rule, create a script to classify the six input vectors. In this script, you will expose the network to all inputs one at a time over many epochs. In the script, plot the inputs and the two weight vectors (with the **quiver** function) associated with each output neuron after each epoch so that the learning process can be visualized (Figure 36.7).

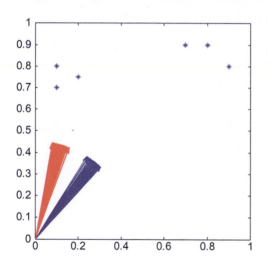

**FIGURE 36.7** The evolution of the two weight vectors (blue and red arrows) associated with the two output neurons during training of a competitive learning neural network. The six two-dimensional inputs are plotted as blue stars.

The results of the trained network should be:

```
>> results =   1   2   1   2   1   2
```

This indicates that the 1st, 3rd, and 5th input vectors activated neuron 1, and the 2nd, 4th, and 6th input vectors activated neuron 2. Of course, because this is an unsupervised neural network, the results could also be the exact opposite:

```
>> results =   2   1   2   1   2   1
```

You may notice that your neural network does not always categorize your inputs into two categories, but sometimes categorizes them instead into one category. This is because one of the weight vectors is accidently very far away from any input and, therefore, never wins the competition. These are called "dead units." You will also notice that the winning weight vector is pointing in between the two classes of inputs. Can you think of a way to solve this problem?

## 36.3.2 Hopfield Network

We will now build a Hopfield neural network to store two four-dimensional patterns. The two column vectors we will store are:

[1   1   −1   −1]' and [ −1   −1   1   1]'

Place those two input vectors into a $4 \times 2$ matrix called **inp** (i.e., the external input to the network). Create a matrix **W** that stores these two patterns using a Hebbian-type learning rule:

```
>>      for ii = 1:size(inp,2)
>>          W = W + inp(:,ii)*inp(:,ii)';
>>      end;
>>      W = W-diag(diag(W));
```

The last line of this code ensures that the diagonals of the weight matrix are zero. This is because the Hopfield network requires that there be no self-connectivity. Now, test to see whether those two patterns were stored in the network. To do this, write code that updates each neuron of the network with a certain probability **p**, which implies that not all neurons will be necessarily updated at the same time (i.e., asynchronous updating). If the neuron is updated, then a simple thresholding operation is performed. If the net input to the neuron is greater than or equal to zero, then the output of the neuron should be set to 1. Otherwise, the output of the neuron should be set to −1.

Create a function called update_hp that takes the net input, the weight matrix W, the current state of the system, and the update probability p, and updates the state of the network.

Let us now feed the trained Hopfield network with one of the two column vectors that we used to build the network: [1 1 -1 -1]'. Let us set the state of the network at t=0 to be all zeros: state(0)=[0 0 0 0]'. The net input at each time point, t, is:

```
>> net_input=[1 1 1 -1]' + W*state;
```

Therefore, at t=0, net_input(0)=[1 1 -1 -1]'. Using the net input, update the state of the network:

```
>> newstate=update_hp(W, state, net_input, p);
```

Repeat this multiple times (e.g., 1000 times) until the state of the system equilibrates. Now feed partial input test patterns, **Test**, and see where the network equilibrates:

```
>> Test = [1   1   0   0; 0   0   −1   −1]';
```

The zeros correspond to missing features in the test patterns. Does the network fill in the missing information?

### 36.3.3 The MATLAB Neural Network Toolbox

MathWorks has developed a specialized toolbox for neural networks. As we have shown previously, everything that constitutes a neural network (inputs, weights, transfer function, and outputs) can be implemented using matrices and matrix operations. However, the Neural Network Toolbox has a rich variety of different types of neural networks that can be easily implemented and have been optimized.

## 36.4 PROJECT

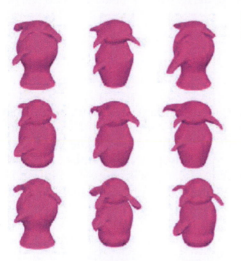

**FIGURE 36.8**  Meet the Greebles. *Image courtesy of Michael J. Tarr, Brown University, http://www.tarrlab.org/.*

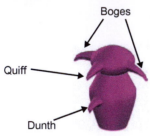

**FIGURE 36.9**  The anatomy of a Greeble. *Image courtesy of Michael J. Tarr, Brown University, http://www.tarrlab.org/.*

Greebles live in dangerous times (Gauthier and Tarr, 1997). Recent events led to the creation of the "Department for Greeble Security." You are a programmer for this recently established ministry, and your job is to write software that distinguishes the "good" Greebles from the "bad" Greebles (Figure 36.8). Researchers in another section of the department have shown that three parameters correlate with the tendency that a Greeble is good or bad. These parameters, identified in Figure 36.9, are: "boges" length, "quiff" width, and "dunth" height (Gauthier, Behrmann, and Tarr, 2004). Specifically, it has been shown that good Greebles have long boges, thin quiffs, and high dunths, while the bad Greebles tend to have short boges, thick quiffs, and low dunths. Of course, this relationship is far from perfect.

A given individual Greeble might have any number of variations of these parameters. In other words, this classification is not as clear-cut and easy as your superiors might want it to be. That's where you come in. You decide to solve this problem with a neural network, since you know that neural networks are well-suited for this kind of problem.

In this project, you will be asked to create two neural networks.

1. The first neural network will be a competitive learning network that distinguishes good Greebles from bad Greebles. Specifically, you should do the following:
   a. Train the network with the training set provided on the companion website (it contains data on Greebles who have been shown to be good or evil in the past, along with their parameters for boges length, quaff width, and dunth height). Plot the training data in three dimensions along with the two weight vectors associated with the good and evil output neurons using **quiver3**.
   b. Test the network with the test set provided on the companion website (it contains parameters on Greebles that were recently captured by the department and suspected of being bad—use your network to determine if they are more likely to be good or bad).
   c. Document these steps, but make sure to include a final report on the test set. Which Greebles do you (your network) recognize as being bad? Which do you recognize as being good?
   d. Qualitatively evaluate the confidence that you have in this classification. Include graphs and figures to this end.

Good luck! The future and welfare of the Greebles rests in your hands.

*Hints:*

- Load the two training populations using the command **xlsread('filename')**. Each file contains measurements of three parameters (in inches): boges length, quaff width, and dunth height. Each row represents an individual Greeble.
- Before you do anything else, you might want to plot your populations in a three-dimensional space (you have three parameters per individual). You can do this by using **plot3**(*param1,param2,param3*). In other respects, **plot3** works just like **plot**.
- Merge the data into a big training vector.
- Create the competitive network.
- Train the competitive network.
- Download the test files, and test the population with your trained network.
- Your program should produce a final list of which Greebles in the test population are good and which are bad. Also graph input weights before and after training.
- *Disclaimer: No actual Greebles were hurt when preparing this tutorial.*

2. The second neural network that you will create is a Hopfield network that will store the prototypical good and bad Greebles. Specifically, you should do the following:

   a. Normalize the features of all the Greebles so that the largest feature value across all Greebles for each of the three features is 1, and the lowest feature value is −1.

   b. Create the prototypical good and bad Greebles by taking the average features of the good and bad Greebles, respectively.

   c. Build a Hopfield network to store the good and bad prototypes (i.e., two feature vectors).

   d. Use the test set to see if the Hopfield network can categorize the suspected Greebles as prototypical good or bad Greebles. Compare these results with the results using the competitive learning network.

The equilibrium state of the Hopfield network should be one of these two vectors:

**[1   −1  1] or [−1  1  −1] for the good and bad Greebles.**

## MATLAB FUNCTIONS, COMMANDS, AND OPERATORS COVERED IN THIS CHAPTER

**rand ('state', number)**

**.**

**quiver**
**quiver3**
**plot3**
**xlsread()**
**diag**

# 37

# Neural Networks Part II: Supervised Learning

## 37.1 GOALS OF THIS CHAPTER

This chapter has two primary goals. The first goal is to be introduced to the concept of supervised learning and how it may relate to synaptic plasticity in the nervous system, particularly in the cerebellum. The second goal is to learn to implement single-layer (technically a two-layer network but the first layer is not always considered a true layer) and multi-layer neural network architectures using supervised learning rules to solve particular problems.

## 37.2 BACKGROUND

### 37.2.1 Single-Layer Supervised Networks

Historically, perceptrons were the first neural networks to be developed, and they happened to employ a *supervised learning rule*. A supervised learning rule is one in which there is a teaching signal that provides the goal of the learning process. Inspired by the latest neuroscience research of the day, McCulloch and Pitts (1943) suggested that neurons might be able to implement logical operations. Specifically, they proposed a neuron with two binary inputs (0 or 1), a summing operation in the inputs, a threshold that can be met or not, and a binary output (0 or 1). In this way, logical operators like AND or OR can be implemented by such a neuron (by either firing or not firing if a threshold is met or not). See Table 37.1 from an example of implementing the logical AND operator with a threshold value of 2.

Later, Frank Rosenblatt (1958) used the McCulloch and Pitts Model in combination with theoretical developments by Hebb to create the first perceptron. It is a modified

**TABLE 37.1**   Perceptron Implementing AND with a Threshold of 2

| Input 1 | Input 2 | Sum | Comparing with Threshold | Output |
|---------|---------|-----|--------------------------|--------|
| 0 | 0 | 0 | <2 | 0 |
| 0 | 1 | 1 | <2 | 0 |
| 1 | 0 | 1 | <2 | 0 |
| 1 | 1 | 2 | = 2 | 1 |

McCulloch and Pitts neuron, with an arbitrary number of weighted inputs. Moreover, the inputs can have any magnitude (not just binary), but the output of the neuron is 1 or 0 depending on whether the total weighted input exceeds the threshold or not. The weight can be different for each input. Setting the weights differently allows for more powerful computations.

Perceptrons are good at separating an input space into two parts (the output). Training a perceptron amounts to adjusting the weights and biases such that it rotates and shifts a line until the input space is properly partitioned. If the input space is higher than two-dimensions, the perceptron implements a hyperplane (with one dimension less than the input space) to partition it into two regions. If the network consists of multiple percep-trons, each can achieve one partition. The weights and biases are adjusted according to the perceptron learning rule:

1. If the output is correct, the weight vector associated with the neuron is not changed.
2. If the output is 0 and should have been 1, the input vector is added to the weight vector.
3. If the output is 1 and should have been 0, the input vector is subtracted from the weight vector.

It works by changing the weight vector to point more towards input vectors categorized as 1 and away from vectors categorized as 0.

Although a real neuron either fires an action potential or not (i.e., it generates a 1 or 0), the response of a neuron is often described by its firing rate (i.e., the number of spikes per unit time), resulting in a graded response (Adrian and Matthews, 1927). To model these responses, we use a different neural network model, called a linear network. The main dif-ference between linear networks and perceptrons lies in the nature of the transfer function. Where the perceptron uses a step function to map inputs to (binary) outputs, a linear net-work has a linear transfer function. The learning rule for the linear network is called the *Widrow-Hoff learning rule*, and is essentially the same as that for the perceptron (Widrow and Hoff, 1960). This rule attempts to minimize the sum squared error between the target, $T$, and the output neurons' (or, more properly, nodes') activations, $O$, by going down the gradient of the error surface in the multi-dimensional weight space. The *sum squared error* is defined as the squared difference between target and output values summed over all the output neurons. The gradient is the derivative of the error with respect to each weight:

$$Error = \sum_k (T_k - O_k)^2 \qquad (37.1)$$

$$\frac{\partial Error}{\partial w_{ij}} = \frac{1}{2}(T_j - O_j)\frac{\partial O_j}{\partial w_{ij}} \tag{37.2}$$

The output node's activation is equal to the net input, *net* (i.e., the weighted sum of the activations of the input neurons, *I*), because the transfer function is linear:

$$O_j = \sum_l w_{lj}I_l \equiv net_j \tag{37.3}$$

$$\frac{\partial O_j}{\partial w_{ij}} = I_i \tag{37.4}$$

$$\frac{\partial Error}{\partial w_{ij}} = \frac{1}{2}(T_j - O_j)I_i \tag{37.5}$$

$$\Delta w_{ij} = \epsilon(T_j - O_j)I_i \tag{37.6}$$

where $\Delta w_{ij}$ is the weight change between input node $i$ and output node $j$, $\varepsilon$ is the learning rate constant, $T_j$ is the target on node $j$, $O_j$ is the activation of output node $j$, and $I_i$ is the activation of input node $i$. Equation 37.6 is the Widrow-Hoff learning rule. Since the error equation above is quadratic, this error function will have one global minimum (if it has any). Hence, we can be assured that the Widrow-Hoff rule will find this minimum by gradually descending into it from the starting point (given by the initial input weights). In other words, we are moving into the minimum of an error surface. MATLAB® has a visual demonstration of that: **demolin1**.

If the output node's transfer function (i.e., activation function) is a differentiable, non-linear function, $f(x)$, the Widrow-Hoff learning rule can be adjusted by changing Equation 37.4:

$$O_j = f(\sum_l w_{lj}I_l) \tag{37.7}$$

$$\frac{\partial O_j}{\partial w_{ij}} = f'(net_j)I_i \tag{37.8}$$

where $f'(net_j)$ is the derivative of the non-linear transfer function evaluated at $net_j$.

$$\frac{\partial E}{\partial w_{ij}} = \frac{1}{2}f'(net_j)(T_j - O_j)I_i \tag{37.9}$$

$$\Delta w_{ij} = \epsilon f'(net_j)(T_j - O_j)I_i \tag{37.10}$$

Let's define the error as seen by output neuron $O_j$ as $\delta_j$:

$$\delta_j \equiv f'(net_j)(T_j - O_j) \tag{37.11}$$

## 37.2.2 Multilayer Supervised Networks

Since perceptrons are vaunted for their ability to implement and solve logical functions, it came as quite a shock when Minsky and Papert (1959) showed that a single layer (technically a two-layer network but the first layer is sometimes not considered a true layer) perceptron could not solve a rather elementary logical function: XOR (exclusive or; see Figure 37.1). This finding also implies that all similar networks (linear networks, etc.) can only solve linearly separable problems. These events caused a sharp decrease in the interest in neural networks until its resurgence during the 1980s.

The revived interest in neural networks occurred in part with the advent of multilayer, nonlinear networks with hidden units, and a learning rule used to train them called *backpropagation*, which is a generalized Widrow-Hoff rule. The power of these networks is that they can approximate any arbitrary nonlinear, differentiable function between the inputs and outputs.

The backpropagation (or backprop, for short) learning rule is a generalization of the Widrow-Hoff learning rule for multilayer, nonlinear networks that adjusts the weight between a pre- and postsynaptic node proportional to the product of the presynaptic activity and a measure of the error registered at the postsynaptic node. For nodes at the output layer, the error is straightforward. It is the difference between the target and the output node's activity multiplied by the derivative of the non-linear activation function as in the single-layer, nonlinear case (see Equation 37.11). But what is the error for a hidden node? By using the chain rule for differentiation, backprop can define such an error as follows:

$$\delta_j \equiv f'(net_j)\sum_k w_{jk}^O \delta_k \qquad (37.12)$$

To derive this, let's assume we have a network with four input nodes, three output nodes, and one hidden layer with two hidden nodes (Figure 37.2). Let's define the input, hidden, and output layers as $I$, $H$, and $O$, and we will use subscripts $m$, $l$, and $k$ to refer to particular nodes in each layer, respectively. Let's also designate the weights feeding into the hidden nodes as $w^H$, and the weights feeding into the output nodes as $w^O$.

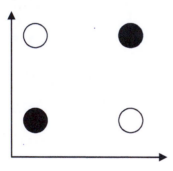

**FIGURE 37.1**    The XOR problem that a single layer network cannot solve. The XOR problem requires that the neuron respond (i.e., white circles) when only one (but not both) of the inputs is on. This is not solvable by a single-layer perceptron or linear network because it is not linearly separable.

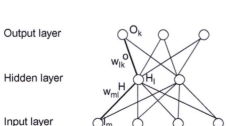

Output layer

Hidden layer

Input layer

**FIGURE 37.2**    A three-layer neural network with one hidden layer.

The partial derivative of the sum squared error with respect to a weight connecting an input node, $i$, to a hidden node, $j$, can be found by the following:

$$\frac{\partial Error}{\partial w_{ij}^H} = \frac{1}{2}\sum_k (T_k - O_k)\frac{\partial O_k}{\partial w_{ij}^H} \tag{37.13}$$

$$O_k = f(net_k) = f(\sum_l w_{lk}^O H_l) \tag{37.14}$$

$$H_l = f(net_l) = f(\sum_m w_{ml}^H I_m) \tag{37.15}$$

$$\frac{\partial O_k}{\partial w_{ij}^H} = f'(net_k)\frac{\partial net_k}{\partial w_{jk}^H} \tag{37.16}$$

$$\frac{\partial net_k}{\partial w_{ij}^H} = \sum_l w_{lk}^O \frac{\partial H_l}{\partial w_{ij}^H} \tag{37.17}$$

$$\frac{\partial H_l}{\partial w_{ij}^H} = f'(net_j)I_i \; if \; l = j, otherwise \; \frac{\partial H_l}{\partial w_{ij}^H} = 0 \tag{37.18}$$

$$\frac{\partial O_k}{\partial w_{ij}^H} = f'(net_k)w_{jk}^O f'(net_j)I_i \tag{37.19}$$

$$\frac{\partial Error}{\partial w_{ij}^H} = \frac{1}{2}\sum_k f'(net_k)(T_k - O_k)w_{jk}^O f'(net_j)I_i = \frac{1}{2}I_i f'(net_j)\sum_k w_{jk}^O f'(net_k)(T_k - O_k) \tag{37.20}$$

$$\Delta w_{ij}^H = \varepsilon I_i f'(net_j)\sum_k w_{jk}^O f'(net_k)(T_k - O_k) \tag{37.21}$$

$$\delta_j \equiv f'(net_j)\sum_k w_{jk}^O \delta_k \tag{37.22}$$

### 37.2.3 Supervised Learning in Neurobiology

Although there is no definitive evidence for neural plasticity that is guided by a "teaching" signal as supervised learning requires, there are some intriguing experimental findings which suggest the possibility that the physiology of the cerebellum may support a form of supervised plasticity. David Marr and James Albus independently proposed that the unique and regular anatomical architecture of the cerebellum could instantiate error-based plasticity that might underlie motor learning. In particular, they proposed that the climbing fibers acting on Purkinje cells from the inferior olive could provide an error signal to modify the synapses between the parallel fibers and the Purkinje cells. Experimental support for this theory first came from Ito and Kano (1982), who discovered that long-term depression could be induced at the synapse between the parallel fibers and the Purkinje cells. By electrically stimulating the parallel fibers and at the same time stimulating the climbing fibers (each of which forms strong synaptic contacts with a particular Purkinje cell), the parallel fiber synapse could be depressed. Because the Purkinje cells are inhibitory on the deep cerebellar nuclei, this synaptic depression would disinhibit the deep cerebellar nuclei.

A number of motor learning experiments have provided additional support for the idea that the cerebellum supports supervised learning via the climbing fibers. Sensory-motor adaptation experiments in which the gain between the movement and its sensory consequences are altered have shown transient increases in climbing fiber input (as measured by the complex spike rate) during learning (Ojakangas and Ebner, 1992, 1994). In addition, classical conditional experiments have suggested that the cerebellum plays a role in learning associations between *unconditioned stimuli* (US) and *conditioned stimuli* (CS) (Medina et al., 2000). For example, learning the association between an air puff (US) generating an eye-blink response and a tone (CS) is disrupted by reversible inactivation to parts of the cerebellum (Krupa, Thompson, and Thompson, 1993). Moreover, it has been shown that the tone enters the cerebellar cortex via the parallel fiber pathway, whereas the air puff (acting like a teacher) enters through the climbing fiber input. In fact, a mathematical formulation of classical conditioning proposed by Rescorla and Wagner (1972) quite closely resembles the Widrow-Hoff error correction learning rule using in supervised neural networks.

## 37.3 EXERCISES

### 37.3.1 Perceptrons

Start by creating a perceptron with two input nodes and one output node by creating a $1 \times 2$ weight matrix $W$ (i.e., 1 output and 2 inputs) initialized to zero, a bias weight $b$ initialized to zero, and a thresholding transfer function for the output unit:

```
>> W = zeros(1,2);
>> b = zeros(1,1);
>> net_inp = W*inp + b;
>> out = sign(sign(net_inp) + 1);
```

If the net input to the output node is larger than 0, the output will be 1; otherwise it will be 0. A bias weight can be thought of as connecting an additional input node whose value is always 1 to the output node. The bias weight changes the point where this decision is made. For example, if the bias is −5, the net sum has to exceed 5 in order for the output to yield 1 (the bias of −5 is subtracted from the net sum; if it is less than 5, it will fall below 0).

Now set the weights from the two input nodes to 1 and −1, respectively. Set the bias weight to zero.

If the sum of the input multiplied by the weight meets or exceeds 0, the network should return 1; otherwise it should return 0. Test it by feeding it some inputs:

**inp1 = [1; 0.5]  %Note: The first, positive weighted input larger than the second**
**inp2 = [0.5; 1]  %Note: The second, negative weighted input is larger than the first**

Your output should be 1 for **inp1** and 0 for **inp2**. The network classified these inputs correctly. You can now feed the network a large number of random numbers and see if they are classified correctly, like this:

**a = rand(2,10)  %Create 20 random numbers, arranged as 2 rows, 10 columns**

If the value in the first input row is larger than the value in the second input row (for a given column), the output value (for that column) should be 1; otherwise it should be 0.

---

### EXERCISE 1

Adjust the bias to some arbitrary value and see how the input/output mapping changes.

---

By adjusting the weights and the bias, it can be shown that any linearly separable problem (a problem space that can be separated by a line or more generally by a hyperplane) can be solved by a perceptron. To verify if this is the case, play around with the interactive perceptron, where you can arbitrarily set the decision boundary yourself. Type **nnd4db** (Figure 37.3). Try to separate the white and the black circles. They represent the inputs. The output is represented by the black line (creating two regions; presumably one region corresponding to an output of zero and another region corresponding to an output of one).

What if you don't know the weights or don't want to find the weights? What if you only know the problem? Luckily, one of the strongest functions of neural networks is their ability to learn—to solve problems like this on their own. We will do this now. The first thing we need is a learning rule, a rule that tells us how to update the weights (and biases), given a certain existing relationship between input and output. The perceptron learning rule is an instance of supervised learning, giving the network pairs of inputs and

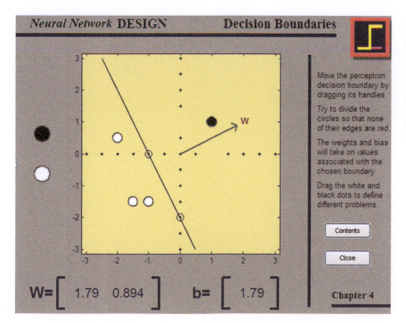

**FIGURE 37.3** An interactive display of the decision boundary of a perceptron provided in the MATLAB Neural Network Toolbox. The display shows how a perceptron creates a linear decision boundary whose slope and intercept can be modified by adjusting the weight vector and bias.

desired (correct) outputs. The perceptron learning rule is essentially equivalent to the Widrow-Hoff learning rule:

```
>> function [Wout bout] = train_perceptron(W,b,inp,out,targ)
>> Wout = W + (targ-out)*inp';
>> bout = b + (targ-out);
>> end
```

To test this, set the weight matrix and bias weight back to zero. Use the random numbers you created as inputs and the correct answer as targets and train the network. You can try this dynamically, in an interactive demo, by typing **nnd4pr** (see Figure 37.4).

### 37.3.2 Linear Networks

Now you will create a linear network with two input nodes and one output node. The only difference between a linear network and a perceptron is the transfer function of the output node. The output node's activation is simply equal to the net input:

**out = net_inp**

Set the weights to 3 and 4 and bias to 0, and feed the network the following input:

**inp = [5; 7];**

The same can be done by typing:

```
>> [3 4] * [5;7]
```

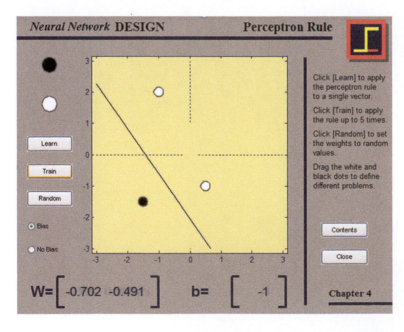

**FIGURE 37.4** An interactive display of the perceptron learning rule provided in the MATLAB Neural Network Toolbox.

In other words, this neural network implements an inner product (dot product).

Of course, we want a more useful neural net than just one that can take the dot product; we can take the dot product without neural nets. A classical function of linear neural networks is the automatic classification of input objects into different categories. In order to achieve that, we need to train the network. To do this, we will use the Widrow-Hoff learning rule again. However, this time we will feed the inputs and targets multiple times (i.e., through multiple epochs), each time incrementing the weights by a small amount scaled by a learning rate parameter $\varepsilon$:

```
>> Wout = W + ε*(targ-out)*inp';
>> bout = b + ε*(targ-out);
```

To assess the quality of learning, we will measure the total sum squared error between the output of the network and the target summed over all output units $k$ and all training samples $p$ after each epoch of learning:

$$Total\ Error = \sum_p \sum_k (T_k^p - O_k^p)^2$$

Why not give it a try? Say we have six two-dimensional inputs from two sets: **set1** and **set2**.

```
>> set1 = [2, 2; -2, 2 ; 0.5, 1.5];
>> set2 = [1, − 2; − 1, 1; − 0.5 − 0.5];
```

Plot them to see what is going on:

```
figure %Opening a new figure
>> plot(set1(:,1),set1(:,2),'*') %Plotting the first set
hold on; %Holding on
>> plot(set2(:,1),set2(:,2),'*', 'color', 'r')
%Plotting the second set in red
>> axis([ − 2.5 2.5 − 2.5 2.5])
>> axis square
```

Looking at the graph, you can see what is going on (Figure 37.5). We also see that the problem is, in principle, solvable; we can draw a line that separates the blue and the red stars. Now create a network that will find this solution, starting at 0 weights and 0 bias.

Next, assign the corresponding targets. We arbitrarily assign 1 to the blue set and 0 to the red set:

```
targets = [1  1  1  0  0  0]
```

Finally, you have to decide on the learning rate, $\varepsilon$, and the number of times you run through the entire training data set (i.e., the number of epochs). Set the learning rate to 0.01 and the number of epochs to 100 (Figure 37.6).

Challenge the trained network with some new input, and see if it correctly classifies it. Pick something in the middle of the red range, like [ −1, 0]. You should get 0.20 as an output. Of course, this should be 0. But this might be the best we can do, given the sparse input.

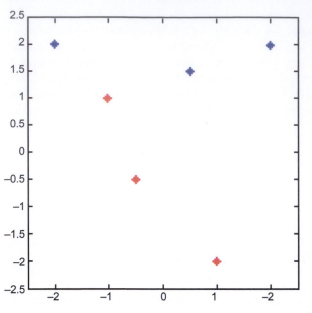

FIGURE 37.5 A plot of six two-dimensional inputs (blue and red stars) that are to be classified.

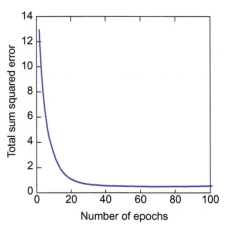

FIGURE 37.6 The sum square error as a function of training epochs.

---

**EXERCISE 37.2**

Explore some points in the space and see what their output is. Is it about what you expect? Does the network make gross errors? What about the initial input vector?

---

A linear classifier allows one to roughly categorize and classify inputs if they are linearly separable. Adjusting the weights and biases amounts to creating a linear transfer function that separates the desired outputs maximally and optimally. Of course, there is only so much that a linear function can do, but it's not too bad.

---

**EXERCISE 37.3**

Use a linear neural network to classify the "good" and "bad" Greebles used in Chapter 36. Initially, set the weights to 1 and the bias to zero.

---

Note that there is some initial classification even without training (Figure 37.7, left side). But it is not very good, and the absolute output values are all over the place. The final classification by the linear network is pretty good (Figure 37.7, right side). Values cluster around 1 and 0, although there is quite a bit of variance. The linear network couldn't separate the clusters any better than this, given the variance in the input and the amount of training. You should always check your network by visualizing the outputs of trained and untrained networks with simple plots like this.

### 37.3.3 Backpropagation

Most of the principles of backpropagation are the same as in the other networks, but we now have to specify the number of layers, the number of input nodes, hidden nodes per hidden layer, and output nodes, and nonlinear transfer functions of the hidden and output nodes. For example, create a three-layer, feedforward network (i.e., one input layer, one hidden layer, and one output layer) with a sigmoidal transfer function:

```
>> Wh = rand(NHIDDEN,      % weight matrix feeding hidden nodes
NINP);
>> Wo = rand(NOUT,         % weight matrix feeding output nodes
NHIDDEN);
>> bh = zeros(NHIDDEN,1);  % bias weights feeding hidden nodes
>> bo = zeros(NOUT,1);     % bias weights feeding output nodes
>> transfer_fn = @(x,alpha) 1./(1 + exp(alpha*x));
```

The last line of code defines a function handle called **transfer_fn** to a sigmoid function whose output ranges from 0 to 1. The steepness of the sigmoid is specified by **alpha**.

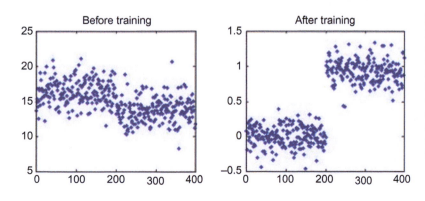

**FIGURE 37.7** The problem of categorizing "good" and "bad" Greebles using a supervised linear network (see Chapter 36). Left: Classification when all the weights are set to 1 (before training). Right: Classification after application of the supervised learning rule (after training).

The backprop learning rule applied to the weights feeding the output nodes is identical to the non-linear Widrow-Hoff learning rule (see Equation 37.10). However, for the weights feeding the hidden units, the rule is specified in Equation 37.21.

---

### EXERCISE 37.4

Solve the Greeble problem with a multi-layer feedforward network using a sigmoid transfer function and the backpropagation learning rule. What are the results?

---

### SUGGESTION FOR EXPLORATION

Try Exercise 37.4 using different transfer functions between the layers. See what can be done. Or wait for the final project to do this.

---

## 37.3.4 Sound Manipulation in MATLAB

As you saw in earlier chapters, MATLAB is not only useful for analyzing data; it is also possible to use it for experimental control and data gathering. Here, you will see that MATLAB can be used to design the stimulus material itself. Before you start, check the volume control to make sure that the loudspeaker of your PC is not muted and that the volume is turned up. There are many MATLAB functions dealing with auditory information; as a matter of fact, there is a whole toolbox devoted to it. Here, we will only handle sound in very fundamental ways.

The first thing you might want to do is to create your sound stimuli. To do so, try this:

```
>> x = 0:0.1:100;
>> y = sin(x);
>> sound(y);
```

Did you hear anything? What about this:

```
>> y = sin(2 * x);
>> sound(y);
>> y = sin(4 * x);
>> sound(y);
```

You know that your code created sine functions of increasing frequency. What you are listening to is the auditory representation of these sine functions, pure sine waves. Of course, most acoustic signals are not that pure. Try this to listen to the sound of randomness, white noise:

```
>> x = randn(1,10001);
>> sound(x)
```

The **sound** function interprets the entries in an array (here, the array "x") as amplitude values, and plays them as sound via your speakers. That means you can manipulate psychological qualities of the sound by MATLAB operations. You already saw how to manipulate pitch (by manipulating the frequency). You can manipulate volume by changing the magnitude of the values in the array. **Sound** expects values in the range 1 to −1. So how does this sound?

```
x = x / 5 ;
sound(x)
```

This should sound much less violent.

Of course, you can manipulate the sound in any way you want; for example, you can mix two signals:

```
>> z = x + y;
>> sound(z)
```

This should sound like the sine wave from before, plus noise.

---

### EXERCISE 37.6

What happens if you add two different frequencies of sine waves and play it?

---

In short, what you hear should sound more complex, and rightfully so. It can be shown that any arbitrarily complex sound pattern (or any signal, really) can be constructed by appropriately adding sine waves. We will use this property later. Speaking of complex sounds, most practical applications will require you to deal with sounds that are much more complex than pure sine waves. So let's look at one. Luckily, MATLAB comes with a complex sound bite:

```
>> load handel
>> sound(y, Fs);
```

You might be confused by the second parameter, **Fs**. It is the sampling rate at which the signal was sampled. It specifies how many amplitude values are played per second. If your array has 10,000 elements and the sampling rate is 10,000, it takes one second to play it as sound.

---

### EXERCISE 37.7

Play it again at a higher/lower sampling rate. What is happening?

---

Why don't we take a look at the structure of the amplitude values in the $y$ matrix? This could help you understand how **sound** works.

```
>> figure
>> plot(y)
```

This will do the trick, and it should look something like Figure 37.8.

If you already listened to it, this will probably not surprise you. As a matter of fact, this information about sound amplitudes is not very powerful in itself. It is much better and more useful to look at the spectral power of a signal over time. To do this, we will use a function called **spectrogram**. It comes with the MATLAB Signal Processing Toolbox.

We won't go into how exactly this function works. It is rather complex, and we could spend a whole chapter on it alone. In principle, let's say that it decomposes the signal into sine waves and plots the power (how much of each frequency is in the signal) over time. A spectrogram of the Handel sounds looks like Figure 37.9.

```
>> spectrogram (y, 256, 'yaxis')
```

The parameter, 256, breaks up the data into 256 segments and applies a Hamming window to each segment. 'yaxis' specifies that the frequency axis should be plotted on the y-axis. Of course, you probably will want to import your own sounds into MATLAB. That can be done using the **wavread** function. Load the appropriate files that we have created, then type:

```
>> y = wavread('Kira1.wav');
```

---

### EXERCISE 37.8

What is the person saying? *Hint*: The signal was sampled at 22050 Hz. You might want to take that into account when playing it. Look at the spectrogram, too.

---

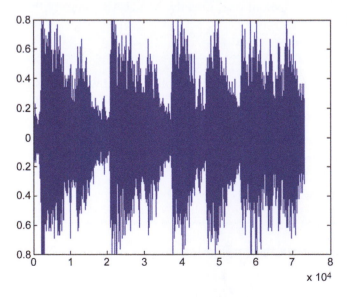

FIGURE 37.8　The raw acoustic pressure amplitude of the Handel sound bite.

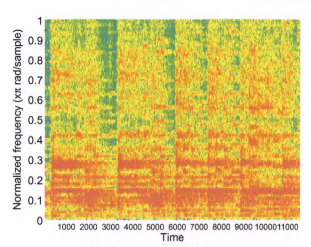

FIGURE 37.9 The spectrogram representation of the Handel sound bite.

---

## SUGGESTION FOR EXPLORATION

Listen to your data, literally. Listen to some of the material you created in previous chapters. How does it sound?

---

In this section, you saw how MATLAB lets you create, manipulate, and analyze acoustic stimuli. You can use it to design sound stimuli that are precisely timed and have very specific properties. Obviously, this is extremely useful for acoustic experiments.

## 37.4 PROJECT

The project is to create a network that correctly classifies the gender of two target speech bites. In order to do this, train the network with a total of six speech bites of both genders. Load these files (called Train_Kira1 to 3 and Train_Pascal1 to 3). Of course, the two speakers differ in all kinds of ways other than just gender (age, race, English as a first/second language, idiosyncratic speech characteristics, lifestyle, etc.). If one were to face this problem in real life, one would have to train the network with a large number of speakers from both genders so that the network can extract gender information abstract from all these irrelevant dimensions. But for our purposes, this will be fine. Specifically, you should do the following:

Create a network that reliably distinguishes the gender of the two target speech bites. Provide some evidence that this is the case and submit the source code.

*Hints:*

- You should implement a network using the backpropagation learning rule.

- Take the spectrogram of the sound files. Use those as the inputs to your neural network. See Figure 37.10 for the amplitude and spectrogram of the sentence "The dog jumped over the fence." The left subplot is a female speaker, and the right subplot is a male speaker.
- **a = spectrogram(b, 256)** will return an array of complex numbers in a. They have an imaginary and a real part. For our purposes, the real part will do. For this, type **c = real(a)**. c will now contain the real part of the complex numbers in a.
- This project is deliberately under-constrained. Basically, you can try whatever you want to solve the problem. As a matter of fact, we encourage this, since it will help you understand neural networks that much better. Don't be frustrated, and don't panic if you can't figure it out right away.
- If you take a spectrogram of the speech patterns with which you are supposed to train the network, it will return 129 rows and multiple columns. A neural network that takes the entire information from this matrix has to have 129 inputs.
- This is obviously rather excessive. If you closely observe the spectrogram (see Figure 37.10, right), you might be able to get away with less. The rows tessellate the frequency spectrum (y axis). Power in a particular frequency band is at a particular row. Power in a particular frequency band at a particular time is in a particular combination of row and column.
- The point is that the left spectrogram has much more power in the upper frequencies than the one on the right. If you properly combine frequency bands (or sample them), you might be able to get away with a neural network that has only five or ten inputs.
- The same applies for time. Your spectrogram will have several thousand columns. This kind of resolution is not necessary to get the job done. Try combining (averaging) the values in 100 or so columns into one. Then feed that to the neural network for training.
- Try using one hidden layer.
- Try having a large number of hidden units (definitely more than 1).

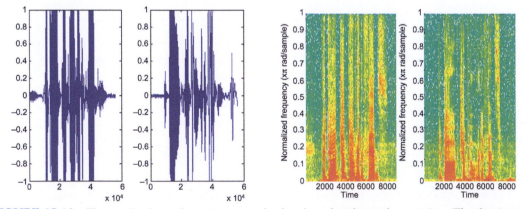

FIGURE 37.10   The amplitudes and spectrograms of a female and male speaker uttering "The dog jumped over the fence."

- You will be using a supervised learning rule. The target is defined by which file the data came from (Pascal or Kira). Create an artificial index, assigning 1 and 0 or 1 and 2 to each.
- Always remember what the rows and columns of the variables you are using represent. Be aware of transformations in dimensions.

## MATLAB FUNCTIONS, COMMANDS, AND OPERATORS COVERED IN THIS CHAPTER

**nnd4db**
**nnd4pr**
**demolin1**
**real**
**sound**
**spectrogram**
**wavread**
**@**

# Creating Publication-Quality Figures

## A.1  INTRODUCTION

While we made a great many figures in the course of working through this book, their appearance was rarely the focus of our discussion. However, as the theoretical neuroscientist Konrad Körding once observed: "Each figure deserves to look good." We agree with this sentiment, and this appendix is written in the spirit of helping you to achieve just that.

Of course, if you do make figures for publication purposes, you should inform yourself about and follow the requirements specified by the journal publishing your work.

Similarly, it is no secret that holy wars are fought about issues of design and style of figures in the scientific community. Some people feel incredibly strongly about buzzwords like "data ink" (Tufte's notion that a figure should maximize the amount of data ink—ink that represents data—and minimize all other ink, as it distracts from data ink, broadly speaking—see http://www-personal.umich.edu/~jpboyd/eng403_chap2_tuftegospel.pdf and Tufte and Graves-Morris, 1983.). One of these controversies revolves around the use of bar graphs. Purists suggest that one should never use them, as the vertical part of the bar (the majority of the figure) represents non-data ink, and only the horizontal piece on top of the bar graph actually represents data. Others point out that using bar graphs hooks into pre-existing cognitive architectures; everyone knows what a bar graph is and how to interpret it. So why not use it?

We do not want to get involved in these controversies about the purity of doctrine. If you do, there is plenty of reading material out there. Rather, we want to provide pragmatic and sensible guidelines on how to make figures that look good, and provide information on how to implement these suggestions (and that's all they are) within MATLAB®.

## A.2  FIGURE MAKEOVERS

Before we get started, we need some data to put into a figure. As usual, if we are in need for a quick data fix, we'll use trigonometric functions. Pretend these data represent the outcomes of two experimental conditions, coarsely sampled, with little to no noise.

Data:

```
>> x = 0:0.2:10;
>> cond1 = sin(x);
>> cond2 = 2.*sin(x);
```

Plotting:

```
>> figure
>> plot(x,cond1)
>> hold on
>> plot(x,cond2)
```

So much for the "before" plot. It does represent the data, but we can do much better than that (see Figure A.1).

---

## SUGGESTION

Don't let any part of the axes obscure or interfere with the representation of the data.

---

As you can see in Figure A.1, the tick marks of the axes are pointed inward, potentially interfering with the representation of the data, e.g., around x = 8 and y = −0.5 and −1. There are several ways to fix this:

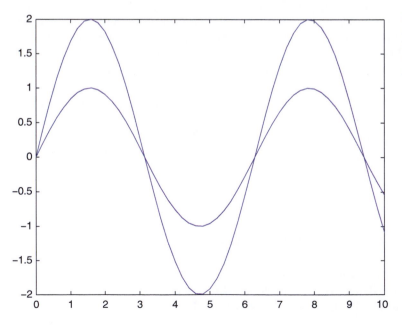

FIGURE A.1 The "before" figure. Note that some of the data is obscured by tick marks. From a data-ink perspective, this is entirely unacceptable.

Use outward-facing tick marks:

>> **set(gca,'TickDir','out')**

Use shorter tick marks (the vector sets the length):

>> **set(gca,'ticklength',[0.005 0.025])**

Make sure the axes are plotted below the data:

>> **set(gca,'layer','bottom')**

The figure should now look something like Figure A.2.

Voila. No more interfering of axes tick marks with data ink. But speaking of data ink, the upper x-axis and right y-axis now really look ridiculous and superfluous. In the spirit of maximizing data ink, we can get rid of them, like so:

>> **box off**

---

### SUGGESTION

Do get rid of superfluous lines in the figure that only detract from the interpretation of the data ("maximizing data ink," sensu Tufte).

---

In this spirit, do use the function **grid** sparingly. Go ahead, try it. Simply typing "grid" toggles a grid that overlays the data on and off. There are cases where the use of a grid

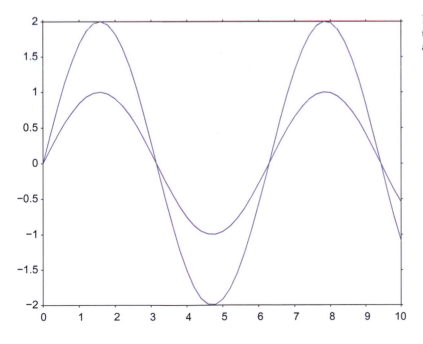

FIGURE A.2 We started the makeover. The tick marks are now facing outwards.

helps with the interpretation of data, but these cases are rare. For the most part, grids are a holdover from the days when people drew figures on grid or graph paper. Today, MATLAB is doing the drawing, so in most cases, it just increases the non-data ink in a figure. Be careful when using it.

While we are at it (and this will depend strongly on the nature of the figure) we could equalize the aspect ratio of the figure. Put differently, if one has a long time series, one might want a figure that is more wide than high. However, in most other figures this is inappropriate, as it can lead to cognitive distortions. For instance, if one makes a scatterplot of measurements before and after an intervention (same units), it is crucial to give the x- and y-axes the same visual weight. In the case of our example figure, it is not strictly necessary, but we'll do it anyway to show how this plays out (see Figure A.3):

>> **axis square**

---

**SUGGESTION**

This one is more than a suggestion. Absolutely make sure that the font properties of your axis labels match the rest of the text in your manuscript. The meaning of a figure is impossible to determine without looking at the figure labels. Thus, there is no point in creating figures where the axis labels are impossible to make out because they are too small. Similarly, they should not clash with the manuscript font. A popular choice is Helvetica oblique, used with the font size of the manuscript text.

---

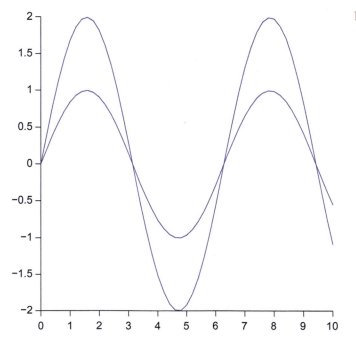

FIGURE A.3   Now with a squared axis.

To implement this suggestion, type

```
>> set(gca,'FontSize',18);
>> set(gca,'FontAngle','italic');
>> set(gca,'FontName','Helvetica');
```

See Figure A.4 for the result.

Getting there. Of course, the visual weight of the axes' tick labels is now way out of proportion with the representation of the actual data.

---

### SUGGESTION

Make sure that data and axes are visually in balance.

---

To restore this balance, we need to scale up the representation of the data (see Figure A.5):

```
>> h1 = plot(x,cond1);
>> h2 = plot(x,cond2);
>> set(h1,'linewidth',2);
>> set(h2,'linewidth',2);
>> set(h1,'marker','.');
>> set(h2,'marker','.');
>> set(h2,'color','r');
>> set(h1,'markersize',25);
>> set(h2,'markersize',25);
```

This will do as an "after" image (not to be confused with an afterimage). If you do want to use a figure like this in a publication, make sure to add suitable axis labels; we left them off here because they are arbitrary labels from made-up data.

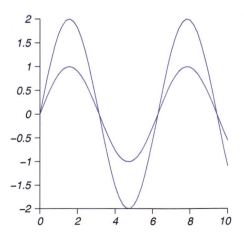

FIGURE A.4   Don't these tick labels look nice now? Note that for a real figure, the axes (and thus the figure) would only be meaningful with specified axis labels. We deliberately left them off here, as we just pulled the data out of a hat (or rather, out of Matlab). But you could imagine that the x-axis represents—for instance, time—and the y-axis voltage. Or something along those lines.

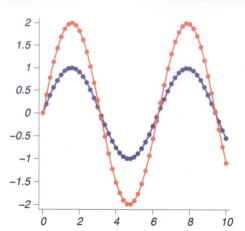

**FIGURE A.5** The "after" image. Compare it with the "before" image to appreciate the power of properly setting figure properties. In many cases, it is worth it. For publications, it is almost always worth it.

---

### SUGGESTION

Don't make a figure too busy. In our experience, it is preferable if each figure makes one easily discernible point. If necessary, use two panels instead. However, if you do this, make sure to use the same axis limits in both panels. Otherwise, they become very hard to compare. This is a common mistake. In our case, you could set the y-axis limits for both subpanels as

>> ylim([min([cond1 cond2]) max ([cond1 cond2])])

---

### EXERCISE A.1

Give your figure a makeover, using the principles outlined in this appendix. Use some of the pre-existing data or figures from the previous chapters to do it. Have your pick.

---

## A.3  SAVING FIGURES IN THE DESIRED FORMAT

Many journals do not accept figures as jpegs or bitmaps. This is understandable. Vector graphics formats scale easily, whereas this is not the case for rasterized images. This can cause problems, as figures might have to get scaled up or down in the printing process.

The key function to use is **print**. It governs the output of a figure both to hardcopy (a printer) and to files. It is quite versatile, but for our purposes, we will focus on saving as PostScript and TIFF files.

For instance, the command

>> **print ('-depsc', '-r300', 'Figure 1')**

saves the current figure as a color Encapsulated PostScript (EPS) file with a resolution of 300 dpi (check with the editors for what is acceptable) and a filename of "Figure 1" in the current folder. The following command:

>> **print ('-dtiff', '-r300', 'Figure 1')**

does the same thing, but the figure is saved as a TIFF file.

## A.4  HOW TO MAKE ANIMATED GIFS

While the controversy on how to pronounce GIF will probably never be settled in a way that makes everyone happy, we will show you here how to make one with MATLAB. The point of this exercise is that animated GIFs are making a comeback. Some journals— for instance, the Journal of Vision—now allow authors to upload an animated GIF that summarizes their experiment. As there are more and more online journals, it is not hard to see how animated figures that illustrate complex experimental procedures could become much more common.

Let's make a simple one, to illustrate the concept. We'll use a quick and dirty sine wave that expands over time. Perhaps this could serve as an illustration of wave propagation or some such. Motion captures attention; why not use it?

```
>> Y = cell(100,1);
>> FR = cell(100,1);
>> for ii = 1:100
>> x = 0:1:ii;
>> Y{ii} = sin(x);
>> end


>> figure
>> xlim([0 100])
>> hold on
>> for ii = 1:100
>> plot(Y{ii})
>> FR{ii} = getframe;
>> pause(0.1)
>> end
```

**getframe** takes a snapshot of the current axes. We put it into the frame repository FR, a cell. These frames can be played back with the command **movie(FR)** within MATLAB. The pause in the previous code is not necessary; it just allows you to see what is going on.

If this is the desired animation (in our case, it is), we can now proceed to convert and save it as an animated GIF:

```
>> for ii = 1:100
>> img = frame2im(FR{ii});
>> [ndx, cmp] = rgb2ind(img,256);
>> if ii == 1
>> imwrite(ndx,cmp,'waveform','gif', 'Loopcount', inf, 'DelayTime', 0.1);
>> else
>> imwrite(ndx,cmp,'waveform','gif', 'WriteMode', 'append', 'DelayTime', 0.1);
>> end
>> end
```

There is a lot going on here. Let's unpack it. First, we open a loop that goes through all 100 frames. Then, we extract the image data from the frames by using the function **frame2im**. Then, we convert the RGB image into an indexed image by using the **rgb2ind** function. Indexing saves a considerable amount of space. The GIF itself is opened if we are in the first frame. A "Loopcount" of **inf** sets the GIF to repeat endlessly. You can also give a finite value as to the number of animation repeats before it should stop. "Delaytime" specifies the delay in seconds before displaying the next image. By setting the "Writemode" to append for all other frame numbers above 1, we add the frames to the existing file. And that's it. The function **imwrite** can be used to write all kinds of other image formats, such as JPG or BMP, as well.

## MATLAB FUNCTIONS, COMMANDS, AND OPERATORS COVERED IN THIS APPENDIX

**grid**
**imwrite**
**print**
**getframe**
**rgb2ind**
**frame2im**

# B

# Relevant Toolboxes

In this appendix, we will give a *brief* overview of some toolboxes that are highly relevant to research in the neural sciences.

## B.1  THE CONCEPT OF TOOLBOXES

One of the great strengths of MATLAB® is that in addition to a large (and growing) number of general purpose functions, MathWorks also sells kits of powerful and highly optimized functions for special purposes. One such kit is called a toolbox. There are toolboxes available for many different fields. For instance, there is a Financial Instruments Toolbox that allows you to design complex financial products. Be very careful when using that one. You wouldn't inadvertently want to cause the downfall of western civilization as we know it. In general, the number of toolboxes is large, growing, and their content is constantly changing. Thus, we will only be able to give a brief overview of some of the most relevant ones here. Yet, if you want to use toolboxes, it is important to keep an open mind. Almost all toolboxes can be useful to neuroscience researchers in some way, e.g. the curve fitting toolbox.

Also, it should be noted that some people oppose the use of toolboxes on philosophical grounds. They cite, for instance, the issue that using a toolbox spoils the purity of MATLAB and makes the user dependent on a sophisticated black box that he doesn't really understand. Others object to the high cost of the individual toolboxes. There is no question that you will understand something better if you code it yourself instead of buying an off-the-shelf solution. This is a balance that you have to strike yourself, and it will largely depend on long term goals versus immediate needs. We advise you to be pragmatic, rather than dogmatic, non-dogmatically.

## B.2  NEURAL NETWORK TOOLBOX

You already encountered neural networks in Chapters 36 and 37 of this book. While we tried to avoid the use of the Neural Network Toolbox in Chapter 36, you did see its utility in Chapter 37. The Neural Network Toolbox features interactive demos of neural network principles, e.g., the perceptron rule (see Figure B.1).

The Neural Network Toolbox also features a large number of functions to implement large and complex neural networks. In principle, you can quickly redo your work from these chapters with these functions. For instance, **newc** creates a competitive network, and allows you to specify the number of input elements and layers. The function **train** starts the training of the neural network, and **sim** classifies the input with the specified network. **newhop** creates a recurrent Hopfield network. If you have a lot of work with complex neural networks, you might want to use the toolbox; but if you worked through Chapter 36, you may appreciate why it is better to code it yourself than to rely on an opaque toolbox.

## B.3 PARALLEL COMPUTING TOOLBOX

This toolbox was developed relatively recently, but will probably play a larger and larger role in coming years as computers with more processing cores proliferate. The advent of multicore processors allows users to run several processing threads in parallel. For some computations, this can speed up time to completion dramatically. For instance, if you have a **for** loop that takes a long time to execute, most of your cores will be idle until an iteration of this loop is complete and another can be started. The Parallel Computing Toolbox introduces the function **parfor**. This is a parallel for loop. As long as iterations of the loop are independent of the outcome of previous iterations, each core of your processor can work on one iteration in parallel. Before you can use **parfor** (or other parallel

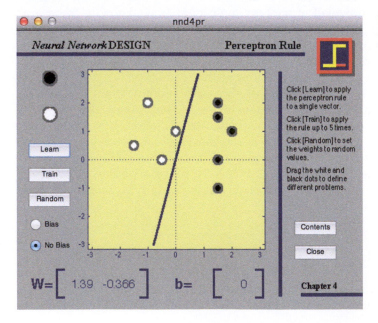

**FIGURE B.1** The perceptron rule demoed with the Neural Network Toolbox.

processing commands), you must start a "pool" of "workers", corresponding to the number of cores you want to use. For instance, the syntax

>> **matlabpool open 3**

opens a worker pool with three parallel workers to execute parallelized code.

If you do use a lot of **for** loops in your code that are, in principle, independent of each other, you might want to look into parallelizing them. It might speed things up a lot.

## B.4 STATISTICS TOOLBOX

MATLAB comes with a number of statistical functions that are sufficient for most basic statistical work, such as **corrcoef**, **mean**, or **std** to calculate the Pearson correlation coefficient, the mean, or the standard deviation of data, respectively. While these are sufficient for many—if not most—purposes, there will come a time when you need to use more advanced statistical functions. Instead of writing them yourself (which you are always welcome to do), you could use the Statistics Toolbox. It contains a large number of statistical functions that are implemented efficiently; for instance, **geomean** returns the geometric mean, **corr** allows you to compute different flavors of rank correlations, and **kurtosis**, aptly, calculates the kurtosis of a distribution. It is trivial to write a bootstrap function, but this toolbox comes with a ready-made one, **bootstrp**. It also includes a fairly large number of probability density functions, e.g., **gampdf** for the gamma distribution. The number of included functions is really quite staggering and there is no way we can cover them here. If you do a lot of statistical computations and don't want to rewrite all of these functions yourself, you might want to consider this toolbox.

## B.5 MATLAB COMPILER

As we covered previously in this book, MATLAB is an interpreted language. The code is interpreted and executed line by line as the script is run. There are many reasons why you might not want to do things that way at some point. For instance, you might want to deploy your data gathering program to machines that don't even have MATLAB installed. For this purpose, MathWorks created the compiler. It compiles your code and builds a standalone executable. This file can then be run on any machine. The key function here is **mbuild**. The first time you run the compiler, you have to run >> **mbuild −setup** to select a compiler. **mcc** invokes the MATLAB compiler, and **deploytool** allows you to do all of this within a GUI. It is worth noting that compiled code usually also executes faster than interpreted code, generally speaking.

## B.6 DATABASE TOOLBOX

If you work with very large and complex datasets, you might want to consider storing them in a relational database, such as MySQL. As a matter of fact, any ODBC/JDBC compliant database will do. Of course, once your data is in this format, the question is how you interface with it via MATLAB. That's where the Database Toolbox comes in. It provides the functionality that allows you to execute MySQL queries in MATLAB, and retrieve them from the database for further processing.

The crucial function in this context is the **database** function. It allows you to create a connection to an existing database, like this:

> `>> conn=database('database_name ', 'username', 'password ')`

This creates a connection object named "conn," which allows you to access the specified database, providing these credentials. Once **conn** exists, it can also be pinged, e.g., by typing

> `>> ping(conn)`

To actually execute a query, you can then type — for instance —

> `>> curs = exec(conn, 'select x from y;')`

which uses the connection object **conn** we just created to query table y in the connected database for x, and puts the whole thing in a "data cursor" variable, which you can then further process in MATLAB. Or you can type

> `>> curs = fetch(conn, 'select x from y;',1000)`

to import data from the database to a structure "curs", which will contain the results of the query. If you retrieve numeric data, you can set the database preferences by typing

> `>> setdbprefs('DataReturnFormat','numeric');`

Executing the command above now returns the data directly to a Matlab array.

You can also skip the command line and use **querybuilder** to access a visual query builder GUI.

If you do have data in a ODBC database and want to use MATLAB to work with it, it is probably prudent to use this toolbox.

## B.7 SIGNAL PROCESSING TOOLBOX

We have already done a lot of signal processing in this book, particularly in Chapters 11–13. Once you know about the Fourier transform, you can — in principle — build most of the signal processing functions yourself. The toolbox provides a large number of signal processing functions that are highly optimized and versatile. For instance, unless you are an expert (or if you want to write one yourself for educational purposes), it will be hard to improve on the **spectrogram** function that comes with this toolbox. Specifically, it provides a large number of industrial strength functions to design, analyze, and implement digital and analog filters. It provides all commonly used window shapes, including the notorious **kaiser** window. Special purpose statistical functions, e.g., those to do cross-correlations, are also included—for instance, **xcorr**. There are also all the usual waveform generations functions, e.g., **sinc** to create a sinc function. Yes, you can write all of these yourself, and many a self-respecting signal processing scientist has done just that. If you don't want to do that, it is neat to have all these little functions (that you do need all the time) ready-made and available. And they come with this toolbox.

## B.8 DATA ACQUISITION TOOLBOX

If you want to collect data via MATLAB, this toolbox might come in handy. It allows you to directly read in signals from devices connected to the computer via a NIDAQ card and define data sources. There are also functions to generate signals and output them to a device. Finally, it allows you to monitor signals via a virtual oscilloscope, invoked by the function **softscope**.

## B.9 IMAGE PROCESSING TOOLBOX

This toolbox might be of particular interest to you if you work on trying to understand the visual system. The toolbox affords all kinds of sophisticated algorithms for image processing and analysis. Those who work with medical imaging, particularly MRIs, will benefit from a direct way to read images from DICOM (digital imaging and communications in medicine) format by using the function **dicomread**. For instance, you could conceivably read MRI images into MATLAB, enhance the contrast, then save them again as DICOM images by using **dicomwrite**. All kinds of image filtering and manipulation algorithms that are familiar from programs like Photoshop are available here, e.g., through the **edge** function, which finds edges in grayscale images or **imhist** which yields a histogram of the luminance of an image.

## B.10 PSYCHOPHYSICS TOOLBOX AND MGL

The Psychophysics Toolbox was *not* developed by MathWorks, and it can be downloaded for free at http://psychtoolbox.org/HomePage. It is most useful for the controlled presentation of visual stimuli, and is thus mostly used by visual psychophysicists. One of the great strengths of this toolbox is its generality, which leads to flexibility. If the hardware and MATLAB allow the display of a stimulus, this toolbox will allow you to display it with high temporal precision. While MATLAB is an interpreted language, the Psychophysics Toolbox achieves high temporal fidelity by utilizing compiled low-level C-code, which is closer to the hardware. MATLAB allows you to call these functions, so-called MEX files (for "MATLAB executable"). With the Psychophysics Toolbox, you create a stimulus as a MATLAB matrix, put it into a frame buffer, and then put that on the screen and wait for a user response. The function **screen** constitutes the core of the Psychophysics Toolbox. After opening a screen with **'OpenWindow'**, you can then put an image represented by a MATLAB matrix on the screen (**'PutImage'**). There are all kinds of other useful functions within this toolbox, e.g., **PsychGamma** for the gamma calibration of the monitor, **Quest** to efficiently calculate thresholds, **getmouse** to retrieve user input from the mouse, etc. Importantly in times of highly multithreaded operating systems, the Psychophysics Toolbox lets you set the priority of the executing program, telling the OS that now is not a good time to run a cleanup process in the background. There is a devoted user base, and the Psychophysics Toolbox is in use by hundreds if not thousands of labs.

A very similar and more recent offering is provided by the "MGL" package, which uses simple and atomic functions (implemented as MEX files) to provide OpenGL functionality. At this point, MGL is only available for the Mac. The advantage of MGL is its full integration with neuroimaging and eye tracking hardware. If anything, the temporal precision is even higher than in the Psychophysics Toolbox. MGL is in active development, is increasingly popular, and can be downloaded here: http://gru.brain.riken.jp/doku.php/mgl/download.

## B.11  CHRONUX

You (presumably) first encountered Chronux in Chapter 23, when analyzing LFP data. Chronux was developed by the Mitra Lab at Cold Spring Harbor Laboratory. It is discussed in detail in *Observed Brain Dynamics*, and it makes extensive use of Slepian functions. It can be downloaded for free at http://chronux.org/.

The large number of efficient and sophisticated signal processing functions afforded by Chronux is beyond the scope of this brief blurb. For instance, Chronux provides spectral analysis functions that inherently make use of multiple tapers—in contrast to those provided by MATLAB—that neatly recover the edges of the time window. If you are serious about the processing of neural signals, you might want to consider giving it a spin.

## B.12  MATHWORKS FILE EXCHANGE

MATLAB has a huge user base. While not everyone has the stamina to build an entire toolbox, lots of people are willing to share individual functions that they think could be useful to the community at large. There are thousands of these functions available on the MathWorks File Exchange, and all of them can be downloaded for free at www.mathworks.com/matlabcentral/fileexchange/.

Many of these functions are modified versions of original MATLAB functions, such as improved subplots, polar plots, etc. Some of them *are* entire toolboxes, for instance the "Circular Statistics Toolbox", which provides functions for the analysis of circular data, for instance circular mean or circular variance, among many others: http://www.mathworks.com/matlabcentral/fileexchange/10676-circular-statistics-toolbox-directional-statistics.

Before you go ahead and reinvent the wheel, you should probably check on the file exchange if someone already did. Of course, you can also make your own toolbox and share it with the world. You can find mine at http://goo.gl/sPxVK.

# References

**Preface References**

Karpicke, J.D., Roediger, H.L., 2008. The critical importance of retrieval for learning. Science 319, 966–968.

**Chapter 1 References**

Genesee, F., 1985. Second language learning through immersion: A review of U.S. programs. Rev. Educ. Res. 55 (4, Winter), 541–561.

Grube, G.M.A., Reeve C.D.C., 1992. Plato: Republic. Hackett Publishing Co., Inc.

Hubel, D.H., Wiesel, T.N., 2004. Brain and visual perception: The story of a 25-year collaboration. Oxford University Press, New York, 707.

Marr, D., 1982. Vision: A Computational Investigation into the Human Representation and Processing of Visual Information. W.H. Freeman and Company, New York.

Whitehead, A.N., 1959. The aims of education. Daedalus 88.1, 192–205.

**Chapter 2 References**

Berry, D.C., Broadbent, D.E., 1984. On the relationship between task performance and associated verbalized knowledge. Q. J. Exp. Psychol. 36A, 209–231.

**Chapter 3 References**

Azad, K., 2011. Math, Better Explained: Learn to Unlock Your Math Intuition. Amazon Digital Services.

Fawcett, T.W., Andrew, D.H., 2012. Heavy use of equations impedes communication among biologists. Proc. Natl. Acad. Sci. 109.29, 11735–11739.

Gabbiani, F., Steven, J.C., 2010. Mathematics for neuroscientists. Academic Press.

Gigerenzer, G., Ulrich, H., 1995. How to improve Bayesian reasoning without instruction: Frequency formats. Psychol. Rev. 102.4, 684–704.

Gladwell, M., 2009. What the dog saw: and other adventures. ePenguin.

"Student". Gosset, W.S., 1908. "The probable error of a mean". Biometrika 6, 1–25.

Hoffrage, U., Gerd, G., 1998. Using natural frequencies to improve diagnostic inferences. Acad. Med. 73.5, 538–540.

MacKay, 2003. Information Theory, Inference, and Learning Algorithms. Cambridge University Press.

**Chapter 6 References**

Donders, F.C., 1868. Over de snelheid van psychische processen. Onderzoekingen gedaan in het Physiologisch Laboratorium der Utrechtsche Hoogeschool. Tweede Reeks II, 92–120, Reprinted in and translated as Donders, F.C. (1969). On the speed of mental processes. Acta Psychologica, 30, Attention and Performance II, 412–431.

Shepard, R., Metzler, J., 1971. Mental rotation of three dimensional objects. Science 171 (972), 701–703.

Treisman, A., Gelade, G., 1980. A feature integration theory of attention. Cognit. Psychol. 12, 97–136.

**Chapter 7 References**

Helmholtz, H., 1867. Handbuch der Physiologischen Optik. Voss, Hamburg.

James, W., 1890. The principles of psychology, vol. 1. Henry Holt, New York.

Posner, M.I., 1980. Orienting of attention. Q. J. Exp. Psychol. 32, 3–25.

## Chapter 8 References

Carpenter, R., Robson, J., 1999. Vision Research: A Practical Guide to Laboratory Methods. Oxford University Press, New York.

Fechner, G.T. 1848. Nanna, oder über das Seelenleben der Pflanzen. Leipzig. Leopold Voss.

Fechner, G.T., 1851. Zend-Avesta oder über die Dinge des Himmels und des Jenseits: Vom Standpunkt der Naturbetrachtung. Leopold Voss, Leipzig.

Fechner, G.T., 1860. Elemente der Psychophysik. Breitkopf und Härtel, Leipzig.

Hecht, S.P., 1942. Energy, quanta, and vision. J. Gen. Physiol. 25, 819–840.

Norton, T.T., Corliss, D.A., Bailey, J.E., 2002. The Psychophysical Measurement of Visual Function. Butterworth-Heinemann, Woburn, MA.

## Chapter 9 References

Smith, S.T., 2006. MATLAB: advanced GUI development. Dog Ear Publishing.

## Chapter 10 References

Fisher, R.A., 1925. Statistical Methods for Research Workers. Oliver and Boyd, Edinburgh.

Green, D.M., Swets, J.A., 1966. Signal Detection Theory and Psychophysics. John Wiley & Sons, Inc, New York.

Rosenthal, R., 1976. Experimenter Effects in Behavioral Research. Irvington, New York.

Ziliak, S.T., McCloskey, D.N., 2008. The Cult of Statistical Significance. How the Standard Error Costs Us Jobs, Justice, and Lives. University of Michigan Press.

## Chapter 11 References

Van Drongelen, W., 2006. Signal Processing for Neuroscientists: An Introduction to the Analysis of Physiological Signals. Academic Press, Burlington, MA.

Hillenbrand, J., Getty, L.A., Clark, M.J., Wheeler, K., 1995. Acoustic characteristics of American English vowels. J. Acoust. Soc. Am. 97, 3099–3111.

Peterson, G.E., Barney, H.L., 1952. Control methods used in a study of the vowels. J. Acoust. Soc. Am. 24, 175–184.

## Chapter 12 References

Van Drongelen, W., 2006. Signal Processing for Neuroscientists: An Introduction to the Analysis of Physiological Signals. Academic Press, Burlington, MA.

## Chapter 13 References

Percival, D., Walden, A., 2000. Wavelet Methods for Time Series Analysis. Cambridge University Press, Cambridge.

Quiroga, R., Nadasdy, Z., Ben-Shaul, Y., 2004. Unsupervised spike detection and sorting with wavelets and superparamagnetic clustering. Neural Comput. 16, 1661–1687.

## Chapter 15 References

Fitzhugh, R., 1961. Impulses and physiological states in theoretical models of nerve membrane. Biophys. J. 1, 445–466.

## Chapter 16 References

Dayan, P., Abbott, L.F., 2001. Theoretical neuroscience. MIT Press, Cambridge, MA.

Lotto, R.B., Williams, S.M., Purves, D., 1999. An empirical basis for Mach bands. Proc. Natl. Acad. Sci. USA 96, 5239–5244.

Ratliff, F., 1965. Mach bands: Quantitative studies on neural networks in the retina. Holden-Day, San Francisco, CA.

Sekular, R., Blake, R., 2002. Perception, 4th Ed. McGraw-Hill, New York.

## Chapter 18 References

Hatsopoulos, N.G., Ojakangas, C.L., Paninski, L., Donoghue, J.P., 1998. Information about movement direction obtained from synchronous activity of motor cortical neurons. Proc. Natl. Acad. Sci. USA 95 (26), 15706–15711.

Optican, L.M., Richmond, B.J., 1987. Temporal encoding of two-dimensional patterns by single units in primate inferior temporal cortex. III. Information theoretic analysis. J. Neurophysiol. 57 (1), 162–178.

Panzeri, S., Senatore, R., Montemurro, M.A., Petersen, R.S., 2007. Correcting for the sampling bias problem in spike train information measures. J. Neurophysiol. 98 (3), 1064–1072.

Richmond, B.J., Optican, L.M., 1987. Temporal encoding of two-dimensional patterns by single units in primate inferior temporal cortex. II. Quantification of response waveform. J. Neurophysiol. 57 (1), 147–161.

Richmond, B.J., Optican, L.M., Podell, M., Spitzer, H., 1987. Temporal encoding of two-dimensional patterns by single units in primate inferior temporal cortex. I. Response characteristics. J. Neurophysiol. 57 (1), 132–146.

Shannon, C.E., 1948. A mathematical theory of communication. Bell Syst. Tech. J. 27, 379–423, 623–656.

## Chapter 19 References

Georgopoulos, A.P., Kalaska, J.F., Caminiti, R., Massey, J.T., 1982. On the relations between the direction of two-dimensional arm movements and cell discharge in primate motor cortex. J. Neurosci. 2 (11), 1527–1537.

Hartline, H.K., 1940. The receptive fields of optic nerve fibers. Am. J. Physiol. 130, 690–699.

## Chapter 20 References

Hatsopoulos, N.G., Xu, Q., Amit, Y., 2007. Encoding of movement fragments in the motor cortex. J. Neurosci. 27, 5105–5114.

Moran, D.W., Schwartz, A.B., 1999. Motor cortical representation of speed and direction during reaching. J. Neurophysiol. 82, 2676–2692.

## Chapter 21 References

Georgopoulos, A.P., Schwartz, A.B., Kettner, R.E., 1986. Neuronal population coding of movement direction. Science 233 (4771), 1416–1419.

Hochberg, L.R., Serruya, M.D., Friehs, G.M., Mukand, J.A., Saleh, M., Caplan, A.H., et al., 2006. Neuronal ensemble control of prosthetic devices by a human with tetraplegia. Nature 442 (7099), 164–171.

Papsin, B.C., Gordon, K.A., 2007. Cochlear implants for children with severe-to-profound hearing loss. N. Engl. J. Med. 357 (23), 2380–2387.

## Chapter 22 References

Brockwell, A.E., Rojas, A.L., Kass, R.E., 2004. Recursive Bayesian decoding of motor cortical signals by particle filtering. J. Neurophysiol. 91 (4), 1899–1907.

Brown, E.N., Frank, L.M., Tang, D., Quirk, M.C., Wilson, M.A., 1998. A statistical paradigm for neural spike train decoding applied to position prediction from ensemble firing patterns of rat hippocampal place cells. J. Neurosci. 18 (18), 7411–7425.

Georgopoulos, A.P., Kettner, R.E., Schwartz, A.B., 1988. Primate motor cortex and free arm movements to visual targets in three-dimensional space. II. Coding of the direction of movement by a neuronal population. J. Neurosci. 8 (8), 2928–2937.

Hochberg, L.R., Serruya, M.D., Friehs, G.M., Mukand, J.A., Saleh, M., Caplan, A.H., et al., 2006. Neuronal ensemble control of prosthetic devices by a human with tetraplegia. Nature 442 (7099), 164–171.

Serruya, M.D., Hatsopoulos, N.G., Paninski, L., Fellows, M.R., Donoghue, J.P., 2002. Instant neural control of a movement signal. Nature 416 (6877), 141–142.

Warland, D.K., Reinagel, P., Meister, M., 1997. Decoding visual information from a population of retinal ganglion cells. J. Neurophysiol. 78 (5), 2336–2350.

Wu, W., Shaikhouni, A., Donoghue, J.P., Black, M.J., 2004. Closed-loop neural control of cursor motion using a Kalman filter. Conf. Proc. IEEE Eng. Med. Biol. Soc. 6, 4126–4129.

## Chapter 23 References

van Drongelen, W., 2007. Signal Processing for Neuroscientists: Introduction to the Analysis of Physiological Signals. Elsevier/Academic Press, Amsterdam.

Katzner, S., Nauhaus, I., Benucci, A., Bonin, V., Ringach, D.L., Carandini, M., 2009. Local origin of field potentials in visual cortex. Neuron 61 (1), 35–41.

Mitra, P., Bokil, H., 2008. Observed Brain Dynamics. Oxford University Press, New York.

O'Leary, J.G., Hatsopoulos, N.G., 2006. Early visuomotor representations revealed from evoked local field potentials in motor and premotor cortical areas. J. Neurophysiol. 96 (3), 1492–1506.

Pesaran, B., Pezaris, J.S., Sahani, M., Mitra, P.P., Andersen, R.A., 2002. Temporal structure in neuronal activity during working memory in macaque parietal cortex. Nat. Neurosci. 5 (8), 805–811.

## *Chapter 24 References*

Aguirre, G.K., Zarahn, E., D'Esposito, M., 1998. The variability of human, BOLD hemodynamic responses. Neuroimage 8, 360–369.

Bandettini, P.A., Wong, E.C., Hinks, R.S., Tikofsky, R.S., Hyde, J.S., 1992. Time course EPI of human brain function during task activation. Magn. Reson. Med. 25, 390–397.

Biswal, B., Yetkin, F.Z., Haughton, V.M., Hyde, J.S., 1995. Functional connectivity in the motor cortex of resting human brain using echo-planar MRI. Magn. Reson. Med. 34, 537–541.

Boynton, G.M., Engel, S.A., Glover, G.H., Heeger, D.J., 1996. Linear systems analysis of functional magnetic resonance imaging in human V1. J. Neurosci. 16, 4207–4221.

Buchel, C., Friston, K.J., 1997. Modulation of connectivity in visual pathways by attention: cortical interactions evaluated with structural equation modelling and fMRI. Cereb. Cortex 7, 768–778.

Buckner, R.L., 2003. The hemodynamic inverse problem: making inferences about neural activity from measured MRI signals. Proc. Natl. Acad. Sci. USA 100, 2177–2179.

Calhoun, V.D., Adali, T., Pearlson, G.D., Pekar, J.J., 2001. A method for making group inferences from functional MRI data using independent component analysis. Hum. Brain Mapp. 14, 140–151.

Calhoun, V.D., Stevens, M.C., Pearlson, G.D., Kiehl, K.A., 2004. fMRI analysis with the general linear model: removal of latency-induced amplitude bias by incorporation of hemodynamic derivative terms. Neuroimage 22, 252–257.

Chumbley, J.R., Friston, K.J., 2009. False discovery rate revisited: FDR and topological inference using Gaussian random fields. Neuroimage 44, 62–70.

Cox, R.W., 1996. AFNI: software for analysis and visualization of functional magnetic resonance neuroimages. Comput. Biomed. Res. 29, 162–173.

Donaldson, D.I., Buckner, R.L., 2003. Effective paradigm design. In: Jezzard, P., Matthews, P.M., Smith, S.M. (Eds.), Functional MRI: An Introduction to Methods. Oxford University Press.

Evans, A.C., Collins, D.L., Mills, S.R., Brown, E.D., Kelly, R.L., Peters, T.M., 1993. 3D statistical neuroanatomical models from 305 MRI volumes. IEEE--Nuclear Science Symposium and Medical Imaging Conference, pp. 1813–1817.

Forman, S.D., Cohen, J.D., Fitzgerald, M., Eddy, W.F., Mintun, M.A., Noll, D.C., 1995. Improved assessment of significant activation in functional magnetic resonance imaging (fMRI): use of a cluster-size threshold. Magn. Reson. Med. 33, 636–647.

Fox, M.D., Snyder, A.Z., Vincent, J.L., Corbetta, M., Van Essen, D.C., Raichle, M.E., 2005. The human brain is intrinsically organized into dynamic, anticorrelated functional networks. Proc. Natl. Acad. Sci. USA 102, 9673–9678.

Friston, K.J., Buechel, C., Fink, G.R., Morris, J., Rolls, E., Dolan, R.J., 1997. Psychophysiological and modulatory interactions in neuroimaging. Neuroimage 6, 218–229.

Friston, K.J., Harrison, L., Penny, W., 2003. Dynamic causal modelling. Neuroimage 19, 1273–1302.

Friston, K.J., Holmes, A.P., Poline, J.B., Grasby, P.J., Williams, S.C., Frackowiak, R.S., et al., 1995. Analysis of fMRI time-series revisited. Neuroimage 2, 45–53.

Genovese, C.R., Lazar, N.A., Nichols, T., 2002. Thresholding of statistical maps in functional neuroimaging using the false discovery rate. Neuroimage 15, 870–878.

Gitelman, D.R., Penny, W.D., Ashburner, J., Friston, K.J., 2003. Modeling regional and psychophysiologic interactions in fMRI: the importance of hemodynamic deconvolution. Neuroimage 19, 200–207.

Handwerker, D.A., Ollinger, J.M., D'Esposito, M., 2004. Variation of BOLD hemodynamic responses across subjects and brain regions and their effects on statistical analyses. Neuroimage 21, 1639–1651.

Henson, R., Rugg, M., Friston, K.J., 2001. The choice of basis functions in event-related fMRI. HBM01 abstract, Neuroimage, 13, 149.

Henson, R.N., Shallice, T., Gorno-Tempini, M.L., Dolan, R.J., 2002. Face repetition effects in implicit and explicit memory tests as measured by fMRI. Cereb. Cortex 12, 178–186.

Huettel, S.A., McCarthy, G., 2001. The effects of single-trial averaging upon the spatial extent of fMRI activation. Neuroreport 12, 2411–2416.

Kwong, K.K., Belliveau, J.W., Chesler, D.A., Goldberg, I.E., Weisskoff, R.M., Poncelet, B.P., et al., 1992. Dynamic magnetic resonance imaging of human brain activity during primary sensory stimulation. Proc. Natl. Acad. Sci. USA 89, 5675–5679.

Logothetis, N.K., Pauls, J., Augath, M., Trinath, T., Oeltermann, A., 2001. Neurophysiological investigation of the basis of the fMRI signal. Nature 412, 150–157.

McIntosh, A.R., Bookstein, F.L., Haxby, J.V., Grady, C.L., 1996. Spatial pattern analysis of functional brain images using partial least squares. Neuroimage 3, 143–157.

McLaren, D.G., Ries, M.L., Xu, G, Johnson, S.C., 2012. A generalized form of context-dependent psychophysiological interactions (gPPI): a comparison to standard approaches. NeuroImage. 61, 1277–1286.

Ogawa, S., Tank, D.W., Menon, R., Ellermann, J.M., Kim, S.G., Merkle, H., et al., 1992. Intrinsic signal changes accompanying sensory stimulation: functional brain mapping with magnetic resonance imaging. Proc. Natl. Acad. Sci. USA 89, 5951–5955.

Pauling, L., Coryell, C.D., 1936. The Magnetic Properties and Structure of Hemoglobin, Oxyhemoglobin and Carbonmonoxyhemoglobin. Proc. Natl. Acad. Sci. USA 22, 210–216.

Raichle, M.E., Mintun, M.A., 2006. Brain work and brain imaging. Annu. Rev. Neurosci. 29, 449–476.

Shulman, G.L., McAvoy, M.P., Cowan, M.C., Astafiev, S.V., Tansy, A.P., d'Avossa, G., et al., 2003. Quantitative analysis of attention and detection signals during visual search. J. Neurophysiol. 90, 3384–3397.

Smith, S.M., 2003a. Overview of fMRI analysis. In: Jezzard, P., Matthews, P.M., Smith, S.M. (Eds.), Functional MRI: An Introduction to Methods. Oxford University Press.

Smith, S.M., 2003b. Preparing fMRI data for statistical analysis. In: Jezzard, P., Matthews, P.M., Smith, S.M. (Eds.), Functional MRI: An Introduction to Methods. Oxford University Press.

Steffener, J., Tabert, M., Reuben, A., Stern, Y., 2010. Investigating hemodynamic response variability at the group level using basis functions. Neuroimage 49, 2113–2122.

Worsley, K.J., Friston, K.J., 1995. Analysis of fMRI time-series revisited—again. Neuroimage 2, 173–181.

## Chapter 25 References

Hodgkin, A.L., Huxley, A.F., 1952. A quantitative description of membrane current and its application to conduction and excitation in nerve. J. Physiol. 116, 500–544.

## Chapter 26 References

del Castillo, J., Katz, B., 1954. The effect of magnesium on the activity of motor nerve endings. J. Physiol (Lond). 124, 553–559.

del Castillo, J., Katz, B., 1954. Quantal components of the end-plate potential. J. Physiol. 124, 560–573.

Fatt, P., Katz, B., 1952. Spontaneous sunthershold activity at motor nerve endings. J. Physiol. (Lond) 117, 109–128.

## Chapter 27 References

Hodgkin, A.L., Huxley, A.F., 1952. A quantitative description of membrane current and its application to conduction and excitation in nerve. J. Physiol. 117, 500–544.

Hodgkin, A.L., Katz, B., 1949. The effect of sodium ions on the electrical activity of the gaint axon of the squid. J. Physiol. 108 (1), 37–77.

## Chapter 29 References

Izhikevich, E.M., 2003. Simple model of spiking neurons. IEEE Trans. Neural. Netw. 14 (6), 1569–1572.

## Chapter 30 References

Murray, J.D., 2002. Mathematical biology I: An introduction. Springer-Verlag, New York.

Strauss, W.A., 1992. Partial differential equations: An introduction. John Wiley & Sons, Inc, New York.

Wilson, H.R., 1999. Spikes, decisions, and actions: Dynamical foundations of neuroscience. Oxford University Press, Oxford.

### Chapter 31 References

Shadlen, M.N., Newsome, W.T., 2001. Neural basis of a perceptual decision in the parietal cortex (area LIP) of the rhesus monkey. J. Neurophysiol. 86 (4), 1916–1936.

Swensson, R., 1972. The elusive tradeoff: Speed vs accuracy in visual discrimination tasks. Percept. Psychophys. vol. 12 (1-A), 16–32.

### Chapter 32 References

Herbst, J.A., Gammeter, S., Ferrero, D., Hahnloser, R.H., 2008. Spike sorting with hidden Markov models. J. Neurosci. Methods 174 (1), 126–134.

### Chapter 33 References

Brown, E.N., Barbieri, R., Ventura, V., Kass, R.E., Frank, L.M., 2002. The time-rescaling theorem and its application to neural spike train data analysis. Neural. Comput. 14, 325–346.

Donald, E.K., 1969. Seminumerical Algorithms, The Art of Computer Programming, vol. 2. Addison Wesley, NJ.

von Neumann, J., 1951. Various techniques used in connection with random digits. Monte Carlo methods. Nat. Bureau Standards 12, 36–38.

### Chapter 34 References

Wilson, H.R., Cowan, J.D., 1972. Excitatory and inhibitory interactions in localized populations of model neurons. J. Biophys. 12, 1–24.

### Chapter 35 References

Bak, P., Chen, K., Tang, C., 1990. A forest-fire model and some thoughts on turbulence. Phys. Lett. A 147, 297–300.

Buice, M., Cowan, J.D., 2007. Field-theoretic approach to fluctuation effects in neural networks. Phys. Rev. E. 75, 051919.

Drossel, B., Schwabl, F., 1992. Self-organized critical forest-fire model. Phys. Rev. Lett. 69, 1629–1632.

Drossel, C., Schwabl, 1994. Crossover from percolation to self-organized criticality. Phys. Rev. E 50, R2399–R2402.

### Chapter 36 References

Anderson, J.A., et al., 1977. Distinctive features, categorical perception, and probability learning: Some applications of a neural model. Psychol. Rev. 84, 413–451.

Bliss, T.V., Lomo, T., 1973. Long-lasting potentiation of synaptic transmission in the dentate area of the anaesthetized rabbit following stimulation of the perforant path. J. Physiol. 232 (2), 331–356.

Gauthier, I., Behrmann, M., Tarr, M.J., 2004. Are Greebles like faces? Using the neuropsychological exception to test the rule. Neuropsychologia 42 (14), 1961–1970.

Gauthier, I., Tarr, M.J., 1997. Becoming a "Greeble" expert: Exploring mechanisms for face recognition. Vision Res. 37 (12), 1673–1682.

Hebb, D.O., 1949. Organization of behavior. John Wiley & Sons, New York.

James, W., 1890. The principles of psychology. Henry Holt & Sons, Inc, New York.

Rumelhart, D.E., Zipser, D., 1985. Feature discovery by competitive learning. Cogn. Sci. 9, 75–112.

### Chapter 37 References

Adrian, E.D., Matthews, R., 1927. The action of light on the eye: Part I. The discharge of impulses in the optic nerve and its relation to the electric changes in the retina. J. Physiol. 63 (4), 378–414.

Albus, J.S., 1971. A theory of cerebellar function. Math. Biosci. 10, 25–61.

Hebb, D.O., 1949. Organization of behavior. John Wiley & Sons, New York.

Ito, M., Kano, M., 1982. Long-lasting depression of parallel fiber-Purkinje cell transmission induced by conjunctive stimulation of parallel fibers and climbing fibers in the cerebellar cortex. Neurosci. Lett. 33 (3), 253–258.

Krupa, D.J., Thompson, J.K., Thompson, R.F., 1993. Localization of a memory trace in the mammalian brain. Science 260 (5110), 989–991.

Marr, D., 1969. A theory of cerebellar cortex. J. Physiol. 202 (2), 437–470.

McCulloch, W.S., Pitts, W., 1943. A logical calculus of the ideas immanent in nervous activity. Bull. Math. Biophys. 5, 115–133.

Medina, J.F., et al., 2000. Mechanisms of cerebellar learning suggested by eyelid conditioning. Curr. Opin. Neurobiol. 10 (6), 717–724.

Minsky, M., Papert, S., 1969. Perceptrons: An introduction to computational geometry. MIT Press, Cambridge, MA.

Ojakangas, C.L., Ebner, T.J., 1992. Purkinje cell complex and simple spike changes during a voluntary arm movement learning task in the monkey. J. Neurophysiol. 68 (6), 2222–2236.

Ojakangas, C.L., Ebner, T.J., 1994. Purkinje cell complex spike activity during voluntary motor learning: Relationship to kinematics. J. Neurophysiol. 72 (6), 2617–2630.

Rescorla, R.A., Wagner, A.R., 1972. A theory of Pavlovian conditioning: Variations in the effectiveness of reinforcement and nonreinforcement. In: Black, A., Prokasy, W.F. (Eds.), Classical conditioning II. Appleton-Century-Crofts, New York, pp. 64–69.

Rosenblatt, F., 1958. The perceptron: A probabilistic model for information storage and organization in the brain. Psychol. Rev. 65, 386–408.

Widrow, B., Hoff, M.E., 1960. Adaptive switching circuits. IRE WESCON Convention Record. IRE, New York, 96–104.

## Appendix A References

Tufte, E.R., Graves-Morris, P.R., 1983. The visual display of quantitative information, Vol. 2. Graphics Press, Cheshire, CT. <http://www-personal.umich.edu/~jpboyd/eng403_chap2_tuftegospel.pdf>.

## Appendix B References

Mitra, P., Hemant, B., 2007. Observed Brain Dynamics. Oxford University Press, USA.

Brainard, D.H., 1997. The psychophysics toolbox. Spat. Vis. 10.4, 433–436.

Cornelissen, F.W., Peters, E.M., John, P., 2002. The Eyelink Toolbox: eye tracking with MATLAB and the Psychophysics Toolbox. Behav. Res. Methods Instrum. Comput. 34.4, 613–617.

# Index

*Note*: Page numbers followed by "*f*" and "*t*" refer to figures and tables, respectively.